Modern Component Families and Circuit Block Design

Nihal Kularatna

Newnes

BOSTON OXFORD AUCKLAND JOHANNESBURG MELBOURNE NEW DELHI

Library of Congress Cataloging-in-Publication Data
Kularatna, Nihal.
  Modern component families and circuit block design / Nihal
Kularatna.
    p.   cm.
  Includes bibliographical references.
  ISBN 0-7506-9992-2 (alk. paper)
    1. Power electronics—Design and construction.   2. Low voltage
systems—Design and construction.   3. Electronic circuit design.
  I. Title.
TK7881.15.K84 1999   99–31348
621.31'7—dc21   CIP

British Library Cataloguing-in-Publication Data
A catalogue record for this book is available from the British Library.

The publisher offers special discounts on bulk orders of this book.
For information, please contact:
Manager of Special Sales
Butterworth–Heinemann
225 Wildwood Avenue
Woburn, MA 01801–2041
Tel: 781-904-2500
Fax: 781-904-2620
For information on all Butterworth–Heinemann publications available, contact
our World Wide Web home page at: http://www.bh.com

10 9 8 7 6 5 4 3 2 1

Printed in the United States of America

Modern Component Families and Circuit Block Design

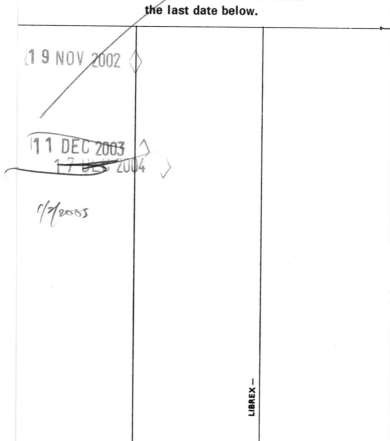

Books are to be returned on or before
the last date below.

To my family. . . Priyani, Dulsha, and Malsha

Contents

Chapter 6 Optoisolators   **241**

Chapter 7   Sensors   **265**

Preface

I began compiling the material for this work when the transistor was celebrating its golden jubilee. It justified me taking a broader look at the world of modern component families without a specific bias toward analog, digital or mixed signal component families.

Nature & human activities being analog, while the modern world of electronic engineering makes use of digital processing, one must be serious about the conversion process too. At the same time designers may have to appreciate a new theory... proposed by Theo A. M. Classen of Philips Semiconductors about the technology component necessary to increase the usefulness of an electronic system that,

$$\text{Usefulness of a product} = \text{Log (Technology)}$$

as cited by Brian Dippert in a recent EDN magazine (EDN, April 15, 1999, page 52).

My intention here is to offer a minute part of the massive world of semiconductors as examples and provide some guidance to designers, academics and students on how to keep balance in the overall design.

I would be thankful to readers & reviewers for feedback on how I have fared in this attempt.

Nihal Kularatna
39A, Sumudu Place
Sri Rahula Road
Katubedda
Moratuwa
Sri Lanka.

6 August 1999

Acknowledgments

Let me commence with an acknowledgement of the co-inventors of the integrated circuit, Jack S. Kilby and Robert Noyce. Their initiative, followed by subsequent efforts at semiconductor fabrication facilities and R&D groups in the global semiconductor industry and academia have given the world a plethora of semiconductor-based consumer products and industrial systems. The world of digital components, such as processors, memories and peripheral ICs opening wider vistas for signal processing, cannot be overlooked, nor should there be less emphasis on analog signal conditioning and processing, data conversion, signal isolation, glue logic, power supplies and battery back-up. Though their development has been slower, modern non-linear devices also provide an important contribution to electronic systems. In modern electronics many function blocks implemented in the form of ICs have become building elements of products or systems. This may remain so for a decade or more in most cases, until the System on a Chip (SoC) concepts are fully matured.

This book provides a broad approach to "system design," embracing many circuit blocks (from the power circuits to processing functions) going through the important mixed signal circuitry. In this work a large amount of published material from US industry and academia has been used and the organizations who deserve acknowledgments are:

(1) Harris Semiconductor, Linear Technology Corporation, Maxim Integrated Products, Motorola and Unitrode for information on power supply circuit blocks, voltage references and power management circuitry etc
(2) Analog Devices Inc., Burr Brown and Apex Microsystems for valuable information on op amps and related techniques in Chapter 2
(3) Analog Devices Inc. for almost 90% of the material used for Chapter 3 on Data Converters

(4) Intel, Zilog, Mips Technologies, Infineon Technologies (former Siemens Microelectronics) and EDN magazine for the information on microprocessors and microcontrollers in Chapter 4

(5) Analog Devices, Inc., and IEEE for information for Chapter 5 on digital Signal processors

(6) Infineon Technologies and International Rectifier for information on optoisolators in Chapter 6

(7) Allegro Microsystems Inc., Analog Devices Inc, Motorola, EG&G Sensors, Microswitch/Honeywell, NIST/IEEE working group on sensor integration and FiS Corporation, Japan for information on modern semiconductor sensors in Chapter 7

(8) Analog Devices Inc. and Bensys Corporation for information in Chapter 8 for non-linear devices

(9) Benchmarq Microelectronics, AER Energy Resources, NEC Moli Energy (Canada) and Unitrode and for information on batteries and battery management ICs in Chapter 9

(10) Texas Instruments for information on programmable logic devices for Chapter 10

I am always indebted to the editorial staff of industry magazines such as EDN, Electronic Design, PCIM, Electronic Engineering and many other IEEE and IEE journals for their valuable reporting on the new developments in the industry. In carrying out this exercise living in a small island in the Indian Ocean, I would not have completed this manuscript within a reasonable period of time, if not for the great support extended by the following personnel from such companies. They either assisted me with valuable publications, component samples, suggestions for chapter contents, or assisted me with copyright permissions.

(a) Dan Sheingold, Walt Kester, Ethan Boudeaux, John Hayes, James Bryant and Derek Bowers and Linda Roche of Analog Devices

(b) Benjamin Actranie and John Tran of Bensys Corporation

(c) Dorein Stein, and Sally Slemons from Infineon Technologies of USA and Simon N.G from Siemens, Singapore respectively.

(d) Patty Smith of Benchmarq Microelectronics and David Heacock of Unitrode Inc.

(e) Karen Bosco of Motorola

(f) Robert Donkins of Linear Technologies

(g) Edward Pawlock of Harris Semiconductors

(h) Mike Postula and Taka Ito of FiS, Japan

(i) Frank M. Harris of AER Energy Resources

(j) Kevin Self and Syd Coppersmith of Dallas Semiconductor, USA and Grace Tao of Dallas Semiconductor, Taiwan

(k) Peter Wood, Jonathan Adams, Kathy Frey and Judy Turner of International Rectifier

(l) Michael Markowitz and Kathy Leonard of EDN Magazine

(m) Kang Lee of NIST/IEEE committee on 1451 sensor standard
(n) Marie G. Riviera of Apex Microtechnolgies, USA

In this exercise I am very thankful for the tireless efforts by the ACCIMT staff, including Mrs. Chandrika Weerasekera, Miss Shayanika Rashikanganie and trainees Arosh Edirisinghe, Kapila Kumara and A.H.S. Chamara in the preparation of the manuscript text and the graphics. Also I am very grateful to engineering undergraduates from University of Moratuwa, Manoj Bandara and Sasiri Yapa, who volunteered to work at my residence and helped with the graphics at early stages. I am also very thankful to Ms. Indrani Hewage who prepared the first draft of the several chapters.

My special gratitude is extended to the Corporate staff of Metropolitan Group, J.J. Ambani, Dinesh Ambani, Erajh Gunaratne, Ivor Maharoof (Thank you, Ivor, for all the bulk copying at lightning speed!!). For my well maintained residential computing resources, Niranjan de Silva, Mohan Weerasuriya and their staff of the same company are gratefully acknowledged.

I am very grateful to Chairman of ACCIMT, Mr. Rohan Weerasinghe, former Chairman Prof. K.K.Y.W. Perera, Director Prof. S. Karunaratne and the Members of the Board of Governors and the management of ACCIMT for their encouragement.

Many friends encouraged me in this work, especially Mr. Keerthi Kumarasena, Sherani Godamunne, Padmasiri Soysa (our ACCIMT librarian who maintains and manages a vast array of industrial documentation) and my friends such as Mohan Kumaraswamy (who helped with my first technical paper in 1977), and many other friends and university colleagues such as Sunil, Kumar, Ranjith, Vasantha, Lakshman, Upali, Kithsiri, Lal, Shantha and Jayantha.

My most sincere appreciation is for Wayne Houser, formerly of Voice of America, for his support for information collection and copyright clearances. A thank you note to Dr. Fred Durant and Dr. Joe Pelton of Arthur C. Clarke Foundation of USA for their support and encouragement for my work. I appreciate the work of the editorial and production staff at Butterworth-Heinemann in Boston, with particular mention of Candy Hall, Pam Chester, Susan Prusak, and Tina Adoniou.

Sir Arthur Clarke, patron of the Clarke Institute . . . my most sincere gratitude to you for your great encouragement of my little works.

Last, but not least, to Priyani for shouldering most difficult family commitments including the workload of the children, Dulsha and Malsha, allowing me free time to complete this work successfully.

Thank you.

Nihal Kularatna
39A Sumudu Place
Sri Rahula Road
Katubedda
Moratuwa
Sri Lanka

21 October 1999.

CHAPTER 1

Voltage References and Voltage Regulators

Coauthor: Dileeka Dias

1.1 Introduction

Almost all electronic systems utilize a regulated power supply as an essential requirement. Most systems need a precision voltage reference as well. In the past, the task of voltage regulation was tediously accomplished with discrete devices. Today, with integrated circuit voltage references and regulators, this task has been significantly simplified. Not only can an extremely high precision be obtained, but also an extremely high degree of temperature stability.

The performance of today's electronic devices such as microprocessors, test and measuring instruments, and sophisticated portable and handheld equipment is directly related to the quality of the supply voltage. This results in the need for tight regulation, low noise, and excellent transient response. The designer now has a wide choice of fixed, adjustable, and tracking voltage regulators, with many also incorporating built-in protection features.

One of the fastest growing markets in the world of power regulation is for switching regulators. These offer designers several important advantages over linear regulators, the most significant being size and efficiency. In addition, the ability to perform step-up, step-down, or voltage inverting functions is an attractive feature.

The old linear regulator is not totally out of business. The proliferation of battery-powered equipment in recent years has accelerated the development and usage of low-dropout (LDO) voltage regulators. Compared to a standard linear regulator, the LDO regulator using PNP transistors can maintain its output in regulation with a much lower voltage across it. While the NPN transistor requires about 2 V of headroom voltage to regulate, the LDO typically will work with less than 500 mV of input-to-output voltage differential. This reduced input voltage requirement is advantageous in battery-powered systems, since it translates directly into fewer battery cells (Simpson, 1996). In low-dropout applications, the efficiency advantage of switching regulators no longer is as great. A linear regulator design on the other hand offers several desirable features, such as low output noise and wide bandwidth, resulting in excellent transient response.

This chapter describes the basics of voltage references, linear and switching regulators, and continues to discuss the state-of-the-art components available, the advantages and disadvantages of different types of devices, their application environments as well as the basics of regulator design using these components.

1.2 Voltage References

1.2.1 Voltage Reference Fundamentals

A wide variety of voltage references are available today. However, all base their performance on the action of either a zener diode or a bandgap cell. Additional circuitry is included to obtain good temperature stability. Although discrete zener diodes are available in voltage ratings as low as 1.8 V to as high as 200 V, with power handling capabilities in excess of 100 W, their tolerance and temperature characteristics are unsuitable for many applications. Therefore, discrete zener diode-based references have additional circuitry to improve performance. The most popular reference is probably the temperature-compensated zener diode, particularly, for voltages above 5 V.

The operation of a bandgap reference is based on specific characteristics of diodes operating at the same current but different current densities. Bandgap references are available with output voltage ratings of about 1.2 to 10 V. The principal advantage of these devices is their ability to provide stable low voltages, such as 1.2, 2.5, or 5 V. However, bandgap references of 5 V and higher tend to have more noise than equivalent zener-based references. This is because, in bandgap references, higher voltages are obtained by amplification of the 1.2 V bandgap voltage by an internal amplifier. Their temperature stability also is below that of zener-based references.

1.2.2 Types of Voltage References

1.2.2.1 Zener-Based Voltage References

Zener diodes are semiconductor PN junction diodes with controlled reverse-bias properties, which make them extremely useful as voltage references. The V-I characteristics of an ideal zener diode is shown in Figure 1-1(a) and a simple regulator circuit based on it in Figure 1-1(b).

The reverse characteristics show that, at the breakdown point, the knee voltage is independent of the diode current. This knee voltage or the zener voltage is controlled by the amount of doping applied in the manufacturing process. In the simple regulator circuit shown in Figure 1-1(b), as long as the zener diode is in its regulating range, the load voltage V_L remains constant and equal to the nominal zener voltage, even when the input voltage and the load resistance varies over a wide range. If the input voltage increases, the diode maintains a

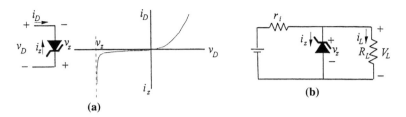

(a)   (b)

FIGURE 1-1 Zener diode and voltage regulator (a) Typical zener characteristics (b) a simple zener diode voltage regulator

constant voltage across the load by absorbing the extra current and keeping the load current constant. If the load resistance decreases, the extra current required to keep the load voltage constant is facilitated by a decrease in the current drawn by the zener diode.

In the preceding simplified analysis, the temperature dependence of the zener voltage was not taken into account. The stability of the output with temperature is a prime requirement of a voltage reference. Not only does the zener voltage vary with temperature, its variation also depends on the type of breakdown that occurs.

A zener diode has two distinctly different breakdown mechanisms: zener breakdown and avalanche breakdown. The zener breakdown voltage decreases as the temperature increases, creating a negative temperature coefficient (TC). The avalanche breakdown voltage increases with temperature (positive TC) (Pryce, 1990). This is illustrated in Figure 1-2. The zener effect and the avalanche effect dominate at low and high currents, respectively.

Although, theoretically, it is possible to select the operating point of a zener diode so that the two temperature coefficients will cancel out each other, in practical IC zener-based voltage references, a conventional forward-biased diode is used in series with a zener operating in the avalanche mode. A forward-biased diode has a negative TC, and this cancels the positive TC of the zener diode.

A simple zener-based voltage reference IC is shown in Figure 1-3. In this circuit, R4 provides the startup current for the diodes, thus setting the positive

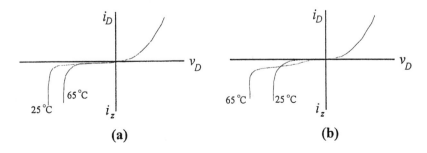

(a)   (b)

FIGURE 1-2 Temperature characteristics of zener diodes: (a) Zener breakdown, (b) Avalanche breakdown

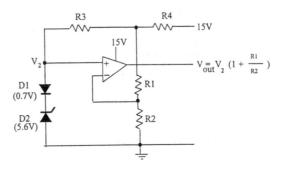

FIGURE 1-3 A simple zener-based voltage reference IC

input of the op amp at V2. R3 sets the desired bias current for the diodes. Manufacturers set the output voltage to a value different from that of V2 through the ratio R1 to R2. By trimming this resistor ratio, the output voltage can be set to the desired accuracy. Also, by trimming R3, the bias current can be optimized to a point where a minimum TC is obtained.

TC specifications as low as 1 ppm/°C are possible with zener-based voltage reference ICs (Pryce, 1990).

1.2.2.2 Bandgap References

Similar to zener-based references, bandgap references also produce the sum of two voltages having opposite temperature coefficients. One voltage is the forward voltage of a conventional diode (the base-emitter junction of a transistor), which has a negative temperature coefficient. The other is the difference between the forward voltages of two diodes with the same current but operating at two current densities. A circuit diagram of a bandgap reference is shown in Figure 1-4.

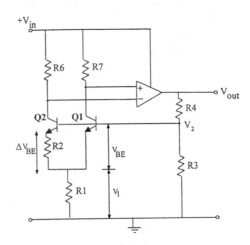

FIGURE 1-4 The circuit diagram of a bandgap reference

Transistors Q1 and Q2 are operating at the same current, but at different current densities. This is achieved by fabricating Q2 with a larger emitter area than Q1. Therefore, the base-emitter voltages of the two transistors are different. This difference is dropped across R2.

Extrapolated to absolute 0, V_{BE} is equal to 1.205 V, the bandgap voltage of silicon, and has a predictable, negative temperature coefficient of -2 mV/°C. By adding a voltage to V_{BE}, which has a positive temperature coefficient, a bandgap reference, at least theoretically, can generate a constant voltage at any temperature.

The base-emitter voltage difference is given by

$$\Delta V_{BE} = \frac{kT}{q} \ln \left(\frac{J_1}{J_2} \right) \tag{1.1}$$

where J_1 and J_2 are the current densities of transistors Q1 and Q2, respectively. Since the sum of the two transistor currents flow through R1, the voltage across R1 can be expressed as

$$V_1 = 2 \left(\frac{R_1}{R_2} \right) \Delta V_{BE} \tag{1.2}$$

Also,

$$V_2 = V_{BE} + V_1 \tag{1.3}$$

Using 1.2 and 1.3,

$$V_2 = V_{BE} + 2 \left(\frac{R_1}{R_2} \right) \Delta V_{BE} \tag{1.4}$$

Therefore, V_2 is the sum of V_{BE} and the scaled ΔV_{BE}. Knapp (1998) shows that, if the emitter areas of the two transistors is eight, the temperature coefficients of V_{BE} and ΔV_{BE} cancel each other. The op amp raises the bandgap voltage V_2 to a higher voltage at the output of the reference.

Bandgap references typically provide voltages ranging from 1.2 to 10 V. The advantage of bandgap references is their ability to provide voltages below 5 V. The greatest appeal of bandgap devices is the ability to function with operating currents from milliamps down to microamps.

IC bandgap references have additional features such as multiple calibrated voltages. Because most bandgap references are constructed in monolithic form, they are relatively inexpensive. However, their temperature coefficient is inferior to that of zener-based references. This is due to the second-order dependencies of ΔV_{BE} on temperature.

1.2.3 Quality Measures of Voltage References

An ideal voltage reference would have the exact specified voltage, and it would not vary with time, temperature, input voltage, or load conditions. However, as it is impossible to fabricate such ideal references, manufacturers provide specifications informing the user of the device's important quality parameters.

1.2.3.1 Output Voltage Error

This is the initial untrimmed accuracy of the reference at 25°C at a specified input voltage. This is specified in millivolts or a percentage. Some references provide pin connections for trimming their initial accuracy with an external potentiometer.

1.2.3.2 Temperature Coefficient

The temperature coefficient of a reference is its average change in output voltage as a function of temperature compared with its value at 25°C. This is specified in ppm/°C or mV/°C.

1.2.3.3 Line Regulation

This is the change in output voltage for a specified change in input voltage. Usually specified in %/V or μV/V of input change, line regulation is a measure of the reference's ability to handle variations in supply voltage.

1.2.3.4 Load Regulation

This is the change in output voltage for a specified change in load current. Specified in μV/mA, %/mA, or ohms of DC output resistance, load regulation includes any self-heating effects due to changes in power dissipation with load current.

1.2.3.5 Long-Term Stability

This is the change in the output voltage of a reference as a function of time. Specified in ppm/1000 hours at a specific temperature, the long-term stability is difficult to quantify. As a result, manufacturers usually provide only typical specifications, based on device data collected during the characterization process.

1.2.3.6 Noise

Although the preceding are the most important quality parameters of a voltage reference, noise is particularly of importance in certain applications such as A/D or D/A converters. In such applications, the noise from the reference should be less than 10% of the LSB value of the converter. Therefore, the higher the resolution of the converter, the lower should be the noise generated from the reference.

Noise depends on the operating current of the reference and generally is specified over a particular bandwidth and for a particular current. The specified bandwidths are 0.1–10 Hz (low-frequency noise) and 10 Hz–10 kHz (high-frequency noise).

1.2.4 Voltage Reference ICs

The levels of sophistication and pricing for voltage references range from simple and inexpensive to complex and costly. Devices are available for almost any conceivable application. Manufacturers of voltage references include National Semiconductor, Motorola, Analog Devices, Linear Technology, SGS-Thompson, Maxim Integrated Circuits, Texas Instruments, Precision Monolithic, and Silicon General.

1.2.4.1 Zener-Based References

Zener-based references usually are used in analog circuits that operate from 12–15 V supplies. Some zener-based voltage references are illustrated in Table 1-1.

A typical high-performance zener diode is the REF101 from Burr-Brown with a reference voltage of 10 V (Burr-Brown, 1989). The combination of its excellent parameters makes this device well suited for use with high-resolution A/D and D/A converters or as a precision calibrated voltage standard. This device has a very high accuracy of 0.005 V and a temperature drift of 1 ppm/°C.

Analog Devices offers a wide range of both zener-based and bandgap precision references as part of its line of data conversion products. The zener-based AD688 is a high-precision +10 and −10 V tracking reference. This device includes the basic reference cell and three additional amplifiers. The amplifiers are laser trimmed for low offset and low drift and maintain the accuracy of the reference. Low initial error and low temperature drift give the AD688 reference absolute ±10 V accuracy performance in monolithic form.

The AD689, an 8.192 V reference, bridges the gap between 5 V and 10 V products. This device is especially useful in data conversion circuits that operate over ±12 V but may swing over a 10% range.

The MAX2700 series (MAX2700/2701/2710) of 10 V references finds typical applications in high-resolution A/D and D/A systems and in data acquisition systems. The MAX2701 in this family is a −10 V reference.

The LTZ1000 from Linear Technology is an ultrastable reference operating at 7.2 V. This includes a heater resistor for temperature stabilization and a temperature sensing transistor, which results in very good temperature stability. Typical applications of this device are in voltmeters, calibrators, standard cells, scales, and low-noise RF oscillators (Linear Technology Corp., 1990).

TABLE 1-1   Illustrative Zener-Based References

Device	Voltage (V)	Operating Current	Accuracy	TC	Noise μV P-P (0.1–10 Hz)
REF101 (Burr-Brown)	10	6 mA max	5 mV	1 ppm/°C	6
AD688 (Analog Devices)	+10 and −10 (tracking)	9 mA	5 mV	2 ppm/°C	6
MAX2701 (Maxim Integrated Circuits)	−10	5 mA	2.5 mV	3 ppm/°C	6
LTZ1000 (Linear Technology)	7.2	5 mA	4 mV	0.05 ppm/°C	1.2
LT1021	5,7,10	0.8 mA	0.05%	2 ppm/°C	3

The LT1021 is a precision reference available in three voltages: 5, 7, and 10. These devices are intended for circuits requiring a precise 5 V or 10 V reference with very low initial tolerance.

1.2.4.2 Bandgap References

Some illustrative bandgap references are illustrated in Table 1-2.

Typical of the lower-cost, general-purpose bandgap references is the LM136 series from National Semiconductor. The LM136 and 336 are bandgap references with an output voltage of 2.5 V and an accuracy of 1–2%. These are particularly useful in obtaining a stable reference from a 5 V logic supply. Typical applications of this series are in digital voltmeters, power supply monitors, and the like (Linear Technology Corp., 1990). The REF-03 from Precision Monolithics is a low-cost, 2.5 V bandgap reference. Silicon General's SG103 series of bandgap references is available in 13 voltage ratings, ranging from 1.8–5.6 V. The LT1019 from Linear Technology is an accurate bandgap reference, available in voltage ratings of 2.5, 4.5, 5, and 10 V. Applications for this device include A/D and D/A converters and precision regulators (Linear Technology Corp., 1990).

Maxim Integrated Circuits produces a wide range of references. One such series, the MAX676/677/678 produces +4.096, 5, and 10 V calibrated, low-drift precision voltage references. One feature of the 4.096 low-dropout reference is that it operates from a 5 V ±10% supply (Maxim Integrated Circuits, 1995). This series of references has excellent line and load regulation in addition to temperature stability. These devices find applications in high-resolution 16-bit A/D and D/A converters, precision test and measurement systems, high-accuracy transducers, and as calibrated voltage reference standards.

Micro-power voltage references, which consume as little as 10 µA operating current, are available, unlike zener-based references, which consume much

TABLE 1-2 Illustrative Bandgap References

Device	Voltage (V)	Operating Current	Accuracy	Temperature Stability	Noise (Typical)
LM136	2.5	400 µA- 10 mA	1%	12 mV max from −55 to +125°C	NA
REF1004-1.2	1.2	10 µA min	4 mV	20 ppm/°C	60 µV RMS (10 Hz–10 kHz)
LT1019-2.5	2.5	0.5 mA	0.02%	5 ppm/°C	ppm P-P (0.1–10 Hz)
LT1034-1.2	1.2	10 µA	1%	20 ppm/°C	4 µV P-P (0.1–10 Hz)
MAX675	5.0	1.4 mA max	0.15	12 ppm/°C max	10 µV P-P (0.1–10 Hz)
MAX872	2.5	10 µA max	0.2	40 ppm/°C max	60 µV P-P (0.1–10 Hz)
MAX676/767/768	4.096, 5, 10	10 mA max	0.02	1 ppm/°C max over −40 to +85°C	1.2 µV P-P (0.1–10 Hz)
LM385-2.5	2.5	8 µA min	1	20 ppm/°C	120 µV RMS (10 Hz–10 kHz)

larger currents. One such example is MAX872. Another micro-power reference, the MAX6120, draws a maximum current of 70 µA and operates over a 2.4–11 V input range. This is ideally suited for battery-powered systems and portable applications such as data acquisition systems (Maxim Integrated Circuits, 1996). The LM385 series of micro-power precision references operate at currents in the range of 15–20 µA.

The LT1034 micro-power precision reference from Linear Technology combines a 1.2 or 2.5 V bandgap reference with a 7 V zener-based auxiliary reference in a single package. The 1.2 V/2.5 V reference is a trimmed, bandgap voltage reference with 1% initial tolerance and guaranteed 20 ppm/°C temperature drift. Operating on only 10 µA, the LT1034 offers guaranteed drift and good long-term stability. The 7 V reference is a subsurface zener device for less demanding applications (Linear Technology Corp., 1990). The REF1004-1.2 and REF1004-2.5 are two terminal micro-power bandgap references designed for high accuracy with outstanding temperature characteristics at low operating currents. The REF1004 is a cost-effective solution when reference voltage accuracy, low power, and long-term temperature stability are required (Burr-Brown, 1993).

1.2.5 Design Basics

Some basic design tips as well as some facilities available in voltage reference ICs are illustrated in Figure 1-5, using the MAX873 as an example (Maxim Integrated Circuits, 1994). Figure 1-5(a) shows a typical application circuit with

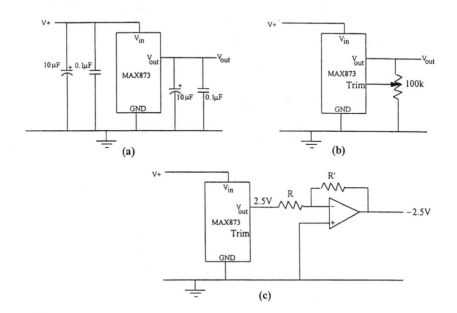

FIGURE 1-5 Designing with voltage references: (a) A basic circuit, (b) An output voltage trimming circuit, (c) Generating a negative reference. (Reproduced with permission from Maxim Integrated Circuits.)

input and output bypassing for best transient performance. Figure 1-5(b) shows an output voltage trimming circuit. Although large adjustments of the output voltage may degrade its temperature performance, adjusting the output over a small range about the nominal output voltage is possible with most reference ICs.

The generation of negative reference voltages is shown in Figure 1-5(c). An op amp in an inverting configuration is used, and the accuracy of the output depends on the matching of the two resistors R and R'.

1.3 Linear Regulators

1.3.1 Linear Regulator Fundamentals

Figure 1-6 illustrates the basic elements of a linear regulator. The output is regulated by controlling the voltage drop across the series-pass element, a power transistor biased in the linear region. The output voltage is maintained at a constant level by changing the voltage drop across this device.

The control circuit detects the output voltage, and changes the on-resistance of the series-pass power transistor by changing its base current to keep the output voltage constant. The power dissipation in the linear regulator is a function of the difference between the input and the output voltages, output current, output driver power, and the quiescent controller power. The power dissipation in the series-pass device contributes largely to lower the efficiency of linear regulators compared to switching regulators. However, this disadvantage is insignificant in low-dropout linear regulators, which find many applications in today's sophisticated electronic and communication equipment.

A major advantage of linear regulators in comparison with switching regulators is their low noise.

1.3.1.1 The Series-Pass Device

The power device selected to provide the pass function must be capable of operating under very low differential input/output voltages while providing

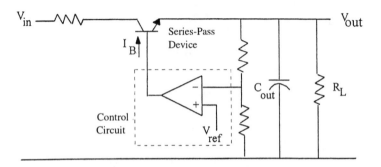

FIGURE 1-6 The basic elements of a linear regulator

reasonable efficiency. Pass devices typically are bipolar transistors or power MOSFETs.

The first linear regulators had NPN Darlington transistors as the series-pass element. However, for low-dropout requirements, PNP transistors are more suitable, as they can maintain output regulation with very little voltage drop across it (Lee, 1989; Simpson, 1996). The dropout voltage of the linear regulator is defined as the input-output voltage differential at which the circuit ceases to regulate against further reduction in input voltage (National Semiconductor Corp., 1987).

As the output requirements of the regulator grow, the gain of suitable PNP power transistors decrease, resulting in excessive base current losses. Therefore, N-channel MOSFETs are a popular choice due to their low drive current, low on-resistance and cost. Recent advances in semiconductor technology have resulted in low on-resistance P-channel devices as well. The low drive current requirement of MOSFETs reduces the quiescent current of the regulator considerably which is a major advantage of these devices. The characteristics of the series-pass device determine what the differential input/output voltage limitations are and how much quiescent power is required by the regulator. Figure 1-7 shows the use of NPN Darlington and PNP transistors as the series-pass element in linear regulators. A comparison of NPN and PNP transistors with several improvements for linear regulators is found in Lee (1989).

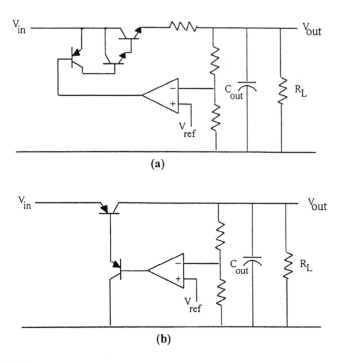

FIGURE 1-7 The basic linear regulator (a) With an NPN Darlington transistor and (b) With a PNP series-pass transistor

1.3.1.2 The Control Circuit

The control circuit samples the output voltage through a resistive divider and uses this feedback signal to control an error amplifier. Here, the regulator output is locked at a constant voltage that is a multiple of the reference voltage as determined by the voltage divider.

Control circuit characteristics directly affect system bandwidth and the achievable DC regulation. The voltage reference is used for comparison of the output voltage in the control circuit and primarily governs the steady-state accuracy of the device.

1.3.1.3 The Output Capacitance

The bulk capacitance maintains the output during transients. The output capacitor is required for the design to meet the specified transient requirements. As with any control system, the voltage loop has a finite bandwidth and cannot respond instantaneously to a change in load conditions. The supply rail for many of today's microprocessors cannot vary more than ±100 mV while handling load transients on the order of 5 A with 20-ns rise and fall times; that is, current slews at 250 A/µs (Goodenough, 1996).

To keep the output voltage within the specified tolerance, sufficient capacitance must be provided to source the increased load current throughout the initial portion of the transient period. During this time, charge is removed from the capacitor and its voltage decreases until the control loop can catch the error and correct for the increased current demand. The amount of capacitance used must be sufficient to keep the voltage drop within specifications. Design considerations in the selection of the capacitor value are detailed in O'Malley (1994).

1.3.2 Linear Regulator ICs

The linear regulator dates back to 1969. The first IC regulators, such as the LM340 or LM317, were NPN devices. Since then, many advances in technology have improved the performance of linear regulators. Regulators are available in a wide output power range. Many additional features, such as reverse-current/overcurrent/overvoltage protection, dual mode (fixed or adjustable) operation, multiple output capabilities, thermal overload protection, and advanced control techniques, have been incorporated into linear regulator ICs since then.

Controller ICs also have been developed, which, together with external pass devices, can be used to implement linear regulators.

The basic parameters of a linear regulator are its accuracy, output current, efficiency, and the dropout voltage. Superior performance with respect to these parameters as well as low quiescent current, wide input range, and fast transient response is essential in today's applications. Special design techniques are used to develop regulators to suit particular environments such as battery-powered equipment, microprocessors, and automotive applications. National Semiconductor, Motorola, Maxim Integrated Circuits, Unitrode, Linear Technology, and Analog Devices are among the major companies producing linear regulators.

General purpose linear regulator ICs as well as those with special features such as high power output, high output current, and low-dropout voltage are available to suit a wide variety of requirements. Adjustable output as well as advanced features such as shutdown facilities to turn off all bias currents, thermal overload protection to limit the overall power dissipation in the device, and current limiting facilities are available.

1.3.2.1 General Purpose Linear Regulators

The LM123 is an example of a general purpose linear regulator, providing 5 V at 3 A, and 30 W output power (Linear Technology Corp., 1990). This and equivalent three-terminal regulators having NPN-Darlington pass transistors commonly are found in on-card regulators, laboratory supplies, and instrumentation supplies.

An example of a high-power linear regulator is the LT1038, a three-terminal, bipolar, adjustable voltage regulator capable of providing current in excess of 10 A over the 1.2–32 V range. This high-power device typically is used in battery chargers and system power supplies (Linear Technology Corp., 1990). The output voltage is adjusted by external resistors.

The LT1036 is a logic-controlled dual linear regulator, one providing 12 V at 4 A and the other 5 V at 75 mA. This device is under the control of a logic shutdown signal.

The LM137/LM337 are adjustable negative regulators, delivering up to 1.5 A of output current over an output voltage range of −1.2 V to −37 V. Table 1-3 compares these regulators.

1.3.2.2 Low-Dropout Linear Regulators

MAX603/604 are dual mode regulators providing either 5 V/3.3 V fixed or adjustable output. Adjustable output from 1.25–11 V may be obtained using external resistors. The P-MOSFET limits quiescent currents to as low as 35 μA. This provides several advantages over similar designs using PNP pass transistors, including longer battery life. A functional block diagram of the MAX603/604 is shown in Figure 1-8(a). Its operation as an adjustable reference is shown in Figure 1-8(b) (Maxim Integrated Circuits, 1996). Typical applications of these

TABLE 1-3 General Purpose Linear Regulators

Device	Manufacturer	Output Voltage	Output Current
LM337	National Semiconductor	−1.2 to −37 V adjustable	1.5 A
LT0136	Linear Technology	12 V 5 V	3 V 75 mA
LT1038	Linear Technology	1.2–32 V adjustable	10 A
LM123	National Semiconductor	5 V	3 A

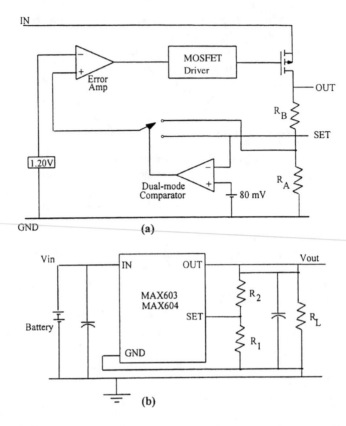

FIGURE 1-8 MAX603/604: (a) Functional block diagram, (b) MAX603/604 in the adjustable mode. [Reproduced with permission from Maxim Integrated Circuits, Inc.]

devices primarily are in battery-powered devices, pagers and cellular phones, and solar-powered instruments.

The ADP330X is a family of precision micro-power low-dropout regulators from Analog Devices. The ADP3302 contains two fully independent regulators. Typical applications of this device are in cellular phones, notebook computers, and portable instruments. MAX687 is a high-accuracy ($\pm 2\%$) linear regulator controller that directly drives high-gain external PNP transistors. The output current can exceed 1 A with a minimum drive current of 10 mA. It has dropout voltages of 40 mV (at 200 mA output current) and 0.8 V (at 4 A output current).

An LDO controller capable of handling ultrafast current transients is the LT1575 from Linear Technology. This device, along with a discrete N-MOSFET is ideally suited for powering today's microprocessors such as the Pentium. The nine versions of the LT1575 range from an adjustable-output controller to controllers with fixed outputs of 1.5, 2.8, 3.3, 3.5, and 5 V (Simpson, 1996). Very low-dropout voltages can be obtained, depending on the external MOSFET's on-resistance.

TABLE 1-4 Low-Dropout Linear Regulators

Device	Manufacturer	Output Voltage	Output Current	Dropout Voltage
MAX603/604	Maxim Integrated Circuits	5/3.3 V or adjustable	500 mA	550 mV max
ADP3302	Analog Devices	3, 3.2, 3.3, 5 V	100 mA	120 mV
MAX687	Maxim Integrated Circuits	3.3 V	>1 A with high β transistor	40 mV typically at 200 mA output current
LT1575	Linear Technology	1.5–5 V fixed or adjustable	1 mA	Very low, depends on external N-MOSFET's on-resistance

The UC3833 from Unitrode is described as a linear regulator controller suited for low-dropout, high-current regulators with a high transient response. This device allows the use of a variety of bipolar and MOSFET power devices. The crux of the design lies in the selection of a pass device. The design of a 3.3 V, 4 A regulator suited for today's microprocessor power supplies is described using this IC with a P-MOSFET in National Semiconductor's *Linear Data Book 1* (1987).

Table 1-4 compares these regulators.

1.4 Switching Regulators

1.4.1 Switching Regulator Fundamentals

Although the linear regulator is a mature technology, due to its low efficiency and other associated disadvantages, this type of power supply tends to be unfit for most of today's compact electronic systems.

The disadvantages of the linear regulator are greatly reduced by the switching regulator. In this technology, the AC line voltage is directly rectified and filtered to produce a raw high-voltage DC. This in turn is fed into a switching element that operates at a high frequency, 20 kHz to 1 MHz, chopping the DC voltage into a square wave. The square wave then is filtered to produce a DC output. The input/output relationship of this DC/DC converter is directly related to the duty cycle of the chopping signal. Regulation is achieved by sampling the output, comparing it with a reference, and modifying the duty cycle of the chopping waveform to compensate for any drifts.

Today, most switchers operate well above 500 kHz, with new magnetics, resonant techniques, and surface mount technology extending this to several MHz. Therefore, the associated components such as transformers and capacitors are much smaller than for linear regulators. In addition, due to the lower power loss, smaller heat sinks may be used. Therefore, the overall size of a switching regulator is smaller than an equivalent linear regulator.

The recent rapid advancement of microelectronics has created a necessity for the development of sophisticated, efficient, lightweight power supplies that have a high power-to-volume (W/in^3) ratio, with no compromise in performance. High-frequency switching power supplies, able to meet these demands, have become the prime power source in a majority of modern electronic systems. The combination of high efficiency and relatively small magnetics results in compact, lightweight switching regulators, with power densities in excess of 100 W/in^3 (Travis, 1996; Goodenough, 1995b) versus 0.3 W/in^3 for linear regulators.

However, a major design concern in such high-frequency switching power supplies is minimization of the EMI pollution generated.

1.4.1.1 Modes of Operation

The DC to DC converter has two major operational modes for switching power supplies: the forward mode and the flyback mode. Although they have only subtle differences between them with respect to component arrangement, their operation is significantly different and each has advantages in certain areas of application.

Forward Mode Converters

Figure 1-9(a) shows a simple forward mode converter. This type of converter can be recognized by an L-C filter section, directly after the power switch (a

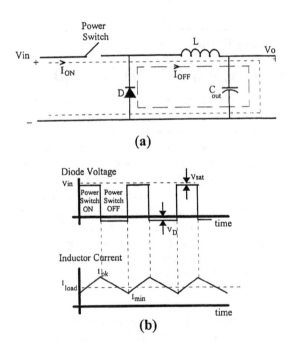

(a)

(b)

FIGURE 1-9 The forward mode converter: (a) The basic circuit, (b) Associated waveforms

power transistor or power MOFSET operating between fully conducting and cutoff modes) or after the output rectifier on the secondary of a transformer.

The operation of the converter can be seen by breaking its operation into two periods:

- *Power switch on period.* When the power switch is on, the input voltage is presented to the input of the L-C section and the inductor current ramps upward linearly. During this period the inductor stores energy.
- *Power switch off period.* When the power switch is off, the voltage at the input of the inductor flies below ground since the inductor current cannot change instantly. Then the diode becomes forward biased. This continues to conduct the current that was formally flowing through the power switch. During this period, the energy that was stored in the inductor is dumped onto the load. The current waveform through the inductor during this period is a negative linear ramp. The voltage and current waveforms for this converter are shown in Figure 1-9(b). The DC output load current value falls between the minimum and the maximum current values and is controlled by the duty cycle. In typical applications, the peak inductor current is about 150% of the load current and the minimum is about 50%.

The advantages of forward mode converters are that they exhibit lower output peak-to-peak ripple voltages and they can provide high levels of output power, up to kilowatts.

Flyback Mode Converters

In this mode of operation, the inductor is placed between the input source and the power switch, as shown in Figure 1-10. This circuit also is examined in two stages:

- *Power switch on period.* During this period, a current loop including the inductor, the power switch, and the input source is formed. The inductor current is a positive ramp, and energy is stored in the inductor's core.
- *Power switch off period.* When the power switch turns off, the inductor's voltage *flies back* above the input voltage, resulting in forward biasing of the diode. The inductor voltage then is clamped at the output voltage. This voltage, which is higher than the input voltage, is called the *flyback voltage*. The inductor current during this period is a negative ramp.

In Figure 1-10, the inductor current does not reach zero during the flyback period. This type of a flyback converter is said to operate in the *continuous mode*. The core's flux is not completely emptied during the flyback period, and a residual amount of energy remains in the core at the end of the cycle. Accordingly, there may be instability problems in this mode. Therefore, the *discontinuous mode* is the preferred mode of operation for flyback mode converters. The voltage and current waveforms are shown in Figure 1-11.

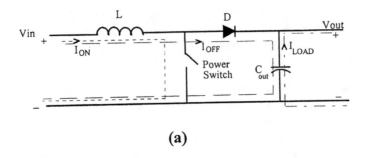

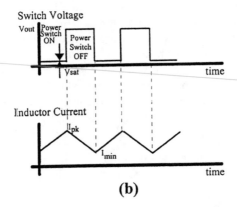

(a)

(b)

FIGURE 1-10   The flyback mode converter: (a) The basic circuit, (b) Associated waveforms

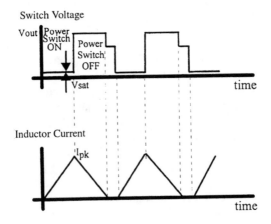

FIGURE 1-11   Voltage and current waveforms for the discontinuous flyback mode converter

The only storage for the load in the flyback mode of operation is the output capacitor. This makes the output ripple voltage higher than in forward mode converters. The power output is lower than in forward mode converters, owing to the higher peak currents generated when the inductor voltage flies back. As they consist of the fewest number of components, they are popular in low- to medium-power applications.

1.4.1.2 A Simplified Analysis of DC/DC Converters

The input/output characteristics of all DC/DC converters can be examined by using the requirement that the initial and the final inductor currents within a cycle should be the same for steady-state operation; that is, the net energy storage within one switching cycle in each inductor should be 0. This leads to the volt-second balance for the inductor, which means that the average voltage per cycle across the inductor must be 0: that is, the volt-second products for the inductor during each switching cycle should sum to 0 (Brown, 1990; Bose, 1992). This can be illustrated using a typical inductor current waveform, as shown in Figure 1-12.

Let the positive slope of the current be m_1 and the negative slope m_2. For the initial and final currents to be the same,

$$m_1 T_{on} - m_2 T_{off} = 0 \qquad (1.5)$$

Since the voltage across an inductor L is given by $L\, di/dt$,

$$V_{Lon} = Lm_1 \qquad (1.6)$$

$$V_{Loff} = -Lm_2 \qquad (1.7)$$

Combining (1.5), (1.6), and (1.7),

$$V_{Lon} T_{on}/L + V_{Loff} T_{off}/L = 0 \qquad (1.8)$$

Equation (1.8) can be written as a function of the duty cycle D:

$$D = T_{on}/(T_{on} + T_{off}) \qquad (1.9)$$

Combining (1.8) and (1.9),

$$V_{Lon} D + V_{Loff}(1 - D) = 0 \qquad (1.10)$$

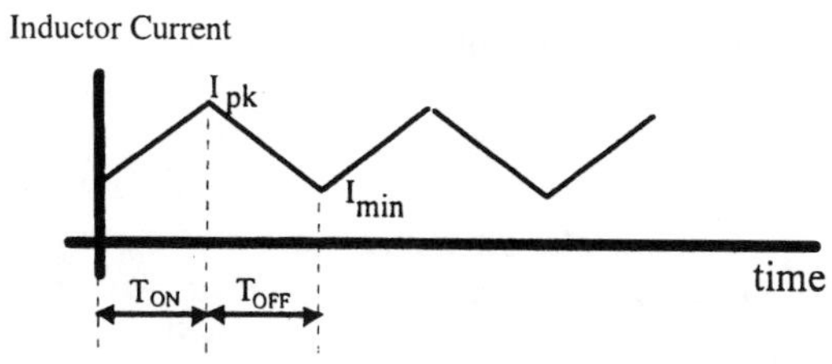

FIGURE 1-12 A typical inductor current waveform

where V_{Lon} and V_{Loff} are the voltages across the inductor during the switch on and switch off periods, respectively, and D is the duty cycle.

The output voltage of forward mode converters is given approximately by $V_{out} \approx DV_{in}$, where D is the duty cycle of the switching waveform. Hence, this mode of operation always performs a *step-down* operation.

The output voltage for the flyback mode of operation is given by $V_{out} = V_{in}/(1 - D)$. Hence, flyback mode converters always are used as *step-up* converters.

1.4.2 Converter Topologies

The converter *topology* is the arrangement of components within the converter. Converter topologies fall into two main categories: transformer-isolated and non-transformer-isolated. Each category contains several topologies, with some available in both forms. An excellent discussion on converter topologies is found in Bose (1992).

1.4.2.1 Non-Transformer-Isolated Converter Topologies

This type of switching converter is used when some external component, such as a 50–60 Hz transformer or bulk power supply, provides the DC isolation and protection. These are simple, hence easy to understand and design. However, they are more prone to failure due to the lack of DC isolation. Therefore, these are used by designers mostly in situations such as distributed power systems, where a bulk power supply provides the necessary DC isolation.

There are four basic non-transformer-isolated topologies, the buck (or step down), the boost (or step up), the buck-boost (or inverting), and the Cük converters. Each of these types is described in detail by Linear Technology Corporation (1990, 1992), Datel Inc. (1991), Siliconix Inc. (1994a), Motorola Inc. (1987), and Unitrode Integrated Circuits (1995–96).

1.4.2.2 Transformer-Isolated Converter Topologies

In non-transformer-isolated converter topologies, only semiconductors provide DC isolation from input to output. Transformer-isolated switching converters rely on a physical dielectric barrier to provide galvanic isolation. Not only can this withstand very high voltages before failure, it also provides a second form of protection in the event of a semiconductor failure. Another advantage is the ease of adding multiple outputs to the power supply without separate regulators for each. These features make transformer-isolated topologies preferred by designers.

The transformer-isolated category includes both forward and flyback mode converter topologies. The isolation transformer now provides a step-up or step-down function. The topologies are called *isolated forward, flyback, push-pull, half-bridge*, and *full-bridge*. Each of these types is described in detail by Linear Technology Corp. (1990, 1992), Datel Inc. (1991), Siliconix Inc. (1994a), Motorola Inc. (1987), and Unitrode Integrated Circuits (1995–96).

1.4.2.3 Selection of a Converter Topology

The converter topology has a major bearing on the conditions in which the power supply can operate safely and the amount of power it can deliver. Cost versus performance trade-offs also are needed in selecting a suitable converter topology for an application.

The primary factors that determine the choice of topology are whether DC isolation is needed, the peak currents and voltages to which the power switches are subjected, the voltages applied to transformer primaries, the cost, and the reliability.

Figure 1-13 illustrates the approximate range of usage for these topologies. The boundaries to these areas are determined primarily by the amount of stress the power switches must endure and still provide reliable performance. The boundaries delineated in Figure 1-13 represent approximately 20 A of peak current.

The flyback configuration is used predominantly for low- to medium-output power (150 W) applications because of its simplicity and low cost. Unfortunately, the flyback topology exhibits much higher peak currents than the forward-mode supplies. Therefore, at higher output power, it quickly becomes an unsuitable choice. For medium-power applications (100–400 W) the half-bridge topology becomes the predominant choice. The half-bridge is more complex than the flyback and therefore costs more. However, its peak currents are about one-third to one-half those exhibited by the flyback. Above 400 W, the dominant topology is the full-bridge, which offers the most effective utilization of the full capacity of

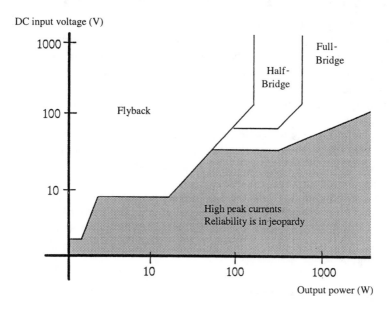

FIGURE 1-13 Industry favorite configurations and their areas of usage. [Source: Brown, 1990.]

the input power source. It also is the most expensive to build, but for those power levels, the additional cost becomes a trivial matter. Above 150 W, the push-pull topology sometimes is used. However, this exhibits some shortcomings, such as core imbalance, that may make it tricky to use (Brown, 1994).

Estimating the major power supply parameters at the beginning of the design and using charts such as Figure 1-13 will lead to the selection of a final topology that is safe, reliable, and cost effective.

Some relative merits of the converter topologies as well as their advantages, disadvantages, and typical applications are summarized in Table 1-5. Mathematical expressions for estimating some of the important parameters, such as peak currents and voltages and output power, are found in Brown (1990, 1994). A comparison of the strengths and weaknesses of the different converter topologies also is found in Moore (1994).

The industry has settled into several primary topologies for a majority of the applications. Due to reasons discussed before, transformer-isolated topologies have become more popular than non-transformer-isolated topologies.

1.4.3 Control Techniques

Two basic methods of control are used in switching regulators: pulse width modulation (PWM) and resonant. In both techniques, the output voltage is sampled and the switching of the transistors is modified in some manner to keep the output constant.

1.4.3.1 Pulse Width Modulation Control Techniques

In a PWM switch mode power supply, a square wave pulse normally is generated by the control circuit to drive the switching transistor on and off. By varying the width of the pulse, the conduction time of the transistor correspondingly is varied, regulating the output voltage. The major function of the control subsection of a PWM supply therefore is to sense any change in the DC output voltage and adjust the duty cycle of the power switches to correct for such changes.

An oscillator sets the basic frequency of operation of the power supply. A stable, temperature-compensated reference is used, to which the output voltage is compared in a high-gain voltage error amplifier. An error voltage-to-pulse width converter is used to adjust the duty cycle.

The PWM control circuit may be *single ended* for driving single-transistor converters, such as the buck or boost topology, or it may be *double ended* to drive multiple-transistor converters, such as the push-pull or half-bridge topology.

Two basic modes of control are used in PWM converters: voltage mode and current mode.

Voltage-Mode Control

This is the traditional mode of control in PWM switching converters. It also is called *single-loop control*, as only the output is sensed and used in the

TABLE 1-5 Comparison of Converter Topologies

Topology	V_{in}/V_{out}	Advantages	Disadvantages	Typical Application Environment	V_{in} (dc) Range (V)	Typical Efficiency (%)	Relative Parts Cost
Buck	D	High efficiency, simple, low switch stress, low ripple	No isolation, potential overvoltage if switch shorts	Small-sized, embedded systems	5–1000	78	1.0
Boost	$1/(1-D)$	High efficiency, simple, low input ripple current	No isolation, high switch peak current, regulator loop hard to stabilize, high output ripple, unable to control short-circuit current	Power-factor correction, battery up-converters	5–600	80	1.0
Buck-boost	$-D/(1-D)$	Voltage inversion without a transformer, simple, high-frequency operation	No isolation, regulator loop hard to stabilize, high output ripple	Inverse output voltages	5–600	80	1.0
Flyback	N_2D/N_1 $(1-D)$	Isolation, low parts count, has no secondary output inductors	Poor transformer utilization, high output ripple, fast recovery diode required	Low output power, multiple output	5–500	80	1.2

continued overleaf

TABLE 1-5   Continued

Topology	V_{in}/V_{out}	Advantages	Disadvantages	Typical Application Environment	V_{in} (dc) Range (V)	Typical Efficiency (%)	Relative Parts Cost
Push-pull	$2N_2D/N_1$	Isolation, good transformer utilization, good at low input voltages, low output ripple	Cross-conduction of switches possible, high parts count, transformer design critical, high voltage required for switches	Low output voltage	50–1000	75	2.0
Half-bridge	N_2D/N_1	Isolation, good transformer utilization, switches rated at input voltage, low output ripple	Poor transient response, high parts count, cross-conduction of switches possible	High input voltage, moderate to high power	50–1000	75	2.2
Full-bridge	$2N_2D/N_1$	Isolation, good transformer utilization, switches rated at input voltage, low output ripple	High parts count, cross-conduction of switches possible	High power, high input voltage	50–1000	73	2.5

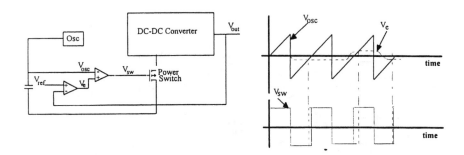

FIGURE 1-14 Voltage-mode control: (a) Block diagram, (b) Associated waveforms

control circuit. A simplified diagram of a voltage-mode control circuit is shown in Figure 1-14.

The main components of this circuit are an oscillator, an error amplifier, and a comparator. The output voltage is sensed and compared to a reference. The error voltage is amplified in a high-gain amplifier. This is followed by a comparator, which compares the amplified error signal with a sawtooth waveform generated across a timing capacitor.

The comparator output is a pulse width modulated signal that corrects any drift in the output voltage. As the error signal increases in the positive direction, the duty cycle is decreased; and as the error signal increases in the negative direction, the duty cycle is increased.

The voltage mode control technique works well when the loads are constant. If the load or the input changes quickly, the delayed response of the output poses a drawback to the control circuit, as it senses only the output voltage. Also, the control circuit cannot protect against instantaneous overcurrent conditions on the power switch. These drawbacks are overcome in current mode control.

Figure 1-15 shows a functional block diagram of the commonly used TL494 voltage mode controller. Control for both push-pull and single-ended operation can be achieved with this chip. The frequency of the oscillator is set by the external resistors R_T and C_T. The typical operating frequency is 40 kHz, and the maximum is 200 kHz.

Output pulse width modulation is accomplished by comparison of the positive sawtooth waveform across the capacitor to either of two control signals. The NOR gates, which drive the output transistors, are enabled only when the flip-flop clock input is in its low state. This happens only during that portion of time when the sawtooth voltage is greater than the control signals. Therefore, an increase in control signal amplitude causes a corresponding linear decrease of output pulse width.

Current-Mode Control

This is a *multiloop* control technique, which has an AC current feedback loop in addition to the voltage feedback loop. The second loop directly controls

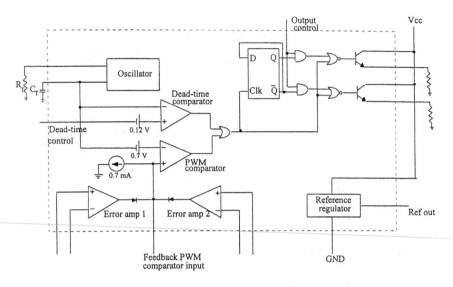

FIGURE 1-15 A functional block diagram of the TL494 voltage-mode controller IC. (Reproduced by permission of Motorola Inc.)

the peak inductor current with the error signal rather than controlling the duty cycle of the switching waveform. Figure 1-16 shows a block diagram of a basic current mode control circuit.

The error amplifier compares the output to a fixed reference. The resulting error signal then is compared with a feedback signal representing the switch current in the current sense comparator. This comparator output resets a flip-flop that is set by the oscillator. Therefore, switch conduction is initiated by the oscillator and terminated when the peak inductor current reaches the threshold level established by the error amplifier output. Thus, the error signal controls the peak inductor current on a cycle-by-cycle basis. The level of the error voltage dictates the maximum level of peak switch current. If the load increases, the voltage error amplifier allows higher peak currents. The inductor current is sensed through a ground-referenced sense resistor in series with the switch.

The disadvantages of this mode of control are loop instability above 50% duty cycle, less than ideal loop response due to peak instead of average current sensing, and a tendency toward subharmonic oscillation and noise sensitivity, particularly at very small ripple current.

However, with careful design as explained by Brown (1990), Cryssis (1989), and others, these disadvantages can be overcome. Therefore, current-mode control becomes an attractive option for high-frequency switching power supplies.

Figure 1-17 shows a functional block diagram of the MAX747 current mode controller from Maxim Integrated Circuits (1994). This is a CMOS step-down controller, which drives external P-channel FETs. The IC operates in a continuous mode under heavy loads but in a discontinuous mode at light loads. Stability

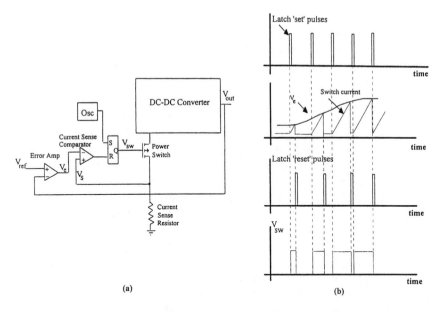

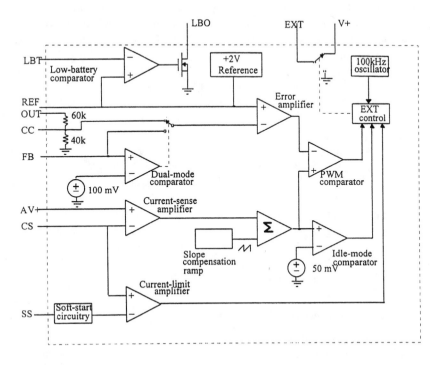

FIGURE 1-16 Current-mode control: (a) Block diagram, (b) Associated waveforms

FIGURE 1-17 A functional block diagram of the MAX747 current-mode controller IC. (Reproduced by permission of Maxim Integrated Circuits.)

of the inner current-feedback loop is provided by a slope-compensation scheme that adds a ramp signal to the current-sense amplifier output. The switching frequency nominally is 100 kHz and the duty cycle varies from 5 to 96%, depending on the input/output voltage ratio. EXT provides the gate drive for the external FET.

1.4.3.2 Resonant Control Techniques

The term *resonance* refers to a continuous sinusoidal signal. Resonant converters process power in a sinusoidal form. Resonant techniques have long been used with thyristor converters, in high-power SCR motor drives, and in UPSs. However, due to its circuit complexity, it had not found application in low-power DC-to-DC converters until recently. With the development of surface mount technology, resonant forms of DC-to-DC converters have gained rapid acceptance recently, harnessing certain advantages inherent in resonant techniques. The thrust toward resonant mode power supplies has been fueled by the industry's demand for miniaturization, together with increasing power densities and overall efficiency, and low EMI.

With available bipolar devices and circuit technologies, PWM converters have been designed to operate with switching frequencies in the range 50–200 kHz. The advent of power MOSFETs enabled the switching frequencies to be increased to several MHz. However, increasing the switching frequency, although allowing for miniaturization, leads to increasing switching stresses and losses, which leads to a reduction in efficiency. The detrimental effects of the parasitic elements also become more pronounced as the switching frequency is increased. With quasi-resonant techniques, higher frequency as well as higher efficiency compared to PWM techniques are achieved.

Resonant circuits in power supplies operate in two modes that define the flow of current in the resonant circuit: continuous and discontinuous. In the continuous mode, the circuit operates either above or below resonance. The controller shifts the frequency either toward or away from resonance, using the slope of the resonant circuit's impedance curve to vary the output voltage. This is a truly resonant technique but is not commonly used in power supplies due to its high peak currents and voltages.

In the discontinuous mode, the control circuit generates pulses having a fixed on-time but at a varying frequency, determined by the load requirements. This mode of operation does not generate continuous current flow in the tuned circuit, is the common mode of operation in a majority of resonant converters, and is called the *quasi-resonant* mode of operation.

The Quasi-Resonant Principle

The quasi-resonant principle is used in power converters by incorporating a resonant LC circuit with the power switch. The power switch is turned on and off in the same manner as in PWM converters, but the tank circuit forces the current through the switch into a sinusoidal form. The actual conduction period

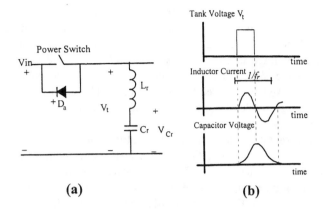

FIGURE 1-18   The resonant principle: (a) The basic circuit, (b) Associated waveforms

of the switch is governed by the resonant frequency f_r of the tank circuit. This basic principle is illustrated in Figure 1-18.

The tank circuit exhibits a relatively fixed ringing period to which the conduction period of the power switch is slaved.

The on-period of the power switch is fixed to the resonance period of the tank elements. The quasi-resonant supply is controlled by changing the number of on-times of the power switch per second.

All resonant control circuits keep the pulse width constant and vary the frequency, whereas all PWM control circuits keep the frequency constant and vary the pulse width. The main advantages of the quasi-resonance techniques arise from the sinusoidal wave shapes of the switching currents and voltages. The switching losses are reduced, leading to higher efficiency and greatly reduced EMI.

The resonant switch consists of a semiconductor switch and resonant LC elements. Because resonant circuits generate sinusoidal waves, designers can operate the power switches either at zero current or at zero voltage points in the resonant waveform. Based on this, there are two types of resonant switches: zero current switches (ZCS) and zero voltage switches (ZVS). The two types of switches are the duals of each other. A description of the operations of these switches is found in Brown (1990).

Comparable converter topologies as with the PWM technique are available with ZCS and ZVS quasi-resonant switching techniques. Circuits for each of these topologies are found in Brown (1990, 1994), Cük and Middlebrook (1994), Steigerwald (1984), and Cük and Maksimovic (1988).

Control Techniques

Control methods for resonant converters are variable frequency ones. Either the on-time is fixed and the off-time is variable or vice versa.

The basic functioning of a resonant mode control circuit is shown in Figure 1-19. The fundamental blocks are a wide-band error amplifier, a voltage-controlled oscillator (VCO), and a temperature-stable one-shot timer.

The output voltage is compared with the reference, and the error voltage is used to drive the VCO. The VCO output triggers the one-shot timer, whose pulse duration is fixed as required by the converter. These control techniques are based on the voltage mode of control and, hence, suffer from poor input transient response characteristics.

Some representative control ICs currently available are the MC34066, LD405, UC1860, and UC3860. One of the first resonant mode controller ICs to appear on the market was the LD405. Subsequently, the CS3805 by Cherry Semiconductors and an improved GP605 by Gennum Corp. were released.

The LD405 and the CS3805 have drive current capabilities of 200 mA and operating frequencies in the range 10 kHz–1 MHz. They have single-ended and complementary outputs for driving power MOSFETs. The dissipation rate is specified at 500 mW at 50°C. The operating range of the GP605 is between 1 kHz and 1.2 MHz.

Unitrode Integrated Circuits introduced its UC3860 family of resonant mode controller chips in 1988. These could supply 800 mA, about four times the drive current, and operate at higher frequencies (up to 2 MHz) than the previously available ICs. Their power dissipation rate is specified at 1.25 W at 50°C.

A newer series of resonant mode control chips by Unitrode Integrated Circuits and their applications are described in Wofford (1990). These are the UC3860, UC1861 (dual ZVS), UC1864 (single ZVS), and UC1865 (dual ZCS).

A functional block diagram of the UC1860 from Unitrode Integrated Circuits (1995–96) is shown in Figure 1-20. The nominal operating frequency for this IC is 1.5 MHz, and the device implements resonant mode, fixed on-time control as well as a number of other power supply control schemes with its various dedicated and programmable features. The IC contains dual high-current totem-pole output drivers that can be programmed to operate alternately or in unison (Unitrode Integrated Circuits, 1995–96).

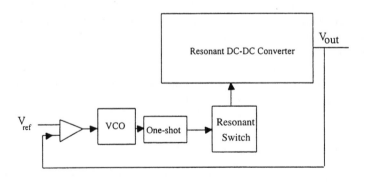

FIGURE 1-19 Resonant mode control

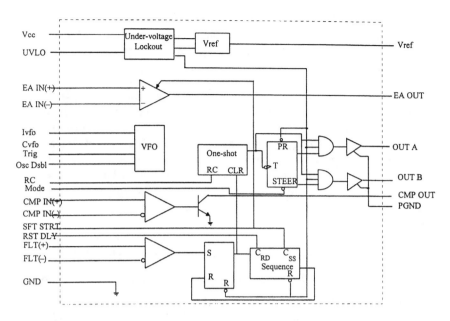

FIGURE 1-20   A functional block diagram of the UC1860 resonant mode controller IC. (Reproduced by permission of Unitrode Integrated Circuits.)

Although varying in detail and complexity, all these resonant mode control chips provide the same basic functions. All contain a VCO that varies the operating frequency, a monostable circuit that establishes the pulse on-time, and a steering circuit that determines the output drive mode (controlled on- or off-times, single-ended, or complementary). The chips also provide many of the same basic protection features such as soft start, undervoltage lockout, and overload protection — although in different ways and with varying degrees of sophistication. Some of these control ICs are also described by Linear Technology Corporation (1990, 1992), Datel Inc. (1991), Siliconix Inc. (1994a), Motorola Inc. (1987), and Unitrode Integrated Circuits (1995–96).

1.4.4 Special Application Requirements

1.4.4.1 Sub-5 V Applications

As IC process lithography becomes finer and the need to reduce overall power dissipation becomes more important, ICs demand lower and better-regulated power supply rails. Power supply rails for modern microprocessors and logic ICs have dropped below 5 V to a 3.3 V pseudo-standard and are fast heading toward sub-2 V and sub-1 V levels.

These low-voltage supply requirements also differ in voltage level from system to system and from version to version of the same IC. If Intel's Pentium processor is taken as an example, some versions just require 3.3 V while others require two supplies at 3.3 V and 3 V. Yet other versions require supplies such as 3.383 V and 3.525 V, being the optimum voltages for maximum speed in the respective versions. Furthermore, the Pentium Pro comes with a 4-bit operating voltage code, which selects 1 of 16 discrete voltages between 2 and 3.5 V (Travis, 1996; Goodenough, 1995b, 1995c, 1996).

Lower voltage levels, however, do not translate to low power. These multimillion-transistor ICs use CMOS technology, which at today's high operating frequencies consume substantial amounts of current although virtually no DC current. The power levels required therefore are rising. Top-of-the-line CPU power is approaching 100 W; and desktop CPU, power 30 W. Systems such as high-performance multiprocessor servers are approaching 400–600 W. The combination of increasing power requirements and reducing voltage rails requires the power source to provide and distribute very high currents within a system. For example, a 300 W system running off a 3 V supply must provide and distribute 100 A of current.

Furthermore, clock rates heading toward 400–600 MHz imply high current transients as devices come out of the sleep mode within a clock cycle of a few nanoseconds.

In addition to microprocessors, high-speed data buses, such as the 60 MHz GTL (Gunning Transceiver Logic) bus, are being recommended for interconnection of processors and peripherals on the motherboard. These require an active 1.2 or 1.5 V terminator at each end, each potentially capable of handling up to 7 A. Other data buses with similar requirements include Futurebus (2.1 V) and Rambus (2.7 V) (Goodenough, 1995c). Therefore, the challenges facing the power supply designer in the sub-5 V range can be summarized as follows:

- Multiple voltage rails.
- Tight tolerances.
- High efficiency while generating high current (low power loss).
- High accuracy.
- High current transients.

The basic design requirements in the face of these challenges are low power loss and higher gain and bandwidth in the control loop to handle tighter voltage regulation and transients.

1.4.4.2 Converters for Battery-Operated Equipment and Battery Chargers

The proliferation of portable computers and handheld communication devices has opened up another branch in DC/DC converter evolution. Users of these portable devices demand features such as compactness, lightness, and low power consumption for extended periods of operation. Corresponding trends in DC/DC converters emphasize excellent conversion efficiency and compactness.

Additional constraints placed on the DC/DC converter include minimum noise intrusion into communication and audio circuits in the device. The increase in the integration of high-speed modems and CD-ROM drives into portable computers has made power supply switching noise an important consideration.

A key requirement for designers of battery-powered products is that they minimize the number of cells used in the product. Products ideally should run off a few high-capacity cells to minimize size and cost. Tiny devices such as pagers run off one or, at most, two cells. So do telephones and PDAs. Converters operating off input sources as low as that of a single-cell alkaline battery (1 V) are needed. Therefore, the voltage required, in most instances, is greater than the voltage available from the cells. Furthermore, for extended battery life, the converter should be able to operate from waning batteries.

Some systems need to operate with input voltages that approach the output voltage. This low-dropout condition requires the DC/DC converter duty cycle to approach 100%. Furthermore, a waning battery may swing the input voltage from a value above the output to one below it.

High efficiency is another prime consideration in battery-operated equipment. This means not only increased operating time on a battery charge but also reduced heat that must be dissipated in or removed from the IC and the device. Low quiescent current is another requirement for extended battery charge.

Therefore, the challenges facing the power supply designer for battery-powered equipment can be summarized as follows:

- Low-dropout voltage.
- Extended use of battery charge.
- Small size.
- Low EMI.

1.4.5 Practical Design Approaches

This section describes some of the approaches taken in DC/DC converter and controller design to address the special requirements of sub-5 V and battery-powered applications. Illustrative examples of some successful solutions are presented.

1.4.5.1 Synchronous Rectification

In a conventional DC/DC converter such as the buck converter, typically an N-channel MOSFET with a low R_{DS}(on) is selected for the switch. However, the diode's forward voltage becomes a limiting factor in improving the converter's efficiency, as the output voltage drops. This has led to the design of synchronous converters; these replace the diode, which normally is a Schottky, with another N-channel MOSFET. This usually is called the *lower MOSFET*, and the switch is called the *upper MOSFET*. The lower MOSFET conducts current during the off-time of the upper MOSFET. Figure 1-21 shows a simplified diagram of a synchronous buck converter. In conventional synchronous converters, a single IC is used for PWM control and the synchronous drive of two external MOSFETs.

Newer process technologies take another approach, where a SynchroFET integrates the two MOSFETs, their drive circuits, and the synchronous control logic. This IC can be used with a conventional PWM control IC to design a

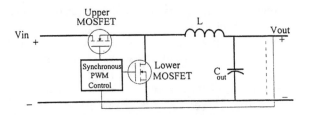

FIGURE 1-21 A synchronous buck converter

converter with features superior to those with discrete MOSFETs (Maxim Integrated Circuits, 1997). Also, this approach allows the SynchroFET to be paired with many different PWM controllers to achieve various performance trade-offs.

HIP5015 and HIP5016 are a widely used pair of SynchroFETs from Harris Semiconductor. A 5 V to 3.3 V DC/DC converter using a generic PWM controller and the HIP5015 SynchroFET is shown in Figure 1-22. This typically is used to derive 3.3 V from a 5 V input. The implementation of a two-output converter using these SynchroFET ICs is described in Goodenough (1996). The HIP5015/5016 is designed to run at over 1 MHz. The efficiency of a 5 to 3.3 V converter running at 400 kHz using this IC is reported to be 85% for load currents between 0.5 and 4.5 A and over 90% for load currents between 0.8 and 2.8 A (Goodenough, 1996). The newer HIP5020 synchronous buck converter (Harris Corp., 1997) is optimized for battery-powered systems with 4.5–18 V input. This buck converter can be operated up to a switching frequency of 1 MHz and demonstrates an efficiency of over 95%.

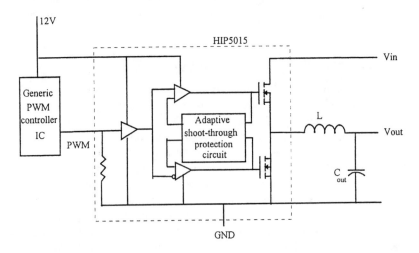

FIGURE 1-22 A PWM synchronous DC/DC converter using a SynchroFET. (Reproduced with permission from Harris Semiconductor.)

1.4.5.2 Increased Gain and Bandwidth Control Loop

In low-voltage synchronous converters, the preferred method of control is the voltage mode. The power dissipated in the current sense resistor in current mode control cuts the efficiency by about 2% and therefore is unsuitable for low-voltage, high-efficiency applications. However, the low bandwidth of the voltage mode control loop reduces the converter's response to dynamic load conditions.

Maxim's three PWM controllers, the MAX796/797/799 designed for portable computer and communication applications, are current mode controllers that drive synchronous rectifiers, primarily in the buck mode. Efficiencies as high as 97% are reported by Maxim Integrated Circuits (1997).

Figure 1-23 shows the MAX797 controller. The special features of this circuit are a proprietary PWM comparator for handling transients, a proprietary idle mode control scheme used at low-load conditions for extended battery life, and the reduction of PWM noise.

This series is designed for output voltages as low as 2.5 V but can provide 1.5 V with external circuitry. To provide the required DC accuracy at the low-output voltage while handling high-speed current transients, these devices require an external op amp in the error amplifier circuit, which operates as an integrator (Maxim Integrated Circuits, 1997).

The heart of this PWM controller is a multi-input open-loop comparator that sums the output voltage error with respect to the reference, the current sense signal, and a slope compensation ramp. This is of the direct summing type and lacks the traditional error amplifier with the associated phase shift. This direct summing configuration approaches the ideal of direct cycle-by-cycle control of the output voltage. This PWM comparator is shown in Figure 1-24.

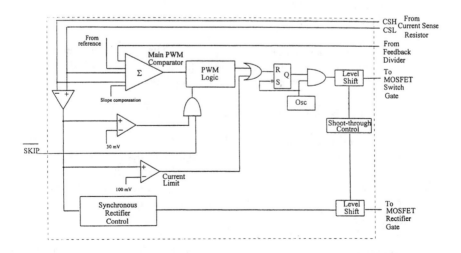

FIGURE 1-23 A block diagram of the MAX797 PWM controller. (Reproduced with permission from Maxim Integrated Circuits.)

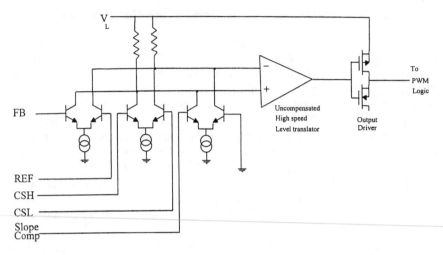

FIGURE 1-24 A block diagram of the main PWM comparator in the MAX797. (Reproduced with permission from Maxim Integrated Circuits, Inc.)

1.4.5.3 Idle Mode Control Scheme

The MAX797's approach for conservation of battery power uses the SKIP input shown in Figure 1-23. At light loads (SKIP = 0) the inductor current fails to exceed the 30 mV threshold set by the minimum current comparator. When this occurs, the minimum current comparator immediately resets the high-side latch at the beginning of the cycle unless the output voltage drops below the reference. This sends the controller into a variable-frequency idle mode, skipping most of the oscillator pulses, to cut back on gate-charge losses.

Operation at a fixed frequency, regardless of load conditions, however, is advantageous in terms of lowering PWM noise interference. Steps can be taken to remove the known emissions at fixed frequencies. Therefore, a low-noise mode is enabled in the MAX797 by making SKIP high. This forces fixed frequency operation by disabling the minimum current comparator.

1.4.5.4 Updated Voltage Mode Control

A newer solution by Linear Technology employs an advanced voltage mode control scheme in its LTC1430 controller. This controller, optimized for high-power, 5 V/3.x V applications, operates at an efficiency greater than 90% in converter designs from 1 A to greater than 50 A output current (Linear Technology Corp., 1995). The LTC1430 uses a synchronous switching architecture with two N-channel output devices.

A block diagram of the LTC1430 is shown in Figure 1-25. The primary control loop is a conventional voltage mode feedback loop. The load voltage is sensed across the output capacitor by the SENSE+ and SENSE− inputs, divided and fed to the feedback amplifier, where it is compared to a 1.26 V internal

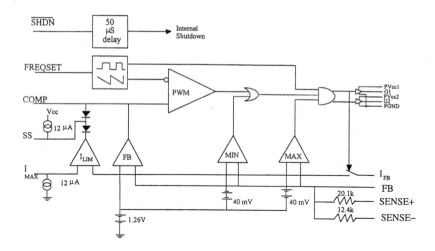

FIGURE 1-25 A block diagram of the LTC1430 controller. (Reprinted with permission by Linear Technology Corporation.)

reference. The error signal then is compared to a sawtooth waveform from the oscillator to generate a pulse width modulated switching waveform.

Two other comparators in the feedback loop, MIN and MAX, provide high-speed fault correction in situations where the feedback amplifier may not respond quickly enough. MIN compares the feedback signal to a voltage 40 mV (3%) below the reference, and MAX compares it to a voltage 3% above the reference. If the output falls below the minimum level, the MIN comparator overrides the feedback comparator and forces the loop to full duty cycle (about 90%). If the output rises above the 3% level, the MAX comparator forces the duty cycle to 0. These two comparators prevent extreme output perturbations in the presence of fast output transients.

Additionally, the controller senses output current across the drain source resistance of the upper N-channel FET, providing an adjustable current limit without an external sense resistor.

1.4.5.5 A Master/Slave Architecture

Power Trends' solution (Travis, 1996) is a master/slave architecture for DC/DC converter modules, which allows the increase of available current in 3 A steps. This design consists of a 5 V/3.3 V converter with a 3 or 8 A master coupled with three 3 A slaves. The slaves add 9 A of current to the master's current output. This approach is an attractive solution, which leaves room for future system upgrades.

1.4.5.6 Improved Process Technologies

Aiming at improved control loop bandwidth, Siliconix developed a high-speed CBiC/D (complementary bipolar/CMOS/DMOS) process that provides vertical PNP transistors with an f_t of 2 GHz. The Si9145 voltage mode controller built with this technology has an error amplifier with a functional closed-loop bandwidth of 100 kHz, helping to handle transient loads provided by today's microprocessors (Goodenough, 1995c).

1.4.5.7 Switched Capacitor Converters

Switched capacitor (charge pump) converters use capacitors rather than inductors or transformers to store and transfer energy. The most compelling advantage of these converters is the absence of inductors. Compared with capacitors, inductors have greater component size, more EMI, greater layout sensitivity, and higher cost.

Compared with other types of voltage converters, the switched capacitor converter can provide superior performance in applications that process low-level signals or require low-noise operation.

Many capacitor-based voltage converters offer extremely low operating current, a useful feature in systems for which the load current is either uniformly low or low most of the time. Thus, for small handheld products, the light-load operating currents can be much more important than full-load efficiency in determining battery life (Williams and Huffman, 1988). The basic operation of switched capacitor voltage converters is shown in Figure 1-26.

When the switch is in the left position, C_1 charges to V_1. The total charge on C_1 is given by $q_1 = C_1 V_1$. When the switch moves to the right position, C_1 discharges to V_2. The total charge on C_1 now is given by, $q_2 = C_1 V_2$. The total charge transfer is given by

$$q = q_1 - q_2 = C_1(V_1 - V_2) \qquad (1.11)$$

If the switch is cycled at a frequency f, the charge transfer per second (the current) is given by

$$I = fC_1(V_1 - V_2) = (V_1 - V_2)/R_{eq} \qquad (1.12)$$

where R_{eq} is given by $1/fC_1$. The reservoir capacitor C_2 holds the output constant. The basic charge pump circuit can be modified for an inverted output by flipping C_1 in the internal switching arrangement. It can be used for doubling, tripling, or halving the input voltage with different external connections and by the use of external diodes.

Figure 1-27 shows a simple switched capacitor IC implementation with on-chip switches. The usage of switched capacitor converters in laptop computers

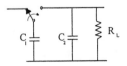

FIGURE 1-26 The principle of operation of switched capacitor converters

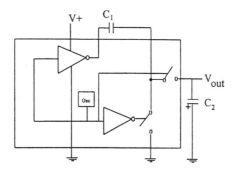

FIGURE 1-27 A simple IC switched capacitor implementation

and small handheld communication devices is increasing as their output current capabilities are increasing and the supply current required by the portable devices is decreasing.

Simple switched capacitor converters such as the MAX660 can generate 100 mA at 3.3 V when powered from a two-cell battery of alkaline, NiCd, or NiMH cells or a single primary lithium cell. A disadvantage of a circuit such as this is the lack of regulation, which is overcome by adding a regulator externally.

Internal regulation in monolithic switched capacitor converter chips is achieved either as linear regulation or as charge pump modulation. Linear regulation offers low output noise. Charge pump modulation controls the switch resistance and offers more output current for a given cost or size because of the absence of a series-pass resistor. Newer charge pump ICs employ an on-demand switching technique that enables low quiescent current and high output current capability at the same time.

The simplicity of switched capacitor circuits, and hence its suitability for miniature equipment, is amply illustrated by Siliconix Inc. (1994b), where the eight-pin Si7660 is used for voltage inversion, doubling, and splitting.

1.4.5.8 SEPIC

The SEPIC (single-ended primary inductance converter) architecture is especially suited for battery-operated equipment and battery charger applications. The SEPIC topology is shown in Figure 1-28.

The two inductors L_1 and L_2 often are two identical windings on the same core. This topology essentially is similar to a 1:1 transformer flyback converter, except for the addition of a capacitor C, which forces identical AC voltages across the two inductors. This is a step-up/step-down topology with no inversion and no transformer. However, the SEPIC topology provides DC isolation from the input to the output through the presence of the capacitor.

The input/output relationship for this converter is given by

$$V_{out} = V_{in}D/(1 - D) \tag{1.13}$$

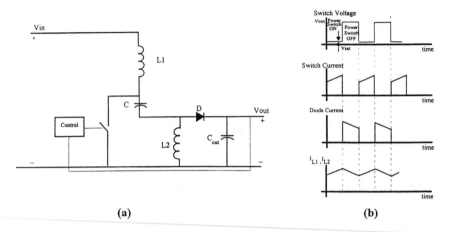

FIGURE 1-28 The SEPIC converter topology: (a) The basic circuit, (b) Associated waveforms

This topology has the advantage of being operable over a wide input voltage range. Due to this, the topology commonly finds applications in battery-powered equipment and battery chargers.

A converter design using the LT1373 current mode switching regulator with a SEPIC converter for generating 3.3 V from a single lithium-ion battery is described in Essaff (1995). In a typical application such as this, the battery voltage at full charge is above the output voltage; and when discharged, it is below the output voltage. This presents special difficulties in the converter design. The SEPIC architecture, with its wide input voltage range, is ideally suited for such applications.

The use of the SEPIC topology in battery charging is illustrated in an application by Linear Technology Corp. (1996). The topology allows charging even when the input voltage is lower than the battery voltage. Further, it allows the current sense circuit to be ground-referenced and completely separated from the battery itself, simplifying battery switching and system grounding problems.

The LT1512 is a constant current controller for SEPIC converters, for charging NiCd and NiMH batteries. It also can provide a constant voltage source for charging lithium-ion batteries (Linear Technology Corp., 1996). A special feature of this controller is an internal low-dropout regulator, which provides a 2.3 V supply for all internal circuitry, allowing input voltages to vary from 2.7 to 25 V. This enables charging batteries from varied sources such as wall adapters, car batteries, and solar cells.

1.4.6 Illustrative DC to DC Converter ICs and Applications

The modularity of design, the wide-ranging power supply requirements, and the need for compactness, portability, and expandability of today's electronic equipment all contribute to the numerous applications that DC/DC converters have found in modern electronic equipment.

The area in which DC/DC converters have found the widest application is in distributed power systems. A distributed power system uses many small regulated power supplies, each located as close as possible to the load (Goodenough, 1995a). The bulk supply develops and distributes via a bus, an arbitrary voltage level. At appropriate points DC/DC converters change this voltage to the levels needed for the local circuitry. A distributed power system can save space and reduce the weight of the system. It also can improve reliability and the quality of the generated power and facilitate modular design and system expansion. Issues relating to distributed power systems are discussed in Ormond (1990, 1992) and Goodenough (1995a).

A typical application of a distributed power system is in telecommunication equipment. A relatively high-voltage battery of 48 V is located in the bottom of each electronic equipment rack and DC/DC converters are used on each rack card cage to step down the voltage to the required level. Another such application is in aircraft power distribution systems.

Modern electronic systems contain both analog and digital circuitry, where analog components require voltage levels such as 9, ±12, and ±15. In mixed mode logic circuits, the conventional 5 V supply and a 3.3 V supply for the latest low-power ICs are required. In such complex cases, DC/DC converters perform the valuable function of generating all required voltages while saving cost and space. A common application of 5–12 V converters is in flash memory programmers, which require a supply of 12 V.

Battery chargers that can be used off diverse power supplies, ranging from the conventional main supply to solar cells, find the step-up/step-down ability of DC/DC converters a useful feature.

Another application area of DC/DC converters is in small, battery-operated, portable equipment such as pagers, cameras, cellular phones, laptop computers, remote data acquisition, and instrumentation systems. The high efficiency and the small size of DC/DC converters become especially useful in such applications.

1.4.6.1 DC/DC Converter ICs

An increasing variety of chip-level DC/DC converters are appearing on the market. These devices not only are changing the way system designers structure their power supplies but are providing solutions to applications that previously required more costly, bulky, and cumbersome approaches.

Integrated circuit DC/DC converters are available in a wide range of power and other capabilities. The lower-power ICs can supply the exact voltage needed for a specific IC board, and higher-power types can simplify the design by reducing the component count (Pryce, 1988).

The key parameters and capabilities of modern low-power DC/DC converters are described in this section using a representative sample from principal manufacturers such as Maxim Integrated Circuits, Datel Inc., Linear Technology Corp., and Harris Semiconductor. For example, output power capabilities range from less than 1 W to more than 300 W. Single, double, or triple output configurations are available. Common output voltages are 3.3, 5, 12, and 15 V, while input voltages are quite varied. Common devices perform 5 to 12 V, 9 to 12 V, and 5 to 3.3 V conversions. Table 1-6 shows key features of some representative DC/DC converters.

Linear Technology Corp. produces a wide range of DC/DC converters, ranging from 0.5 W micro-power devices to high-efficiency 5 A devices. Simple voltage doublers and dividers as well as circuits that can be used in many configurations are available. The LT1073, a gated oscillator mode IC, can operate from a supply of 1 V and typically is used to generate 5 or 12 V from a single cell.

The XWR (wide input range) series by Datel Inc. (1991) is composed of high-efficiency current mode converters with typical efficiencies of about 85%. Common applications of these are in telecommunication, automotive, avionic, and marine equipment and in portable battery-operated systems. In addition to the XWR series, Datel also has the LP series of converters having 3, 4.5, 5, and 10 W outputs.

Maxim Integrated Circuits also developed a wide variety of DC/DC converters, operating in current mode and resonant mode. The MAX632 is a typical low-power device, which can operate from an input voltage of 5 V to produce a 12 V output. This type of converter is ideal for powering low-power analog circuits from a 5 V digital bus. These converters are relatively inexpensive and require very few external components.

TABLE 1-6 Some Representative DC/DC Converter ICs

Manufacturer	Model	Output Power (W)/ Output Current (A)	Output Voltage (V)	Input Voltage (V)
Linear Technology Corp.	LT1070	5 A	Circuit dependent	3 to 60
	LT1073	0.5 W	5, 12	1 to 12
	LT1074/76	5/2 A	5	7 to 45/64
	LT1173	0.5 W	5, 12	2 to 12 (step down) 2 to 30 (step up)
Datel Inc.	XWR series	3, 10, 20 W	3.3, ±5, ±12, ±15	4.6 to 13.2, 4.7 to 7, 9 to 18, 18 to 72
Maxim Integrated Circuits	MAX632	0.025 A	12	2 to 12.6
	MAX742	1 A	5, 12, 15, 28	2 to 16.5
	MAX756	0.3 A	3.3, 5	1.1 to 5.5
Siliconix Inc.	Si9100	0.350 A	$V_{out} < V_{in}$	10 to 70
	Si9102	0.250 A	$V_{out} < V_{in}$	10 to 120
	Si7660	Up to 100 mA depending on V_{in}/V_{out}	$-V_{in}$, $2V_{in}$, $V_{in}/2$	1.5 to 10
National Semiconductor	LM3578	0.750 A	$V_{out} < V_{in}$ $V_{out} > V_{in}$	2 to 40
Harris Corp.	HIP5020	3.5 A	$V_{out} < V_{in}$	4.5 to 18
Analog Devices	ADP3603	50 mA	-3 mA	4.5 to 6

The LM3578 from National Semiconductors is a low/medium power converter which can be used in buck, boost, and buck-boost configurations. This can supply output currents as high as 750 mA.

Siliconix's Si9100 and Si9102 can handle high input voltages. These chips typically are used in transformer-coupled flyback and forward converter applications, such as ISDN and PABX equipment and modems. Due to their high input voltage ratings, the Si9100 can operate directly from the -48 V telephone line supply and the Si9102 from the -96 V double-battery telecom power supplies.

Harris Semiconductor's high-efficiency synchronous buck converter controller with integrated MOSFETs has typical applications in notebook computers, portable instruments, and portable telecommunication equipment.

The switched capacitor converter ADP3603 from Analog Devices (1997–98) provides a -3 V regulated output from a 4.5–6.0 V supply and provides 50 mA of current. A 150 mA version, the ADP3604, also is available. Among the typical applications of these are negative voltage regulators, computer peripherals and add-on cards, pagers and radio control receivers, disk drives, and mobile phones.

Further details of these ICs as well as other similar products, application notes, and the like are found in Linear Technology Corp. (1990, 1992, 1996, 1997), Datel Inc. (1991), Siliconix (1994a, 1994b), Motorola Inc. (1987), Unitrode Integrated Circuits (1995–96), Malinaik (1995), Travis (1996), Goodenough (1995b, 1995c, 1996), Sherman and Walters (1996), Maxim Integrated Circuits (1997), Sherman (1988), Pflasterer (1997), "Trends in Battery Power" (1997), Williams and Huffman (1988), Essaff (1995), Harris Corp. (1997), and Analog Devices Inc. (1997–98).

References

Analog Devices Inc. *Designers' Reference Manual*, revision C.2, DES-CD-98-INT-DISK. Analog Devices Inc., Winter 1997–98.

Bose, B. K. (Editor). *Modern Power Electronics: Evolution, Technology and Applications.* Piscataway, NJ: IEEE Press, 1992.

Brown, Marty. *Practical Switching Power Supply Design.* Academic Press, 1990.

Brown, Marty. *Power Supply Cookbook.* Boston: Butterworth-Heinemann, 1994.

Burr-Brown. *Integrated Circuits Data Book*, vol. 33. Burr-Brown, USA, 1989.

Burr-Brown. "1.2 V and 2.5 V Micropower Voltage Reference." At http://www.burr-brown.com/download/DataSheets/REF1004.pdf., October 1993.

Chryssis, George C. *High Frequency Switching Power Supplies*, 2d ed. New York: McGraw-Hill, 1989.

Cük, S. "Basics of Switched-Mode Power Conversion: Topologies, Magnetics and Control." *Powerconversion International* (August-October 1981).

Cük, S., and Dragan Maksimovic. "Quasi-Resonant Cük DC-to-DC Converter Employs Integrated Magnetics-Part I." *PCIM* (December 1988), pp. 16–24.

Cük, S., and R. D. Middlebrook. "A New Optimum Technology Switching DC-to-DC Converter." In *IEEE Power Electronics Specialist Conference Record*, Palo Alto, CA, June 14–16, 1994, pp. 160–179.

Datel Inc. *Datel Databook*, vol, 4: *Power*. Datel Inc., Mansfield, MA: USA, 1991.

Essaff, Bob. "250 kHz, 1 mAI_Q Constant Frequency Switcher Tames Portable Systems Power." Linear Technology Design Note 108. Linear Technology Corp., Milpitas, CA: USA, July 1995.

Goodenough, Frank. "System Designers vs. Power Supplies." *Electronic Design* (May 15, 1995a), p. 22.

Goodenough, Frank. "Power Supply Rails Plummet and Proliferate." *Electronic Design* (July 24, 1995b), pp. 51–55.

Goodenough, Frank. "Fast LDOs and Switchers Provide Sub-5v Power." *Electronic Design* (September 5, 1995c), pp. 65–74.

Goodenough, Frank. "Driver ICs Create 1 or 2 Processor Power Rails." *Electronic Design* (September 16, 1996a), pp. 65–74.

Goodenough, Frank. "LDO Controller Handles 250 A/µs Load Transients." *Electronic Design* (November 18, 1996b), pp. 162–166.

Harris Corporation. *Harris Design Resource CD.* CD-001. Harris Corporation, Palm Bay, FL: USA, September 1997.

Kerridge, Brian. "Microamps Sustain Stable Sources." *EDN* (March 30, 1992), pp. 53–60.

Knapp, Ron. "Selection Criteria Assist in Choice of Optimum Reference." EDN (February 18, 1988), pp. 183–192.

Kularatna, N. *Power Electronics Design Handbook: Low Power Components and Applications.* Boston: Butterworth-Heinemann, 1998.

Lee, Fred C., Wojciech A. Tabisz, and Milan M. Jovanovic. "Recent Developments in High Frequency Quasi-Resonant and Multi-Resonant Converter Technologies." In *Proceedings of the 1988 European Power Electronic Conference*, Aachen, Germany, 1989, pp. 401–410.

Lee, Mitchell. "Linear PNP Regulator Outperforms NPN Types." PCIM (May 1989), pp. 36–42.

Linear Technology Corporation. *1990 Linear Databook.* Linear Technology Corporation, Milpitas, CA: USA, 1990.

Linear Technology Corporation, *1992 Databook Supplement.* Linear Technology Corporation, Milpitas, CA: USA, 1992.

Linear Technology Corporation. "LTC1430 High Power Step-Down Switching Regulator Controller." At http://www.linear-tech.com/pdf/lt1430.pdf, July 1995.

Linear Technology Corporation. "LT1512 SEPIC Constant Current/Constant Voltage Battery Charger." At http://www.linear-tech.com/pdf/1512f.pdf., October 1996.

Maliniak, David. "Modern DC-DC Converter Sends Power Density Soaring." *Electronic Design* (August 21, 1995), pp. 59–63.

Maxim Integrated Circuits. *New Releases Data Book*, vol. 3. Maxim Integrated Circuits, Sunnyvale, CA: USA, 1994.

Maxim Integrated Circuits. *New Releases Data Book*, vol. 4. Maxim Integrated Circuits, Sunnyvale, CA: USA, 1995.

Maxim Integrated Circuits. *New Releases Data Book*, vol. 5. Maxim Integrated Circuits, Sunnyvale, CA: USA, 1996.

Maxim Integrated Circuits. "Step Down Controllers with Synchronous Rectifier for CPU Power." At http://209.1.238.250/arpdf/1190.pdf, November 1997.

Mitchell, Daniel M. *DC-DC Switching Regulator Analysis.* New York: McGraw-Hill, 1988.

Moore, Bruce D. "Step-up/Step-down Converters Power Small Portable Systems." *EDN* (February 3, 1994), pp. 79–84.

Motorola Inc. *Linear/Switchmode Voltage Regulator Handbook.* Motorola Inc., USA, 1987.

National Semiconductor Corporation. *Linear Data Book 1.* National Semiconductor Corporation, USA, 1987.

National Semiconductor Corporation. "National P/N LM236 - 2.5 Voltage Reference Diode." At http://www.national.com/pf/LM/LM236-2.5.html.

O'Malley, Kieran. "Understanding Linear Regulator Compensation." *Electronic Design* (August 22, 1994), pp. 123–128.

Ormond, Tom. "Distributed Power Schemes Simplify Design Tasks." *EDN* (December 6, 1990), pp. 132–136.

Ormond, Tom. "Distributed Power Schemes Put Power Where You Need It." *EDN* (July 6, 1992), pp. 158–164.

Pflasterer, Jim. "DC-DC Converter ICs Address the Needs of Battery Powered Systems." *PCIM* (April 1997).

Pryce, Dave. "Growing Array of One-Chip DC/DC Converters Provides Power for Diverse Applications." *EDN* (February 18, 1988), pp. 73–80.

Pryce, Dave. "Voltage References." *EDN* (January 18, 1990), pp. 121–126.

Shah, Rajesh J. "Power Supply Basics: Voltage Regulators." PCIM (April 1996), pp. 20–41.

Sherman, Jeffrey D., and Michael M. Walters. "Synchronous Rectification: Improving the Efficiency of Buck Converters." *EDN* (March 14, 1996).

Sherman, Leonard H. "DC/DC Converters Adapt to the Needs of Low-Power Circuits." *EDN* (January 7, 1988).

Siliconix Inc. *Power Products Data Book.* Siliconix Inc., 1994a.

Siliconix Inc. *Analog Integrated Circuits Data Book.* Siliconix Inc., 1994b.

Simpson, Chester. "LDO Regulators Require Proper Compensation." *Electronic Design* (November 4, 1996), pp. 99–104.

Steigerwald, Robert L. "High Frequency Resonant Transistor DC-DC Converters." *IEEE Transactions on Industrial Electronics* IE-31, no. 2 (May 1984), pp. 181–191.

Strassberg, Dan. "A Surfeit of Power Supply Voltages Plagues Designs of Compact Products." *EDN* (May 12, 1994), pp. 55–62.

Travis, Bill. "Low-Voltage Power Sources Keep Pace with Plummeting Logic and µP Voltages." *EDN* (September 26, 1996).

"Trends in Battery Power Recharge DC/DC Converter Advances." *AEI* (February 1997).

Unitrode Integrated Circuits. *Product and Application Handbook.* Unitrode Integrated Circuits, Merrimack, NH: USA, 1995–96.

Unitrode Integrated Circuits. *Unitrode Applications Handbook.* Unitrode Integrated Circuits, Merrimack, NH: USA, April 1997.

Williams, Jim, and Brian Huffman. "Precise Converter Designs Enhance System Performance — Part I." *EDN* (October 13, 1988), pp. 175–185.

Williams, Jim, and Brian Huffman. "Switched Capacitor Networks Simplify DC/DC Converter Design." *EDN* (November 24, 1988), pp. 171–175.

Wofford, Larry. "A New Family of Integrated Circuits to Control Resonant Mode Power Converters." *PCIM* (April 1990), pp. 26–35.

Zendzian, D. "A High Performance Linear Regulator for Low Dropout Applications." In *Applications Handbook.* Unitrode Integrated Circuits, Merrimack, NH: USA, April 1997.

CHAPTER 2

Operational Amplifiers

2.1 Introduction

The world is analog. With the invention of the transistor, electronics engineers gained competence in the digital processing of analog signals. However, in all signal processing systems, designers have to receive, amplify or attenuate, or process the analog signals using varieties of analog techniques. The operational amplifier (op amp) is one of most commonly used analog components for such tasks.

Bob Widlar, working at Fairchild, designed the first successful op amp back in 1965. The first commercial monolithic device was µA709. The developments that continued from this experience provided several successful op amps, such as the µA741 and the LM301. With advancements in semiconductor manufacturing technologies, many advanced op amps entered the market. Today op amps are available in a wide variety of technologies, specifications, prices, and package styles from a multitude of vendors. When a circuit requires an op amp, the designer is confronted with a bewildering number of devices from which to choose. These varieties may be identified as basic voltage feedback, current feedback, micro-power, video, and chopper stabilized types.

This chapter provides a designer's viewpoint on the use of op amps, identifying their characteristics and limitations in a practical sense and discussing the use of different kinds. Many excellent references (Horowitz and Hill, 1996; Analog Devices, 1987, 1990, 1992, 1995; Dostal, 1993) support this chapter with more detailed fundamentals.

2.2 Introduction to Amplifiers

There are four general classes of amplifiers, characterized by transfer characteristics expressed in volts per volt, amperes per ampere, volts per ampere, and amperes per volt. These four cases are illustrated in Figure 2-1.

These basic amplifiers are the building blocks required to synthesize larger electronic systems. Although some simple electronic devices are voltage amplifiers (the triode vacuum tube), current amplifiers (the bipolar transistor), and transconductance amplifiers (the field effect transistor), presently no electronic devices demonstrate the intrinsic characteristics of a transimpedance amplifier. Hence transimpedance amplifiers must be synthesized from voltage, current, or transconductance amplifiers. Characteristics of the ideal amplifiers are shown in Table 2-1 and the amplifier equivalent circuits are shown in Figure 2-2.

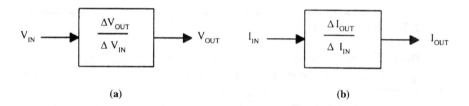

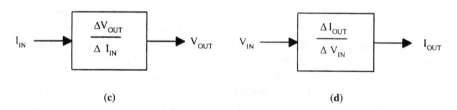

FIGURE 2-1 Types of amplifiers: (a) Voltage amplifier, (b) Current amplifier, (c) Transimpedance amplifier, (d) Transconductance amplifier

TABLE 2-1 Ideal Amplifier Characteristics

	Voltage	*Current*	*Transconductance*	*Transimpedance*
Transfer Function	$V_0 = AV_{in}$	$I_0 = AI_{in}$	$I_0 = G_M V_{in}$	$V_0 = R_M I_{in}$
R_{in}	$\propto$	0	$\propto$	0
R_{out}	0	$\propto$	$\propto$	0

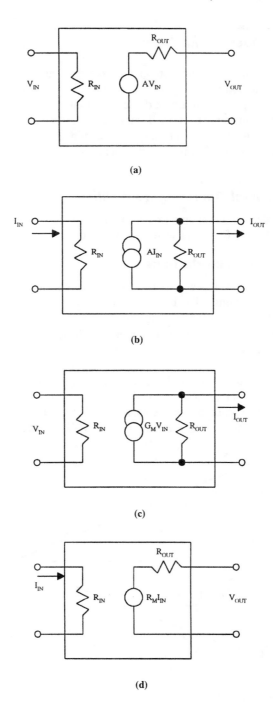

FIGURE 2-2 Amplifier equivalent circuits: (a) Voltage amplifier, (b) Current amplifier, (c) Transconductance amplifier, (d) Transimpedance amplifier

2.3 Basic Operational Amplifier

The ideal op amp has differential inputs, an infinite input impedance, a single-ended output, and infinite gain at all frequencies. The ideal op amp always must be considered as a four-terminal device, the fourth terminal being the return path for the output current. In most designs, it is assumed that the amplifier is ideal and that no current flows into the input terminals, there is no input offset voltage, and no power is required for its operation. Figure 2-3 shows the ideal op amp.

2.3.1 The Real Operational Amplifier

Figure 2-4 depicts a real op amp based on a voltage amplifier. Let us briefly discuss the input imperfections and output obstacles.

2.3.1.1 Input Imperfections

The actual characteristics of real op amps are considerably more complicated. Each input contains a DC current source (I_B, the bias current), and a DC voltage source (V_{OS}, the offset voltage) in series with the inputs. The amplifier has differential and common mode input impedances ($Z_{in(DIFF)}$ and $Z_{in(CM)}$, respectively), which usually are complex and consist of a resistor and a capacitor in parallel.

Also, there are three uncorrelated noise sources: two current sources (I_N) and a voltage source (V_N) that appear differentially. Finally, the amplifier has gain with regard to common mode signals, which the ideal amplifier does not have, and so its common mode rejection ratio (CMRR) needs to be specified.

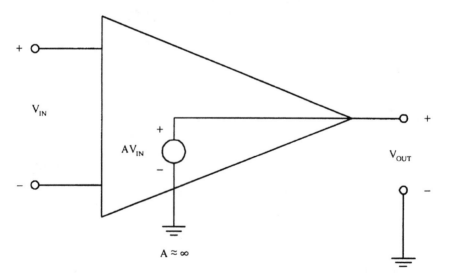

FIGURE 2-3 The ideal op amp

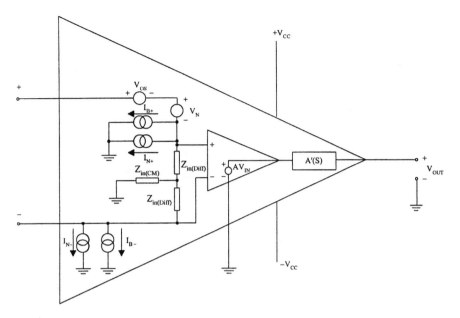

FIGURE 2-4 A practical op amp

2.3.1.2 Output Stage

The output side of the model also is not ideal. There is an output impedance (R_0) in series with the voltage sources. The gain (A, infinite in the ideal model) is both finite and a function of frequency in a real amplifier, which also has a finite slew rate (the rate of rise of output voltage per microsecond) and limited output voltage and current capabilities.

2.3.1.3 Differential to Single-Ended Conversion

One fundamental requirement of a simple op amp is that an applied signal, which is fully differential at the input, must be converted to a single-ended output; that is, with respect to the often neglected fourth terminal. To see how this can lead to difficulties, look at Figure 2-5. The signal flow illustrated by Figure 2-5 is used in several popular integrated circuit families, such as the 101, 741, 748, 503, and other integrated circuit amplifiers.

The circuit first transforms a differential input voltage into a differential current. This input stage function is represented by PNP transistors in Figure 2-5. The current then is converted from differential to single-ended form by a current mirror connected to the negative supply rail. The output from the current mirror drives a voltage amplifier and power output stage, which is connected as an integrator. The integrator controls the open-loop frequency response, and its capacitor may be added externally, as in the 101, or self-contained, as in the 741. Most descriptions of this simplified model do not emphasize that the integrator has

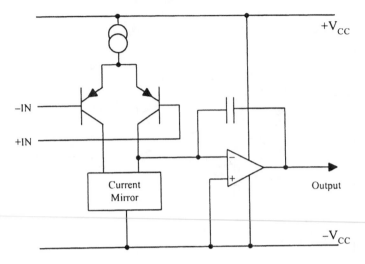

FIGURE 2-5 A simplified real op amp showing a single-ended output and differential input

a differential input, of course. It is biased positive by a couple of base-emitter voltages, but the noninverting integrator input is referred to the negative supply.

It should be apparent that most of the voltage difference between the amplifier output and the negative supply appears across the compensation capacitor. If the negative supply voltage is changed abruptly, the integrator amplifier will force the output to follow the change. When the entire amplifier is in a closed-loop configuration the resulting error signal at its input will tend to restore the output, but the recovery will be limited by the slew rate of the amplifier. As a result, an amplifier of this type may have outstanding low-frequency power supply rejection, but the negative supply rejection is fundamentally limited at high frequencies. Since the feedback signal to the input causes restoration of the output, the negative supply rejection will approach 0 for signals at frequencies above the closed-loop bandwidth. This means that high-speed, high-level circuits can "talk" to low-level circuits through the common impedance of the negative supply line. This phenomenon demands some special consideration in decoupling and grounding; details are discussed in Brokraw (Analog Devices, AN-202).

2.3.2 Amplifier Specifications

2.3.2.1 Offset Voltage

Offset voltage (V_{OS}) is defined as the voltage that must be applied to the input to cause the output to be 0. Offset voltage is the result of a mismatch in the base-emitter voltages of the differential input transistors (the gate-source voltage mismatch in FET-input amplifiers) and is indistinguishable from a DC input signal. This offset can be trimmed to 0 with a potentiometer, which adjusts the balance of the operating currents in the input stage until V_{BE1} and V_{BE2} (or V_{GS1}

and V_{GS2}) are equal. Even if the offset voltage is trimmed to 0 at one particular temperature, it will vary with the temperature. When a bipolar transistor op amp is trimmed for minimum offset, it is trimmed for minimum temperature drift, but this is not the case for FET-input op amps.

2.3.2.2 Input Bias Current

Another DC parameter of op amps is input bias current (I_B). If an op amp uses bipolar transistors in its input stage, a base current must be supplied from somewhere to bias them into their active operating region. Since Kirchhoff's law tells us that current must flow in circles, this current also must return to its origin through a DC path. Therefore, operational amplifiers cannot be used with input signal sources that are not referred to the same power source as the amplifier itself. Although FETs do not require a base current, they nevertheless have a leakage current from their gate junction diode, which results in an input bias current. In many applications, the errors due to bias currents actually are less than the errors caused by the mismatch of the bias currents on the two inputs. This difference between the bias currents, called the *input offset current*, usually is specified along with the bias current.

Like the input offset voltage, bias currents also vary with temperature. In an amplifier with a bipolar input stage, the bias current decreases with increasing temperature because the transistors' β increases, and since their emitter current remains constant, the base current decreases. In FET input amplifiers, the bias current is the gate leakage current of an FET, which is the leakage current of a reverse-biased junction diode. Such leakage currents double for every 10°C rise in junction temperature.

An ideal op amp has no current going into its input terminals; real op amps approach this by reducing bias currents to femtoamp levels. As a category, one can consider low bias current op amps as those with less than 1 nA bias currents.

2.3.2.3 Open-Loop Voltage Gain

Another op amp parameter that distinguishes a real amplifier from an ideal amplifier is the open-loop gain. The open-loop gain of an ideal op amp is assumed to be infinite. The same assumption occasionally is made of real amplifiers, with unfortunate results. Op amps generally have around 20 V of output swing and gains of over 1 million — the input therefore would need to be on the order of 1 µV, and it is very hard to handle such signals without unacceptable errors due to thermoelectric potentials. Special circuits are necessary to measure the open-loop voltage gain.

2.3.2.4 Frequency Response

Most operational amplifiers have a very simple frequency response. The gain is constant at DC and very low frequencies, then has a single-pole roll-off, falling at 6 dB/octave (-20 dB/decade). It is obvious that, throughout the region where this single-pole frequency response applies, the product of gain and frequency is a constant, known as the *gain-bandwidth product* of the amplifier, and a measure of its high-frequency performance. In the majority of op amps, this single-pole response continues past the point where the gain has dropped to unity (such amplifiers are known as *internally compensated* or *unity gain stable amplifiers*). Some amplifiers have a more complex response, and at a gain of something less than 10, a second pole appears. These amplifiers are not stable at low-closed loop gains but generally have better high-frequency performance than the internally compensated types. Op amps are considered wideband fast settling if their bandwidth is greater than 5 MHz, they slew at more than 10 V/μs, and they settle to 0.1% in 1 μs or less.

2.3.2.5 Slew Rate

The slew rate of an amplifier is the rate at which the output voltage can change when high drive is applied to the input. When the circuit is used to measure slew rate, the input signal is a fast square wave of sufficient amplitude to drive the output of the device under test (DUT) to saturation. An oscilloscope is used to observe the slew rate of the DUT output. See Figure 2-6 for slew rate and settling time.

2.3.2.6 Common Mode Rejection Ratio

The ideal operational amplifier has only differential gain and is insensitive to the absolute voltage on the inputs. A real amplifier has several nonideal characteristics associated with input levels. First of all, the range of input voltage is limited. Few IC op amps will operate when the voltages on the input terminals

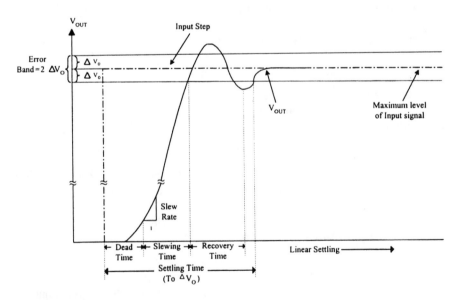

FIGURE 2-6 Amplifier settling time and slew rate

are outside the supply voltages. The second, and perhaps more subtle, characteristic is the common mode rejection ratio. The CMRR is the ratio of a change in common mode voltage to the change in differential input voltage that would produce the same change in output. It often is convenient to specify this parameter in decibels.

2.3.2.7 Settling Time

Settling time is an important op amp specification, especially when the amplifier must handle rapidly changing signals. In applications such as multiplexers, sample/hold amplifiers, and amplifiers used with A/D and D/A converters, the amplifier settling time often determines the maximum data rate for a specified accuracy.

Settling time is the time that elapses between the application of a step input to the time at which the amplifier output enters, and remains within, a specified error band symmetrical to the final output value.

Settling time is determined by both linear and nonlinear characteristics of the amplifier. It varies with the input signal level and is greatly affected by impedances external to the amplifier. For these reasons, extrapolation of settling times from one set of operating conditions to another becomes virtually impossible. Settling time cannot be predicted from open-loop specifications, such as slew rate or small signal and bandwidth, as it is a closed-loop parameter. The best way to know how fast an amplifier will settle in a particular application is to measure it.

2.3.2.8 Noise

In addition to noise present on the input signal, op amp circuits have noise due to external interference and the inherent noise of the circuit itself. Interference noise originates from sources not related to the actual circuit; such noises include ground and power supply noise, stray electromagnetic pickup, contact arcing in switches and relays, and transients due to switching in reactive circuits. Even mechanical vibration can create noise in high-impedance amplifier circuits, either by piezoelectric pickup due to the use of piezoelectric plastic material in cables or circuit boards or by capacitance variation as the circuit vibrates. External interference often can be eliminated, once the interfering source is identified and appropriate action taken. The inherent noise of the op amp circuit itself cannot be totally eliminated, since it is caused by components within the circuit. The best that can be accomplished is to minimize the noise in a specific bandwidth of interest. Four types of noise are commonly encountered in operational amplifiers: popcorn noise, flicker noise, shot noise, and Johnson noise. Popcorn noise is well understood and of less importance in modern components. Flicker noise is the dominant noise at low frequencies. It has a power spectral density inversely proportional to frequency (hence the term *1/f noise*). The noise voltage spectral density therefore is inversely proportional to the square root of frequency. In modern op amps, it is rarely significant above 50 Hz.

Thermal excitation of the electrons in conductors causes random movement of charge. In a resistor, this random current causes a noise voltage, known as *Johnson noise,* whose amplitude is given by the formula

$$E_N(\text{rms}) = \sqrt{4KTRB} \tag{2.1}$$

where

$$K = \text{Boltzman's constant } (1.38 \times 10^{-23} \text{ J/}^\circ\text{K});$$

$$T = \text{temperature (K)};$$

$$R = \text{resistance (Ohms, } \Omega);$$

$$B = \text{bandwidth (Hertz)}.$$

At room temperature (25°C), this may be simplified to

$$E_N(\text{rms}) \approx \tfrac{1}{8}\sqrt{RB} \tag{2.2}$$

or

$$e_n \approx 4\sqrt{R} \tag{2.3}$$

where

$$E_N = \text{total noise } (\mu\text{ V rms});$$

$$R = \text{resistance (k } \Omega);$$

$$B = \text{bandwidth (kHz)};$$

$$e_n = \text{spectral density } (\text{nV/}\sqrt{\text{Hz}}).$$

Johnson noise is a fundamental property of resistance and important in designing low-noise circuitry. It could be reduced only by reducing the temperature, the resistance, or the working bandwidth. As a reference point, it is useful to remember that at room temperature a 1 kΩ resistor has 4 nV/$\sqrt{\text{Hz}}$ of white noise. This is equivalent to 128 nV rms noise in a 1 kHz bandwidth.

Shot, or Schottky, noise is caused by the statistical variations in the rate of electron flow; these manifest themselves as a noise current in semiconductors, where the current consists of a flow of electrons or holes. Shot noise I_N is given by the formula

$$I_N = 5.7 \times 10^{-4}\sqrt{I_j B} \tag{2.4}$$

where

$$I_N = \text{noise current (picoamps rms)};$$

$$I_j = \text{junction current (picoamps)};$$

$$B = \text{bandwidth of interest}.$$

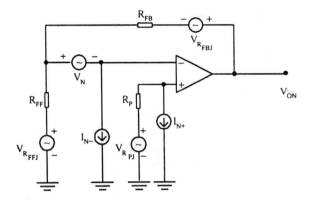

FIGURE 2-7 Op amp noise model

Shot noise is important only when the operational bandwidth is large or the noise current is an appreciable factor of the total current.

Figure 2-7 shows a generalized noise model applicable to all op amps, including current feedback types. In addition to the input noise currents and voltages, the Johnson thermal noise voltages associated with the three resistors are included.

Equation (2.7) is given for calculating the effective integrated output noise voltage in a given bandwidth. The appropriate values for I_{N-}, I_{N+}, and V_N are taken from the current and voltage noise spectral densities given on the data sheet. Usually the noninverting input noise current I_{N+} is neglected because the noninverting input almost always is grounded, bypassed, or driven from a low-impedance source. In making noise performance comparisons among op amps, it often is useful to convert the output noise voltage (V_{ON}) to an effective integrated input noise voltage (V_{IN}). This is accomplished easily by dividing the integrated output noise voltage by the noise gain; that is, $1 + R_{FB}/R_{FF}$.

As applied to Figure 2-7, the gain of stage is

$$G = -\frac{R_{FB}}{R_{FF}} \tag{2.5}$$

$$4KT = 0.01656\frac{(nv)^2}{Hz \cdot \Omega} \text{ at } 25°C \tag{2.6}$$

$$V_{ON} = \sqrt{BW}[I_{N-}^2 \cdot R_{FB}^2 + I_{N+}^2 \cdot R_p^2(1 - G)^2 + V_N^2(1 - G)^2 4KTR_{FB}$$
$$+ 4KTR_{FF}G^2 + 4KTR_p(1 - G)^2]^{1/2} \tag{2.7}$$

where

$$V_{ON} = \text{effective integrated output noise;}$$

$$V_{IN} = \text{effective integrated input noise;}$$

$$V_{IN} = \frac{V_{ON}}{1 - G} \tag{2.8}$$

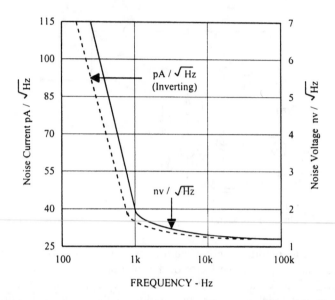

FIGURE 2-8 Equivalent input noise for the AD9617. (Reproduced by permission of Analog Devices Inc.)

Figure 2-8 shows the noise current and voltage spectral density for a practical op amp such as AD9617 from Analog Devices.

Both the noise current and noise voltage are a function of frequency but flatten out above 1 kHz, where the 1/f current noise no longer is significant. To determine the effective noise over a bandwidth where the curves are not flat, the designer must determine the areas under the respective curves for the bandwidth of interest.

Figure 2-9 shows a plot of the effective integrated output noise for the AD844 current feedback op amp from Analog Devices. Note that, at low closed-loop gains, the predominant noise source is the input current noise that flows through the feedback resistor (1000 Ω). For closed-loop gains greater than 15, however, the effects of the "gained-up" input voltage noise begin to dominate the output noise.

Low-noise op amps can be considered those with less than 15–20 nV/$\sqrt{\text{Hz}}$. In the most recent low-noise devices, typical noise voltages of 2–4 nV/$\sqrt{\text{Hz}}$ and noise currents of 2–4 nA/$\sqrt{\text{Hz}}$ are common.

2.3.2.9 Linearity and Distortion

An ideal amplifier produces an exact scaled replica of its input signal at its output. To do this, the slope of its transfer characteristic must be constant. Altering the shape of the input signal between the input and the output is referred to as *distorting* it. Distortion is the result of processing a signal in a nonlinear system. A remarkable property of feedback amplifiers is their ability to improve

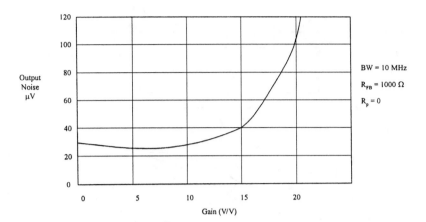

FIGURE 2-9 Equivalent integrated output noise vs. gain for the AD844. (Reproduced by permission of Analog Devices Inc.)

linearity through the use of feedback. The actual mechanism by which this is accomplished is not obvious but may be approached by using a combination of analytical and graphic techniques.

The effect of adding feedback to improve linearity may be treated mathematically. These methods of analysis become increasingly more important as the total harmonic distortion (THD) in an amplifier is used as a criterion for selecting it.

It is clear from Figure 2-10(a) that feedback improves linearity. As per Figure 2-10(b),

$$\frac{V_o}{V_s} = \frac{G}{(1 + GH)} = A \qquad (2.9)$$

$$\frac{dA}{dG} = \frac{1}{1 + GH} \qquad (2.10)$$

It is apparent from equation (2.10) that increasing the product GH reduces the sensitivity of the overall amplifier to a variation of G and hence reduces nonlinearity. It is difficult to discern the presence of distortion on a sinusoidal waveform visually by using an oscilloscope; our eyeballs just are not calibrated to detect slightly distorted sinusoids (it sometimes is less difficult to perceive distortion on noncurvy waveforms such as pulses and square, triangular, and sawtooth waves). The most effective means of measuring distortion is to use a spectrum analyzer and measure the harmonically related components resulting from the nonlinear characteristics of an amplifier.

The most common form of distortion is "limiting" or "clipping," which occurs when the required output voltage from an amplifier is larger than the maximum voltage the amplifier actually can provide (the output is said to exceed the amplifier's "headroom" or to "limit"). Symmetrical clipping introduces high levels of odd harmonics into a waveform, asymmetrical clipping introduces even

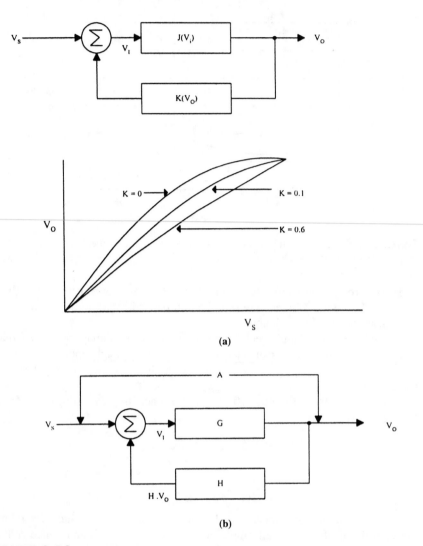

FIGURE 2-10 Feedback and nonlinearity: (a) Feedback improving linearity, (b) Non-linearity gain sensitivity

harmonics as well. The cures are to either reduce the gain of the amplifier or increase its supply voltages.

Although clipping may be considered a gross form of nonlinearity, the term normally is reserved for phenomena that cause the central part of the amplifier transfer characteristic to deviate from a straight line. Consider a 12-bit system with a full-scale range of 10 V. Specifying the system to be linear to 12 bits implies that the maximum deviation from the ideal transfer curve is 1 LSB (2.44 mV).

The two most common types of nonlinearity are square law and logarithmic. Square law nonlinearity occurs when there is a quadratic term in the transfer characteristic of the amplifier and can be produced by a field effect transistor with a resistive load. Logarithmic nonlinearity is produced by a logarithmic (inverse exponential) term in the transfer characteristic and can be caused by bipolar transistors and diodes. A rigorous treatment of op amp characteristics and specifications is given by Analog Devices (1987, 1990).

2.4 Different Types of Operational Amplifiers and Application Considerations

In a modern approach, practical op amps available from the component manufacturers could be grouped into several categories, such as

1. Voltage feedback op amps (the most common form of op amp, similar to the 741 type).
2. Current feedback op amps.
3. Micro-power op amps.
4. Single supply op amps.
5. Chopper stabilized op amps.
6. Wideband, high-speed, high-slew-rate op amps.

As many excellent references describe the design techniques for voltage feedback op amps, the discussion here is limited to special types that have appeared on the market during the last ten years. For details on voltage feedback op amps, see Horowitz and Hill (1996) and Analog Devices (1987, 1990).

2.4.1 Voltage Feedback Op Amps

The voltage feedback operational amplifier (VOA) is used extensively throughout the electronics industry, as it is an easy-to-use, versatile analog building block. The architecture of the VOA has several attractive features, such as the differential long-tail pair, high-impedance input stage, which is very good at rejecting common mode signals. Unfortunately, the architecture of the VOA provides inherent limitations in both the gain-bandwidth trade-off and the slew rate. Typically, the gain-bandwidth product (f_T) is a constant and the slew rate is limited to a maximum value determined by the ratio of the input stage bias current to the dominant-pole compensation capacitor. As a comprehensive discussion on the characteristics and applications of VOAs are well documented, only few examples of applications that highlight the importance of some parameters are described here.

For example, in many applications, bias current can cause offset errors and errors in the transfer function of the circuit. For example, Figure 2-11(a) shows a photodiode amplifier application. Photodiodes generate a current output proportional to the light intensity falling on the photodiode. These currents, however,

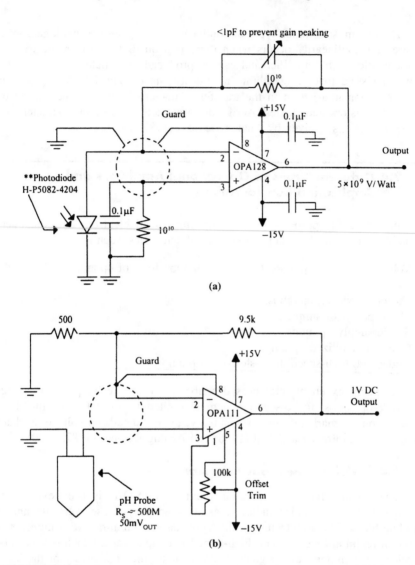

FIGURE 2-11 Some application examples showing the importance of VOA parameters: (a) Sensitive photodiode amplifier, (b) High-impedance amplifier (10^{14} Ω), (c) Low-noise/low-distortion RIAA preamplifier, (d) Low-droop positive peak detector. (Reproduced by permission of Burr-Brown Inc.)

usually are in the subnanoamp range. The circuit of Figure 2-11(a) is simply a current-to-voltage converter. An ideal op amp would cause all the current from the photodiode to flow through the feedback resistor, causing the output voltage of the circuit to be linearly proportional to the current from the photodiode. A real op amp would lose some of the photodiode current into its input terminal. If the photodiode current is 10 nA, then an op amp with a 1 nA bias current would

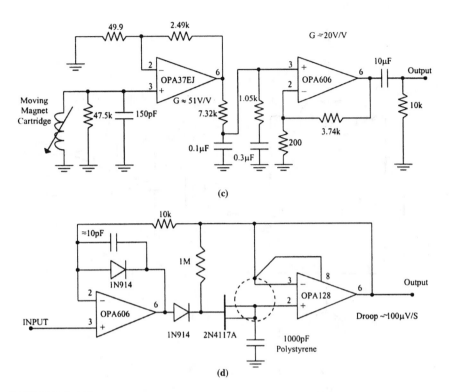

FIGURE 2-11 Continued

cause only 9 nA of signal to pass, causing a 10% error. If the amplifier has a 75 fA bias current, then the error caused by the amplifier is only 0.00075%.

In a pH transducer application such as Figure 2-11(b), a low bias current amplifier is used for two reasons: The OPA111 is an FET input device, thereby giving it an extremely high input impedance. Since the pH probe is a high-impedance element, the input impedance must be even higher. Low bias current also is necessary, since with a source impedance of 500 MΩ, a 1 nA bias current would cause an offset error of 500 mV. The signal output of the probe is only 50 mV; offset error is ten times the signal.

In critical applications where noise can contribute significant errors, such as data acquisition system front ends or audio applications, op amps with a low voltage noise density are crucial. Figure 2-11(c) shows an audio application. Preamplifiers for audio deal with small signal levels; noise that is negligible in large signal applications suddenly becomes a concern when the signal is only an order of magnitude above the noise (an S/N ratio of 20 dB).

2.4.1.1 Closed-Loop Gain and Bandwidth of VOA

Equivalent open-loop voltage gain-related components of VOAs are shown in Figure 2-12(a). With $A = g_m Z_z$, where g_m is the transconductance of the input

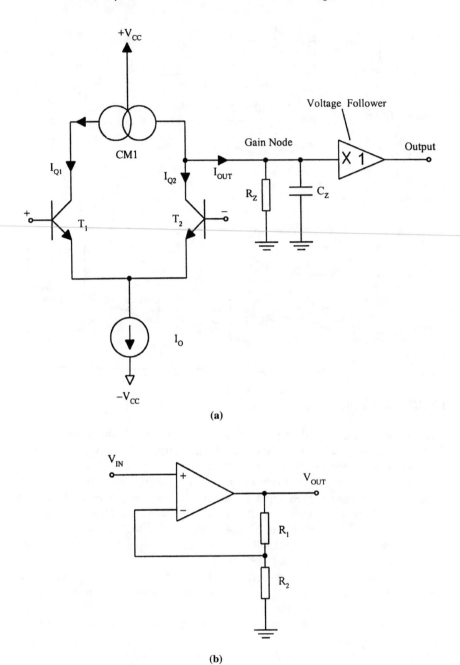

FIGURE 2-12 Simplified VOA and noninverting amplifier: (a) Architecture, (b) Noninverting amplifier

stage and Z_z is the parallel impedance formed by R_z and C_z. The transfer function of the closed-loop gain of the amplifier, configured for the noninverting case (Figure 2-12(b)) is given by

$$\frac{V_{out}}{V_{in}} = G\left(\frac{A}{A + G + \dfrac{R_2}{r_i}}\right) \tag{2.11}$$

where

$$G = 1 + \frac{R_2}{R_1} \tag{2.12}$$

$$A = \text{open-loop gain} = g_m Z_z; \tag{2.13}$$

$$r_i = \text{differential input impedance.}$$

For a VOA, $r_i \rightarrow \infty$ and the equation (2.11) can be simplified to

$$\frac{V_{out}}{V_{in}} = G\frac{(A)}{(A + G)} \tag{2.14}$$

It is important to stress at this stage that, since the VOA is a high differential input impedance device, when negative feedback is applied in this way, voltage is sampled at the output and fed back as a voltage signal to the input. The VOA generally is internally compensated to give a low-frequency dominant pole, f_0, and so the open-loop gain is $A = A_0[1 + j(f/f_0)]$, where $f_0 \approx 1/(2\pi R_z C_z)$ and A_0 is the low-frequency open-loop voltage gain. Substituting this into equation (2.14) gives

$$\frac{V_{out}}{V_{in}} = \frac{GA_0}{(A_0 + G)}\left(\frac{1}{1 + j\dfrac{fG}{f_T}}\right) \tag{2.15}$$

But $A_0/(A_0 + G) \approx 1$, since $A_0 \gg G$, and therefore

$$\frac{V_{out}}{V_{in}} = G\left(\frac{1}{1 + j\dfrac{fG}{f_T}}\right) \tag{2.16}$$

where $(A_0 + G)f_0 \approx A_0 f_0 = f_T$, since $A_0 \gg G$, and this is the gain-bandwidth product of the VOA, which is constant. The closed-loop bandwidth (f_p) of the amplifier is $f_p = f_T/G$; therefore increasing G results in a decrease in the closed-loop bandwidth, while a decrease in G leads to an increase in f_p. This is the "classical" gain-bandwidth trade-off exhibited by a voltage amplifier with a single dominant-pole frequency response. The maximum closed-loop bandwidth of f_T will be obtained when 100% feedback is applied; that is, when $G = 1$.

Generally, VOAs are internally compensated for resistive feedback to guarantee stable operation for any value of G, including $G = 1$.

The equivalent circuit for a voltage feedback op amp is shown in Figure 2-13(a). Figure 2-13(b) shows the classic gain versus frequency response obtained with dominant pole compensation. The 6 dB/octave roll-off is the optimum choice for best phase margin and fastest settling time. The product of the closed-loop gain and the closed-loop bandwidth (gain-bandwidth product) is constant for fixed compensation. The closed-loop bandwidth therefore varies inversely with the closed-loop gain.

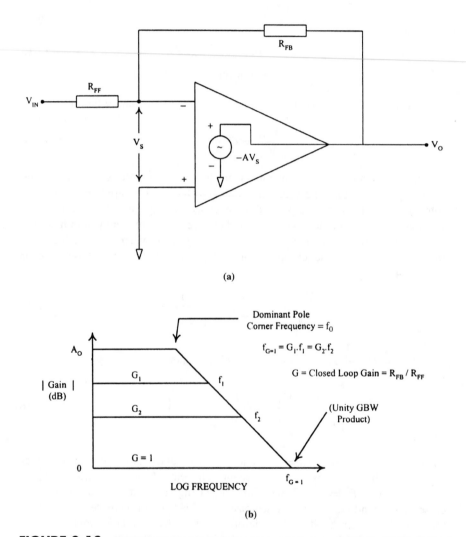

FIGURE 2-13 Equivalent circuit and frequency response of VOA: (a) Equivalent circuit, (b) Frequency response

2.4.1.2 High-Speed Voltage Feedback Op Amps

Within the last decade, many op amps with high-speed capability have entered the market to supply the demand for high-speed A/D converters, video signal processing, and other industrial needs. Many of these op amps combine high speed and precision. For example, the devices introduced by Analog Devices in the late 1980s, such as the AD840, 841, and 842 devices, are designed to be stable at gains of 10, 1, and 2 with typical gain-bandwidths of 400, 40, and 80 MHz, respectively.

A simplified schematic of high-speed voltage feedback is shown in Figure 2-14. For purposes of discussion, the amplifier is shown in the inverting mode. The amplifier consists of a differential input stage (common emitter), a voltage amplification stage, and a class AB (push-pull) output driver.

An example of such a device is the AD847 from Analog Devices. The AD847 is fabricated on Analog Devices' proprietary complementary bipolar (CB) process, which enables the construction of PNP and NPN transistors with similar values of f_T in the 600–800 MHz region. The AD847 circuit (Figure 2-15) includes an NPN input stage followed by fast PNPs in the folded cascade intermediate gain stage. The CB PNPs also are used in the current-amplifying output

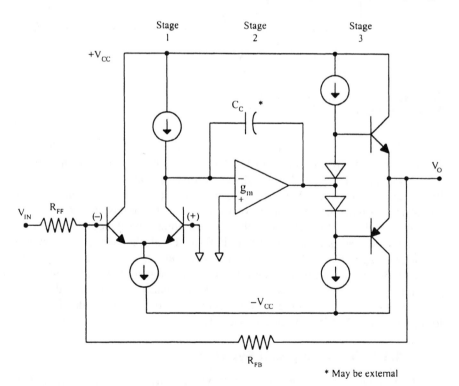

FIGURE 2-14 A simplified schematic of a high-speed VOA. (Reproduced by permission of Analog Devices, Inc.)

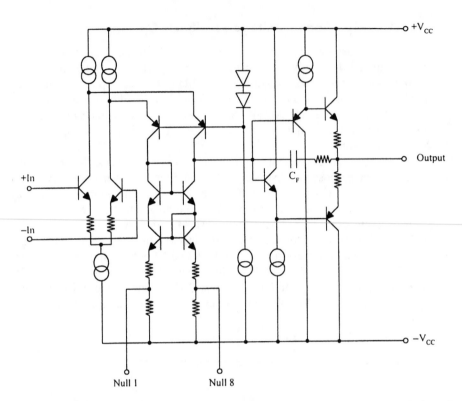

FIGURE 2-15 A simplified schematic of the AD847. (Reproduced by permission of Analog Devices, Inc.)

stage. The internal compensation capacitance that makes the AD847 unity gain stable is provided by the junction capacitance of transistors in the gain stage.

The capacitor, C_F, in the output stage mitigates against the effect of capacitive loads. At low frequencies and with low capacitive loads, the gain from the compensation node to the output is very close to unity. In this case, C_F is bootstrapped and does not contribute to the compensation capacitance of the part. As the capacitive load is increased, a pole is formed with the output impedance of the output stage. This reduces the gain, and therefore, C_F is completely bootstrapped. Some fraction of C_F contributes to the compensation capacitance, and the unity gain bandwidth falls. As the load capacitance is increased, the bandwidth continues to fall and the amplifier remains stable. For more details related to high-speed VOAs, see Analog Devices (1990).

2.4.1.3 Grounding and Bypassing

In designing practical circuits with devices such as AD847, remember that, whenever high frequencies are involved, some special precautions are in order. Circuits must be built with short interconnection leads. A large ground plane should be used whenever possible to provide a low-resistance, low-inductance circuit path as well as minimizing the effects of high-frequency coupling. Sockets should be avoided because the increased interlead capacitance can degrade bandwidth.

Feedback resistors should be of low enough value to assure that the time constant formed with the capacitance at the amplifier summing junction will not limit the amplifier performance. Resistor values of less than 5 kΩ are recommended. If a larger resistor must be used, a small (<10 pF) feedback capacitor in parallel with the feedback resistor, R_F, may be used to compensate for the input capacitance and optimize the dynamic performance of the amplifier. Power supply leads should be bypassed to ground as close as possible to the amplifier pins. Ceramic disc capacitors of 0.1 µF are recommended.

2.4.2 Current Feedback Op Amps

The current feedback operational amplifier (CFOA) is a relatively new arrival to the analog designer's tool kit. The first monolithic device was produced by Elantec Inc. in 1987.

Current feedback operational amplifiers were introduced primarily to overcome the bandwidth variation, inversely proportional to closed-loop gain, exhibited by voltage feedback amplifiers. In practice, current feedback op amps have a relatively constant closed-loop bandwidth at low gains and behave like voltage feedback amplifiers at high gains, when a constant gain bandwidth product eventually results. Another feature of the current feedback amplifier is the theoretical elimination of slew-rate limiting. In practice, component limitations do result in a maximum slew rate, but this usually is much higher (for a given bandwidth) than with voltage feedback amplifiers.

The current feedback concept is illustrated in Figure 2-16. The input stage now is a unity-gain buffer, forcing the inverting input to follow the noninverting input. Thus, unlike a conventional op amp, the latter input is at an inherently low (ideally zero) impedance. Feedback always is treated as a current and, because of the low impedance inverting terminal output, R_2 always is present, even at unity gain. Voltage imbalances at the inputs cause current to flow into or out of the inverting input buffer. These currents are sensed internally and transformed into an output voltage. The transfer function of this transimpedance amplifier is $A(s)$; the units are in ohms.

It can be shown that, if $A(s)$ is high enough (like the open-loop gain of a conventional op amp), very little current flows in the inverting input at balance. The overall closed-loop transfer function becomes

$$\frac{V_{\text{out}}}{V_{\text{in}}} = \left(1 + \frac{R_2}{R_1}\right) \tag{2.17}$$

which is the same as a conventional op amp.

However, if the dominant pole is created by feeding the current imbalances into the compensation capacitor, the time constant will be set by the product of

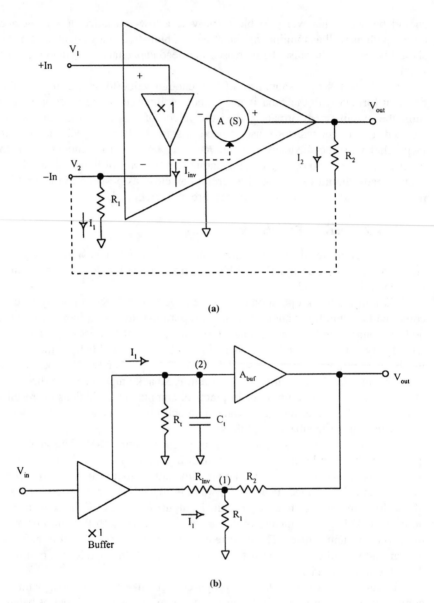

FIGURE 2-16   Current feedback op amp model: (a) Simplified version, (b) Analytical model

this capacitor (C_t) and the feedback resistor R_2. The closed-loop bandwidth now is given by

$$BW = \frac{1}{2\pi R_2 C_t} \qquad (2.18)$$

and is independent of closed-loop gain. A more complete mathematical analysis can be obtained using the representative model shown in Figure 2-16(b). Notice that the input buffer has been given a finite output impedance (R_{inv}) to model practically realizable buffers. The error current from the buffer is mirrored and fed into a transimpedance stage consisting of R_t and C_t, where the current-to-voltage conversion takes place. The voltage generated here is buffered by another unity-gain stage and fed to the main amplifier output. Because the value of the small-signal transresistance, R_t, is very high (often in the megaohm range), only minute error currents are needed to change the voltage at node 2 by several volts. Consequently, the amount of current that must flow into or out of the inverting input terminal under steady-state conditions is extremely small. The feedback network, even though it may be formed from quite low-value resistors, therefore presents a very light effective load on the output of the input buffer.

Applying Kirchhoff's law to nodes of circuit in Figure 2-16(b), the overall transfer function and the like can be derived (Toumazou, Lidgey, and Haigh, 1990, Chapter 16). It can be shown that, if the product of R_t and A_{buf} is large, the closed-loop gain approaches $[1 + (R_2/R_1)]$ as the transconductance approaches infinity, which is the same result as a VOA with large open-loop gain. The AC behavior would appear to be somewhat less intuitive; however, if the product of R_t and A_{buf} is large enough, the closed-loop pole frequency can be closely approximated by

$$f_{pole} = \frac{A_{buf}}{2\pi \left[R_2 + \left(1 + \dfrac{R_2}{R_1} \right) R_{inv} \right] C_t} \tag{2.19}$$

This result indeed is very different from the constant gain-bandwidth product of a voltage feedback amplifier. At low gains, when $R_2 \gg R_1$, and assuming $R_{buf} \ll R_2$ (which always is true in practice), the closed-loop pole frequency is dictated predominantly by the compensation capacitor and the feedback resistor, R_2, and is substantially independent of the exact gain setting. Therefore, the choice of feedback resistor is of somewhat more importance than in the case of voltage feedback amplifiers, and most manufacturers will state a suggested minimum to avoid oscillation problems.

At high gains (in practice 50 or more could be considered high), the R_{inv} term becomes dominant and the amplifier asymptotically assumes a constant gain-bandwidth product given by

$$GBW = \frac{A_{buf}}{2\pi R_{inv} C_t} \tag{2.20}$$

The gain of the output buffer, A_{buf}, also plays its part in determining the closed-loop pole frequency. As the main amplifier output is loaded, the gain drops below unity and causes a reduction in closed-loop bandwidth predicted by equation (2.19). This also tends to make practical amplifiers more stable when heavily loaded. With a constant load, the bandwidth reduction can be compensated for by reducing the value of the feedback resistor.

This theory, although useful, in practice (as always) is compounded by second-order poles and zeroes and stray capacitive and parasitic inductive effects. Such limitations manifest themselves principally as peaking in the small signal response, overshooting and ringing in the transient response, and a sensitivity of overall AC behavior to loading conditions.

The most significant advantages of the CFOAs are

- The slew-rate performance is high, where typical values could be in the range of 500–2500 V/μs (compared to 1–100 V/μs for VOAs).
- The input referred noise of CFOA has a comparatively lower figure.
- The bandwidth is less dependent on the closed-loop gain than with VOAs, where the gain-bandwidth is relatively constant.

2.4.2.1 Practical High-Speed Current Feedback Amplifiers

The equivalent circuit for a current feedback op amp is shown in Figure 2-17. An ideal current feedback amplifier has no input impedance at its inverting input ($R_{in} = 0$), infinite input impedance at its noninverting input, and the voltage at the inverting input is held at that of the noninverting input by a unity gain buffer. The transfer function of this amplifier (V_{out}/I_{in}) is a dimensional quantity with the dimension of a resistance, not a ratio, as in the case of voltage feedback op amps. Because of the very low (ideally zero) inverting input impedance, the current feedback op amp has a bandwidth more or less independent of closed-loop gain for a fixed feedback resistor R_{FB}. This implies that the product of

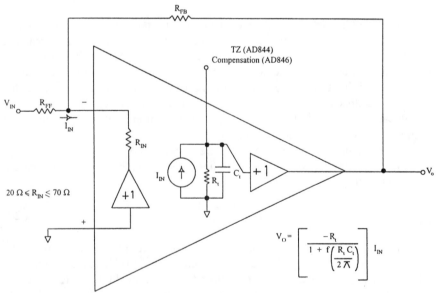

FIGURE 2-17 Equivalent circuit of current feedback op amp

the closed-loop gain and the closed-loop bandwidth is not a constant; hence, it is inappropriate to apply the term *gain-bandwidth product* to current feedback amplifiers. Examples of practical devices are AD844 and AD846 from Analog Devices and HA5004 from Harris Semiconductors. Figure 2-18 compares the frequency response of CFOA with VOA.

The practical current feedback amplifier is based on a common base input stage, as shown in Figure 2-19. This configuration is characterized by a high impedance noninverting input, which drives the inverting input via a unity gain buffer. The open-loop inverting input impedance is approximately equal to any resistance in series with the emitter (R_E) plus the dynamic input impedance of the grounded base transistor (r_E). The output voltage is controlled by the input current and is related to it by the transimpedance gain expressed in ohms. Because of the low-impedance inverting input common base stage, a stepped input voltage ΔV_{in}

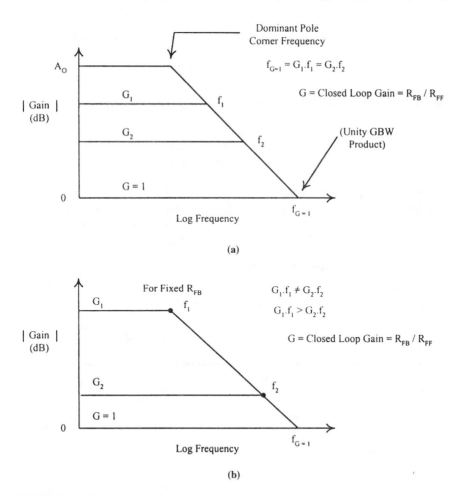

FIGURE 2-18 A comparison of frequency response: (a) VOA, (b) CFOA

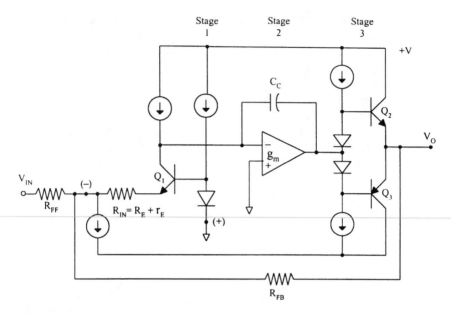

FIGURE 2-19 Simplified current feedback op amp. (Reproduced by permission of Analog Devices, Inc.)

will produce a stepped current in the emitter and collector of Q1. This will yield a slew rate of I_{in}/C_c. Therefore, the rise and fall times of the output essentially are constant regardless of the output voltage swing. This attribute provides for exceptional full-power bandwidth. Bandwidth characteristics for several current feedback op amps are shown in Table 2-2. For further details on applications and design techniques on CFOAs, see Analog Devices (1990), Lidgey and Hayateleh (1997), Toumazou et al. (1990, Chapter 16), and Mancini and Lies (1995).

2.4.3 Single-Rail Op Amps

Single-supply operation is becoming an increasingly important requirement as systems get smaller, cheaper, and more portable. Portable systems rely on

TABLE 2-2 Bandwidth Comparison of Few CFOA. (Reproduced by permission of Analog Devices, Inc.)

Device	R_{FB}	R_{FF}	Gain	Closed-Loop Bandwidth (CLBW)	Gain-Bandwidth Product (GBW)
AD846	1000 Ω	1000 Ω	−1	46 MHz	46 MHz
	1000 Ω	100 Ω	−10	20 MHz	200 MHz
AD844	500 Ω	500 Ω	−1	60 MHz	60 MHz
	500 Ω	50 Ω	−10	33 MHz	330 MHz
AD9617	400 Ω	400 Ω	±1	185 MHz	185 MHz
	400 Ω	27 Ω	±15	105 MHz	1575 MHz
AD9611	1000 Ω	1000 Ω	−1	300 MHz	300 MHz
	1000 Ω	50 Ω	−20	200 MHz	4000 MHz

a battery as the primary power source. Consequently, power consumption and operating time per battery charge are high on the designers' priority list. This makes low-voltage operation and low power consumption critical.

Most general purpose op amps are designed to operate from ±15 V supplies. Trying to make them work at low voltages may require special attention because few are specified to operate at lower voltages. Many even will not function at 5 V or less. In modern products designed to operate from batteries (such as two 1.5 V cells), designers must consider some aspects that are less serious in high rail voltage-based systems. Some important ones follow:

1. A low signal swing compresses signal-to-noise performance.
2. The noise floor tends to rise at low currents.
3. Choosing the ground reference becomes important.
4. The bandwidth suffers as the supply current drops.

2.4.3.1 Effect of Reduced Signal-to-Noise Performance

The most immediate effect on an amplifier circuit running on single supply at a reduced voltage is the reduction in signal swing. Even if the noise floor remains constant (highly unlikely), the signal-to-noise performance will drop as the signal amplitude decreases.

Most op amps that are designed to operate from ±15 V usually have no problem with output headroom: 3–4 V is sufficient for most applications. However, if the supply voltages drop to, say +12 V, a 3 V headroom requirement (at each rail) would limit the signal swing to 6 V P-P (from +3 V to +9 V). Dropping the voltage further to +5 V will render these op amps inoperable, as no output swing capability is left.

For this reason and others, low-voltage single-supply op amps are designed so that their output stage swings as closely to both rails as possible. This helps to restore some of the lost signal-to-noise performance. It is useful if the output can swing to the negative rail. Many op amps have this capability. Single-rail op amps are designed to require 1 or 2 V at most as headroom. Some special designs could have even less headroom, allowing them to operate at +5 V or less. The drawback is that most of these op amps are not designed to source much output current, usually 5 mA or less.

In addition to being compressed at the high-level end by signal reduction, the signal-to-noise performance generally is squeezed from the bottom end as well, as noise floor tends to rise. This is because the single supply usually accompanies an inevitable drive toward lower supply current consumption, which tends to increase noise. The designer must decide how these issues affect the final performance target of the system and make the necessary trade-offs.

2.4.3.2 Ground Reference

Most dual-rail op amps use the signals referenced to 0 V, which is the midpoint of ±V_{cc} rail. This is the most convenient, as there is ample supply headroom to work with. Such is not the case for single-supply circuits. Ground reference can be chosen anywhere within the supply range of the circuit. There is no standard to follow. Indeed, the choice of ground reference depends on the types of signals used. To illustrate this point, choosing the negative rail as the ground reference may optimize the signal dynamic range of an op amp designed to swing to 0 V. On the other hand, the signal may have to be level shifted to be compatible with another device that is not designed to operate at 0 V input because the signal no longer can be inverted. These limitations can force an inefficient design, increasing the cost of the system.

Choosing the negative rail as the ground reference is the simplest and most natural choice for setting the ground reference. Since the negative rail is the return of the supply, its impedance is very low and it makes an ideal ground reference. All signals can be returned to this point without concern about its ability to sink the current, assuming the current is within the rated capability of the supply. Nevertheless, the voltage drop in negative rail ground returns can be a problem, and their resistance must be kept low.

Setting the ground reference at the negative rail usually allows the maximum signal dynamic range, as it establishes the limit of one end of the signal range. However, this may present a problem interacting with other devices, as their inputs or outputs are not designed to operate near the negative rail. Therefore, it is important to know what types of devices are needed for the application, knowing that the signal will swing to the negative rail.

Not all single-supply circuits work well with the 0 V ground system. For example, audio or video signals may be best handled using a false-ground system by biasing the amplifiers to the midpoint of the supply. Then the signal can be AC coupled through the amplifier chain. This removes the necessity of level shifting at each amplifier stage, as otherwise would be required if the 0 V ground system is used. It also saves the cost of additional components to do the level shifting.

It often is assumed that a false-ground circuit need be only a simple buffer amplifier without the bypass capacitors and compensation. In some cases, one can get by without them as long as the false-ground node sees no dynamic or transient load changes. These can occur when driving the ground pin of a D/A or an A/D converter. In these applications, the false-ground node must hold its voltage constant with minimum perturbation. In the presence of reference "bounce," conversion error may result or noise may be injected into the circuit.

The choice of the quality of the false-ground rests on whether the circuit is sensitive to false-ground perturbation, both in DC and AC terms. Not all applications necessarily require a high-quality, low-impedance false-ground. In fact, in some cases, it may be sensible to use both false-ground and negative-rail ground references in different parts of the circuit. One needs to observe the level shifting requirement at interfaces of the two; it depends on what works best.

A solid false-ground can be implemented easily using a voltage divider or reference voltage buffered by an amplifier. The choice of the op amp and the implementation are critical to a good reference node with no reference "bounce." An example is shown in Figure 2-20, and further details may be found in Analog Devices (1992).

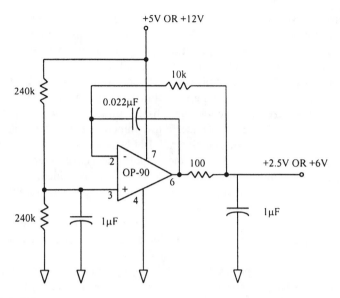

FIGURE 2-20 A supply splitter reference generator for OP-90. (Reproduced by permission of Analog Devices Inc.)

2.4.3.3 Handling 0 V Input Signals

Dual-rail op amps, which generally are designed with NPN input transistor pairs, usually cannot handle signals that are at or near the negative rail, due to the requirement of 2–3 V headroom at either rail. Operating at or near the negative rail would cause the amplifier to go nonlinear as saturation is approached. Even worse, it may cause the output to reverse its phase.

Op amps designed for single-supply operation typically use PNP input pairs to allow the input to operate linearly at or near the negative rail. Another input architecture that can operate at 0 V uses MOSFET transistors in a CMOS amplifier. An example is OP-80 from Analog Devices, which has a P-channel input pair.

2.4.3.4 Handling 0 V Output Signals

Ideally, a single-supply op amp's output should be capable of swinging to the negative rail, especially if it is feeding into the input of the next stage that can operate at 0 V. Beware that, as the output swings very near the negative rail, the output accuracy can drop off rapidly, as a function of the (output sink) load current. This is because a single-supply amplifier's output stage usually has either an active transistor pull-down or it may require a resistor to pull-down to the negative rail. Whatever the means, the pull-down's finite resistance develops a voltage drop that prevents the output from swinging fully to the negative rail. If accuracy is required at or near 0 V, the load current must be kept as small as practical. Always keep in mind that the total load current also includes the current flow in the feedback resistor. As mentioned previously, some op amps have an active output drive to the negative rail. Others require a resistor pull-down to achieve a complete swing to 0 V. Figure 2-21 shows both topologies.

Figure 2-21(a)'s OP-80 output stage relies on the internal on-resistance of Q_{OUT2} to pull the output down to the negative rail. Including a source resistance, the total pull-down resistance typically is 400 Ω. As long as the load current is less than 1μA, its output will swing to within 1 mV of the negative rail.

The OP-90 utilizes a conventional push-pull emitter-follower output stage, as shown in Figure 2-21(b). This provides a low-impedance drive except for output voltages less than 0.5 V to the negative rail. Below this voltage, the bottom output transistor Q_{OUT2} saturates, clamping the output at a base-emitter voltage above the negative rail. If no external pull-down resistor is provided, the top output transistor Q_{OUT1} turns off completely, as its base voltage tends to pull less than its base-emitter voltage, cutting off the base drive. However, a pull-down resistor would keep the Q_{OUT1} active, and therefore linear operation would continue as the output swings to the negative rail. For applications and other design data about these examples, see the device data sheets.

Op amps whose output is designed to swing from one rail to the other are ideally suited for single, reduced supply operation. Theoretically, the maximum

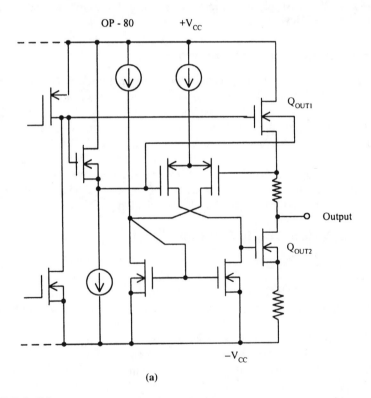

(a)

FIGURE 2-21   Zero-volt swing output stage designs: (a) OP-80 output stage, (b) OP-90 output stage. (Reproduced by permission of Analog Devices Inc.)

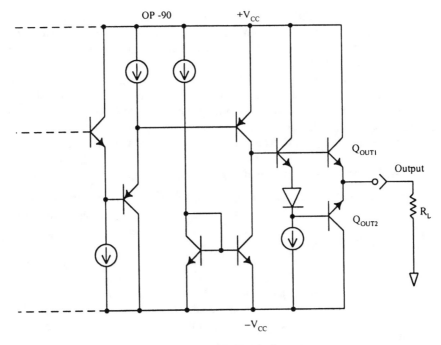

Note : R_L Required

(b)

FIGURE 2-21 Continued

signal-to-noise performance for a given supply voltage can be achieved. One such rail-to-rail device is the OP-295 dual operational amplifier (Analog Devices, 1992).

Most industry standard precision op amps are designed to operate with dual, higher voltage (± 15 V) supplies. Their precision performance is optimized for these supply voltages. As such, trying to operate them at lower voltages may be inadvisable, as their specifications may no longer apply. Indeed, many do not function even if their input is forced to the negative rail or if they are powered by a single +5 V supply.

Very few single-supply op amps exhibit sufficiently low input offset and offset drift to qualify them as precision op amps. The OP-90, OP-290, and OP490 come close to precision performance yet can operate off a single supply.

For a comprehensive practical discussion on single-supply operation, devices, applications, and restrictions, see Analog Devices (1992).

2.4.4 Micro-Power Op Amps

Numerous op amps on the market perform well at sup
500 µA–1 mA range, but certain applications require devices

lower currents. For example, applications that rely on batteries or solar cells need to keep current drain to a minimum. Low-current operation also is essential for minimizing power dissipation in equipment containing large quantities of tightly packed active components.

Micro-power op amps can meet these needs. Although definitions of the term vary, all micro-power devices perform at currents lower than the 500 μA minimum of "low-power" devices. Op amps that operate below about 250 μA supply currents generally can be categorized as micro-power devices. Table 2-3 indicates a representative group of micro-power op amps and their characteristics.

2.4.5 Chopper-Stabilized Op Amps

The best bipolar op amps may have offsets as low as 10 μV. When a design requires an extremely low input offset voltage (V_{os}) with virtually no offset-voltage drift over time or temperature, chopper-stabilized devices are available from many manufacturers.

Low offset-voltage drift probably is the most attractive characteristic of chopper-stabilized op amps. Applications that take advantage of this feature include strain-gauge amplifiers, thermocouple amplifiers, and precision data collection in environments where periodic adjustments are either impractical or impossible. Most often these applications involve low-frequency signals (less than 10 Hz).

Figure 2-22 shows the basic operation of a chopper-stabilized op amp. The device consists of a clock generator, a main and a nulling amplifier, and two holding capacitors, which can be either internal or external. The main amplifier has fixed connections to the input and output pins.

TABLE 2-3 Representative Micro-Power Op Amps

Manufacturer	Type Number	Supply Voltage (V)	Supply Current (μA)	Input Offset Voltage (mV)	Input Bias Current (nA)	Typical Gain× Band-width (kHz)	Typical Slew Rate (V/μs)
Analog Devices	AD548	±4.5 to ±18	200	2	0.02	1000	1.8
Harris	HA7711	±2 to ±8	200	0.25	0.02	800	0.45
Semiconductor	HA7712	±2 to ±8	25	0.25	0.02	100	0.04
Linear Technology	LT1077	5	60	0.06	11	230	0.08
Maxim Integrated Circuits	MAX951	+2.4 V to 7 V	7 μA	1	0.003	20	0.125
Precision Monolithics	OP-282 (dual)	±15	250 (per op amp)	2	0.1	4000	9
SGS-Thomson	TS-271	4 to 10	15	10	0.15	100	0.04
Siliconix	L144 (triple)	±1.5 to ±15	133 (per op amp)	10	250	600	0.4
xas ruments	TL251C	1.4 to 16	20	10	0.6	100	0.04

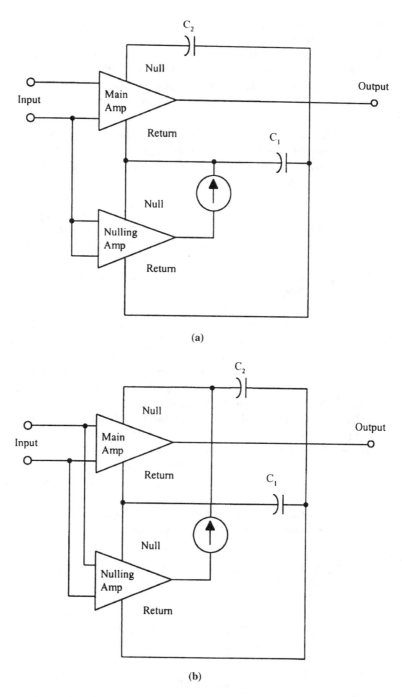

FIGURE 2-22 Basic operation of a chopper-stabilized op amp (a) Storing of offset of nulling amplifier (b) Storing of offset of main amplifier

The nulling amplifier alternately performs two tasks. First, it uses a holding capacitor (C_1) to store a correction for its own offset, which is measured while the amp's input pins are shorted. The nulling amp then measures the offset of the main amplifier and stores a correction voltage in another capacitor (C_2). The amplifier alternates between these tasks at the rate of the clock. Because the op amp is constantly correcting its offset, offset voltages are low and drift is virtually nonexistent.

Practical examples of such op amps are ICL7650S from Intersil or Harris Semiconductor and MAX430 from Maxim Integrated Circuits. Typical specifications of the ICL76505 type devices are 5 µV of V_{os} and an input bias current of 10 pA at 25°C. Offset drift averages 0.02 µV/°C within the range of −25 to 80°C.

The ICL76505 requires external holding capacitors to operate, but devices such as MAX430 have 0.1 µF chip capacitors bonded to the lead frame within the device. Such devices that operate from ±15 V power supplies allow you to directly replace conventional LM741 or LM108 devices.

Early chopper-stabilized amplifiers actually used relays to chop the signal, but today such amplifiers are monolithic and use MOS switches to do the job. Early monolithic amplifiers of this type had high noise at the chopping frequency (which may be between a few hundred Hz and a few tens of kHz). This high-frequency noise is less of a problem in the latest devices, which may contain quite effective filters at the chopper frequency (although careful layout and supply decoupling still are important with these parts), but switching noise remains the major problem with chopper-stabilized op amps. Because of the technology used to manufacture them (frequently CMOS), many chopper-stabilized op amps have a voltage noise of several µV in the band 0.1–10 Hz, and it therefore is necessary to integrate their output for several seconds or tens of seconds to obtain the low offset of which they are potentially capable. Not all systems allow such long integration times, and so a compromise becomes necessary between offset and speed. Table 2-4 compares the precision op amps versus chopper stabilized amplifiers.

For amplifiers that have no built-in capacitors, such as the Intersil 7650S and the 420 series from Intersil and Maxim, you must select the hold capacitors

TABLE 2-4 Comparison of Precision Op Amps vs. Choppers. (Reproduced by permission of Analog Devices Inc.)

Parameter	Precision (Bipolar) Op Amps	Chopper
Offset voltage	10–50 µV	<5 µV
Offset drift	0.1 µV/°C	~0 µV
Open-loop gain	10^7	10^7
Noise: HF glitch	None	>100 mV P-P
Noise: 0.1–10 Hz	<0.2 µV P-P	>1 µV P-P
Cost	Lower	Higher
External components	None	Some require 2 capacitors
Saturation recovery	10–20 µs	>100 ms to seconds

carefully. Noise really is not the critical issue; any low-leakage capacitor is adequate in this regard. But settling time is an issue; capacitors with a high dielectric absorption, such as ceramic capacitors, can take several seconds to settle after power first is applied. Therefore, you should use Mylar or polypropylene capacitors if you need fast initial settling.

When a chopper-stabilized amplifier is overloaded due to saturation, capacitors can acquire excessive charges and it could take a long time to recover when the overload is removed. To avoid the recovery time problem, many chopper-stabilized op amps use clamp circuits that prevent the amplifier from reaching saturation during an overload. Figure 2-23 shows a circuit that prevents the op amps from reaching saturation. When V_{OUT} approaches either rail, the appropriate clamp transistor begins conducting. Connecting the clamp pin to the amplifier's inverting input pin puts that transistor in parallel with the gain resistor. As the transistor conducts, it reduces the gain of the amplifier, thus preventing saturation.

But using the clamp has two drawbacks. First, the available output range of the amplifier is reduced by as much as 1 V from each rail. Second, the leakage of the clamp transistors shows up as additional bias current at the input and reduces accuracy. Typical leakage currents range from 1 to 10 pA.

The sampling techniques that chopper-stabilized op amps use to eliminate drift can result in intermodulation as well as clock noise. Interaction between the input signals and the clock can generate intermodulation products in the form of sum and difference signals. If the input signal's frequency is close to the rate of the clock, the difference product shows up as additional offset error. You

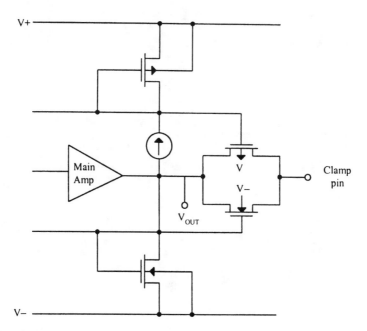

FIGURE 2-23 A technique for preventing saturation of chopper-stabilized op amps

can eliminate this error by filtering the input signal to keep its frequency range well below the sampling clock frequency. To avoid intermodulation problems, many chopper amps allow the designer, using an external signal, to set the clock frequency of the amplifier. For further details, see Quinnell (1989) and Harris Semiconductor (1993–94).

2.4.6 High-Voltage Power Op Amps

For power amplifier applications, special components are available from a limited number of manufacturers. Today, high-voltage amplifiers with total (rail-to-rail) voltages of 1200 are capable of driving around 75 mA to a load. The PA 89 hybrid IC from Apex Microtechnology Corporation is an example. Similarly, high-current amplifiers, which can handle as high as 30 A, with 150 V rail to rail, are available from the some manufacturers. The PA03 power operational amplifier from Apex is an example. Some applications of such special power op amps are sonar transducer drivers, piezoelectric transducer drivers, high-voltage instrumentation, programmable power supplies, and linear and rotary motor drivers. Figure 2-24 shows some of these hybrid devices.

In these types of op amps, the designer has to consider many special situations, such as power supply performance, thermal management, safe operating area, and stability. A discussion of these topics is beyond the limits of this chapter.

FIGURE 2-24 Power operational amplifiers. (Reproduced by permission of Apex Microtechnology Corporation.)

Application notes in Apex Microtechnology Corp. (1996, pp. D1–D105) provide an excellent discussion.

References

Analog Devices. Linear Design Seminar. Analog Devices Inc., USA, October 1987.

Analog Devices. High Speed Design Seminar. Analog Devices Inc., USA, 1990.

Analog Devices. *Amplifier Applications Guide*. Analog Devices Inc., USA, 1992.

Analog Devices. Linear Design Seminar. Analog Devices Inc., USA, 1995.

Apex Microtechnology Corp. Apex Power Integrated Circuits, vol. 7. Apex Microtechnology Corp., USA, 1996.

Baker, Bonnie, and Jerald Graeme. "Systematic Approach Makes Op Amp Circuits Resist Radiated Noise." *EDN* (July 20, 1995), pp. 93–100.

Brokraw, P. "An IC Amplifier User's Guide to Decoupling, Grounding and Making Things Go Right for a Change." Application note AN-202. Analog Devices Inc., USA, ● ● ●.

Dostal, J. *Operational Amplifiers*, 2d ed. Boston: Butterworth–Heinemann, 1993.

Fleming, T. "Monolithic Op Amps." *EDN* (September 3, 1987), pp. 118–132.

Graeme, Jerald. *Optimizing Op Amp Performance*. New York: McGraw-Hill, 1997a.

Graeme, Jerald. *Photodiode Amplifiers: Op Amp Solutions*. New York: McGraw-Hill, 1997b.

Goodenough, Frank. "Amps Put out Fast, High Voltage Linear Power." Electronic Design (March 23, 1989), pp. 99–102.

Harold, Peter. "Current-Feedback Op Amps Ease High Speed Circuit Design." *EDN* (July 7, 1988), pp. 84–94.

Harold, Peter. "Micropower Op Amps Hit New Laws." *EDN* (April 27, 1989), pp. 117–129.

Harris Semiconductor. Linear and Telecom IC Data Book. Harris Semiconductor, USA, 1993–94, pp. 2-694–2-705.

Horowitz, P., and W. Hill. *The Art of Electronics*, 2d ed. Cambridge: Cambridge University Press UK, 1996.

Lidgey, F. J., and K. Hayateleh. "Current Feedback Operational Amplifiers and Applications." *Electronics and Communications Engineering Journal* (August 1997), pp. 176–182.

Mancini, Ronald, and Jeffrey Lies. "Current Feedback Amplifier Theory & Applications"; Application note AN-9420.1. Harris Semiconductor, USA, April 1995.

Pryce, Dave. "Micropower Op Amps: Low Current Devices Offer High Performance." *EDN* (September 17, 1990), pp. 79–84.

Quinnell, Richard A. "Chopper Stabilized Op Amps: Improvements Make Precision Amps Easier." *EDN* (January 19, 1989), pp. 91–96.

Schweber, Bill. "Choosing an Op Amp: It's No Longer a Trying Task." *EDN* (May 25, 1995), pp. 38–52.

Steffes, Michael. "Embedded Gain Supercharges FET-Transimpedance Amplifier." *EDN* (May 22, 1997), pp. 129–143.

Swager, Anne Watson. "High Speed Monolithic Op Amps: Low-Cost Op Amps Break Speed Barriers." *EDN* (January 2, 1992), pp. 53–64.

Toumazou, C., F. J. Lidgey, and D. G. Haigh (Eds.). *Analogue IC Design: The Current-Mode Approach*. IEE Circuits and Systems Series 2. London: Peter Peregrinus, 1990.

CHAPTER **3**

Data Converters

3.1 Introduction

Modern design trends use the power and precision of the digital world of components to process analog signals. However, the link between the digital/processing world and the analog/real world is based on analog-to-digital and digital-to-analog converter ICs, which generally are grouped as the data converters.

Until about 1988, engineers have had to stockpile their most innovative A/D converter (ADC) designs, because available manufacturing processes simply could not implement those designs onto monolithic chips economically. Prior to 1988, except for the introduction of successive approximation and integrating and flash ADCs, the electronics industry saw no major changes in monolithic ADCs. Since then, manufacturing processes caught up with the technology and many techniques such as subranging flash, self-calibration, delta/sigma, and many other special techniques have been implemented on monolithic chips.

High-speed ADCs are used in a wide variety of real-time digital signal processing (DSP) applications, replacing systems that used analog techniques alone. The major reasons for using DSP are that the cost of the processors has gone down, their speed and computational power have increased, and they are reprogrammable, allowing for system performance upgrades without hardware changes. DSP offers practical solutions that cannot be easily achieved in the analog domain; for example, V.32 and V.34 modems.

This chapter provides an overview of design concepts and application guidelines for systems using modern analog/digital and digital/analog converters implemented on monolithic chips.

3.2 Sampled Data Systems

To specify intelligently the ADC portion of the system, one must first understand the fundamental concepts of sampling and quantization and their effects on the signal.

Let us consider the traditional problem of sampling and quantizing a baseband signal whose bandwidth lies between DC and an upper frequency of interest, f_s. This often is referred to as *Nyquist* or *sub-Nyquist sampling*. The topic of super-Nyquist sampling (sometimes called *undersampling*), where the signal of interest falls outside the Nyquist bandwidth (DC to $f_s/2$) is treated later. Figure 3-1 shows key elements of a baseband sampled data system.

3.2.1 Discrete Time Sampling of Analog Signals

Figure 3-2 shows the concept of discrete time and amplitude sampling of an analog signal. The continuous analog data must be sampled at discrete intervals, t_s, which must be carefully chosen to ensure an accurate representation of the original analog signal. It is clear that, the more samples taken (faster sampling rates), the more accurate the digital representation; and if fewer samples are taken (slower sampling rates), a point is reached where critical information about the signal actually is lost.

To discuss the problem of losing information in the sampling process, it is necessary to recall Shannon's information theorem and Nyquist's criteria. Shannon's information theorem:

- An analog signal with a bandwidth of f_a must be sampled at a rate of $f_s > 2f_a$ to avoid loss of information.
- The signal bandwidth may extend from DC to f_a (baseband sampling) or from f_1 to f_2, where $f_a = f_2 - f_1$ (undersampling, or super-Nyquist sampling).

Nyquist criteria:

- If $f_s < 2f_a$, then a phenomenon called *aliasing* will occur.
- Aliasing is used to advantage in undersampling applications.

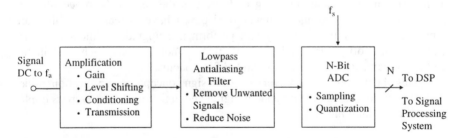

FIGURE 3-1   Key elements of a baseband sampled data system

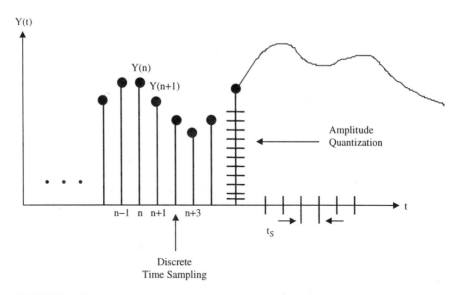

FIGURE 3-2 Sampling and quantizing of an analog signal

3.2.1.1 Implications of Aliasing

To understand the implications of aliasing in both the time and frequency domains, first consider the case of a time domain representation of a sampled sine wave signal shown in Figure 3-3. In Figures 3-3(a) and 3-3(b), it is clear that an adequate number of samples have been taken to preserve the information about the sine wave. Figure 3-3(c) represents the ambiguous limiting condition where $f_s = 2f_a$. If the relationship between the sampling points and the sine wave is such that the sine wave is being sampled at precisely the zero crossings (rather than at the peaks, as shown in the illustration), then all information regarding the sine wave would be lost. Figure 3-3(d) represents the situation where $f_s < 2f_a$, and the information obtained from the samples indicates a sine wave having a frequency lower than $f_s/2$. This is a case where the out-of-band signal is aliased into the Nyquist bandwidth between DC and $f_s/2$. As the sampling rate is further decreased and the analog input frequency f_a approaches the sampling frequency f_s, the aliased signal approaches DC in the frequency spectrum.

Let us look at the corresponding frequency domain representation of each case. From each case of frequency domain representation, we make the important observation that, regardless of where the analog signal being sampled happens to lie in the frequency spectrum, the effects of sampling will cause either the actual signal or an aliased component to fall within the Nyquist bandwidth between DC and $f_s/2$. Therefore, any signals that fall outside the bandwidth of interest, whether they be spurious tones or random noise, must be adequately filtered before sampling. If unfiltered, the sampling process will alias them back within the Nyquist bandwidth, where they will corrupt the wanted signals.

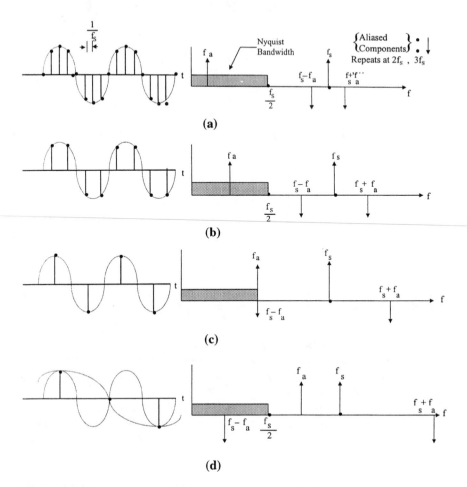

FIGURE 3-3 Time and frequency domain effects of aliasing: (a) $f_s = 8f_a$, (b) $f_s = 4f_a$, (c) $f_s = 2f_a$, (d) $f_s = 1.3f_a$

3.2.1.2 High-Speed Sampling

Now let us discuss the case of high-speed sampling, analyzing it in the frequency domain. First, consider the use of a single-frequency sine wave of frequency f_a sampled at a frequency f_s by an ideal impulse sampler (see Figure 3-4(a)). Also assume that $f_s > 2f_a$ as shown. The frequency domain output of the sampler shows aliases or images of the original signal around every multiple of f_s; that is, at frequencies equal to

$$| \pm Kf_s \pm f_a|, \qquad \text{where } K = 1, 2, 3, 4, \ldots \qquad (3.1)$$

The Nyquist bandwidth, by definition, is the frequency spectrum from DC to $f_s/2$. The frequency spectrum is divided into an infinite number of Nyquist zones, each having a width equal to $0.5f_s$, as shown.

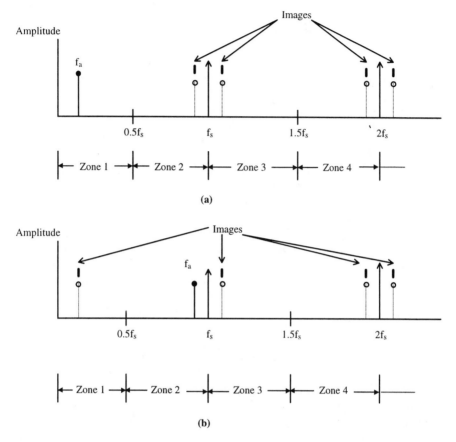

FIGURE 3-4 Analog signal at frequency f_a sampled at f_s: (a) Signal lying within $f_s/2$, (b) Signal lying between $f_s/2$ and f_s

Now consider a signal outside the first Nyquist zone, as shown in Figure 3-4(b). Notice that, even though the signal is outside the first Nyquist zone, its image (or alias), $f_s - f_a$, falls inside. Returning to Figure 3-4(a), it is clear that, if an unwanted signal appears at any of the image frequencies of f_a, it also will occur at f_a, thereby producing a spurious frequency component in the Nyquist zone. This is similar to the analog mixing process and implies that some filtering ahead of the sampler (or ADC) is required to remove frequency components that are outside the Nyquist bandwidth but whose aliased components fall inside it. The filter performance will depend on how close the out-of-band signal is to $f_s/2$ and the amount of attenuation required.

3.2.1.3 Antialiasing Filters

Baseband sampling implies that the signal to be sampled lies in the first Nyquist zone. It is important to note that, with no input filtering at the input of the ideal sampler, any frequency component (either signal or noise) that falls outside the Nyquist bandwidth in any Nyquist zone will be aliased back into the first Nyquist zone. For this reason, an antialiasing filter is used in almost all sampling ADC applications to remove these unwanted signals.

Properly specifying the antialiasing filter is important. The first step is to know the characteristics of the signal being sampled. Assume that the highest frequency of interest is f_a. The antialiasing filter passes signals from DC to f_a while attenuating signals above f_a. Assume that the corner frequency of the filter is chosen to be equal to f_a. The effect of the finite transition from minimum to maximum attenuation on system dynamic range (DR) is illustrated in Figure 3-5.

Assume that the input signal has full-scale components well above the maximum frequency of interest, f_a. The diagram shows how full-scale frequency components above $f_s - f_a$ are aliased back into the bandwidth DC to f_a. These aliased components are indistinguishable from actual signals and therefore limit the dynamic range to the value on the diagram, which is shown as DR.

The antialiasing filter transition band therefore is determined by the corner frequency f_a, the stop band frequency ($f_s - f_a$), and the stop band attenuation DR. The required system dynamic range is chosen based on our requirement for signal fidelity.

Filters have to become more complex as the transition band becomes sharper, all other things being equal. For instance, a Butterworth filter gives 6 dB attenuation per octave for each filter pole. Achieving 60 dB attenuation in a transition region between 1 and 2 MHz (1 octave) requires a minimum of ten poles. This is not a trivial filter and definitely a design challenge. Therefore, other filter types generally are better suited to high-speed applications where the requirement is for a sharp transition band and in-band flatness coupled with linear phase response. Elliptic filters meet these criteria and are a popular choice.

From this discussion, we can see how the sharpness of the antialiasing transition band can be traded off against the ADC sampling frequency. Choosing

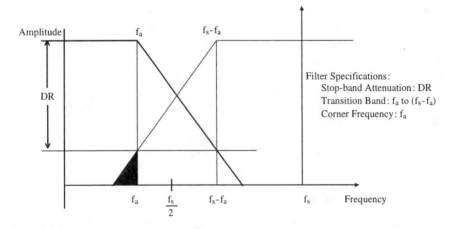

FIGURE 3-5   Effects of antialiasing filters on system dynamic range

a higher sampling rate (oversampling) reduces the requirement on transition band sharpness (hence, the filter complexity) at the expense of using a faster ADC and processing data at a faster rate. This is illustrated in Figure 3-6, which shows the effects of increasing the sampling frequency while maintaining the same analog corner frequency, f_a, and the same dynamic range, DR, requirement.

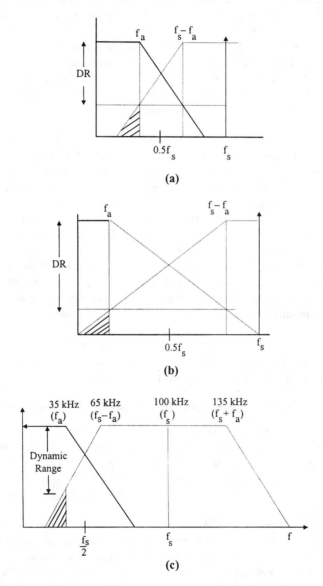

FIGURE 3-6 The relationship between sampling frequency and the antialiasing filter requirement: (a) Low sampling rate with a sharper filter, (b) High sampling rate with a relaxed filter specification, (c) A numerical example

Based on this discussion one could start the design process by selecting a sampling rate of two to four times f_a. Filter specifications could be determined from the required dynamic range based on cost and performance. If such a filter is not realizable, a high sampling rate with a faster ADC will be required.

The antialiasing filter requirements can be relaxed somewhat if it is certain that there never will be a full-scale signal at the stop band frequency, $f_s - f_a$. In many applications, it is improbable that full-scale signals will occur at this frequency. If the maximum signal at the frequency $f_s - f_a$ will never exceed X dB below full scale, the filter stop band attenuation requirement is reduced by that amount. The new requirement for stop band attenuation at $f_s - f_a$ based on this knowledge of the signal now is only $(DR-X)$ dB. When making this type of assumption, be careful to treat any noise signals that may occur above the maximum signal frequency, f_a, as unwanted signals that also alias back into the signal bandwidth.

Properly specifying the antialiasing filter requires a knowledge of the signal's spectral characteristics as well as the system's dynamic range requirements. Consider the signal in Figure 3-6(c), which has a maximum full-scale frequency content of $f_a = 35$ kHz sampled at a rate of $f_s = 100$ kS/s. Assume that the signal has the spectrum shown in Figure 3-6(c) and is attenuated by 30 dB at 65 kHz $(f_s - f_a)$. Observe that the system dynamic range is limited to 30 dB at 35 kHz because of the aliased components. If additional dynamic range is required, an antialiasing filter must be provided to provide more attenuation at 65 kHz. If a dynamic range of 74 dB (12 bits) at 35 kHz is desired, then the antialiasing filter attenuation must go from 0 dB at 35 kHz to 44 dB at 65 kHz. This is an attenuation of 44 dB in approximately one octave; therefore, a seven-pole filter is required. (Each filter pole provides approximately 6 dB attenuation per octave.)

One must consider that broadband noise may be present with the signal, which also can alias within the bandwidth of interest. This is especially true with wideband op amps that provide low distortion levels.

3.2.2 ADC Resolution and Dynamic Range Requirements

Having discussed the sampling rate and filtering, we next discuss the effects of dividing the signal amplitude into a finite number of discrete quantization levels. Table 3-1 shows relative bit sizes for various resolution ADCs, for a full-scale input range chosen as approximately 2 V, which is popular for higher-speed ADCs. The bit size in determined by dividing the full-scale range (2.048 V) by 2^N.

The selection process for determining the ADC resolution should begin by determining the ratio between the largest signal (full-scale) and smallest signals you wish the ADC to detect. Convert this ratio to dB and divide by 6. This is your minimum ADC resolution requirement for DC signals. You actually will need more resolution to account for extra signal headroom, since ADCs act as hard limiters at both ends of their range. Remember that this computation is for DC or low-frequency signals and that the ADC performance will degrade as the input

TABLE 3-1 Bit Sizes, Quantization Noise, and Signal-to-Noise Ratio (SNR) for 2.048 V Full-Scale Converters

Resolution (N Bits)	1 LSB = q	%FS	rms Quantization Noise, $q/\sqrt{12}$	Theoretical Full-Scale SNR (dB)
6	32 mV	1.56	9.2 mV	37.9
8	8 mV	0.39	2.3 mV	50.0
10	2 mV	0.098	580 μV	62.0
12	500 μV	0.024	144 μV	74.0
14	125 μV	0.0061	36 μV	86.0
16	31 μV	0.0015	13 μV	98.1

signal slew rate increases. The final ADC resolution actually will be dictated by dynamic performance at high frequencies. This may lead to the selection of an ADC with more resolution at DC than is required.

Table 3-1 also indicates the theoretical rms quantization noise produced by a perfect N-bit ADC. In this calculation, the assumption is that quantization error is uncorrelated with the ADC input. With this assumption, the quantization noise appears as random noise spread uniformly over the Nyquist bandwidth, DC to $f_s/2$, and it has an rms value equal to $q/\sqrt{12}$. Other cases may be different, and some practical explanation is given in Analog Devices (1995).

3.2.3 Effective Number of Bits of a Digitizer

Table 3-1 shows the theoretical full-scale SNR calculated for the perfect N-bit ADC, based on the formula

$$\text{SNR} = 6.02N + 1.76 \text{ (dB)} \tag{3.2}$$

Various error sources in the ADCs cause the measured SNR to be less than the theoretical value shown in equation (3.2). These errors are due to integral and differential nonlinearities, missing codes, and internal ADC noise sources (some of which are discussed later).

In addition, the errors are a function of the input slew rate and therefore increase as the input frequency gets higher. In calculating the rms value of the noise, it is customary to include the harmonics of the fundamental signal. This sometimes is referred to as the *signal-to-noise-and-distortion*, $S/(N + D)$ or SINAD, but usually simply as SNR.

This leads to the definition of another important ADC dynamic specification, the effective number of bits (ENOB). The effective bits are calculated by first measuring the SNR of an ADC with a full-scale sine wave input signal. The measured SNR (SNR$_{\text{actual}}$ or SINAD) is substituted into the equation for SNR, and the equation is solved for N as shown next:

$$\text{ENOB} = \frac{\text{SINAD} - 1.76 \text{ dB}}{6.02} \tag{3.3}$$

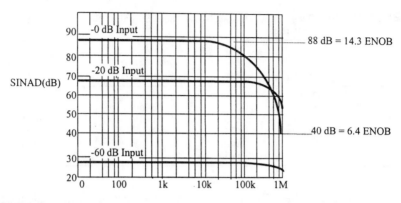

FIGURE 3-7 SINAD and ENOB for the AD676. (Reproduced by permission of Analog Devices Inc.)

For a typical ADC, the AD676 from Analog Devices (a 16-bit ADC) is shown in Figure 3-7.

For this device, the SNR value of 88 dB corresponds to approximately 14.3 effective bits (for 0 dB input), while it drops to 6.4 ENOB at 1 MHz. The methods for calculating ENOB, SNR, and other parameters are described in Analog Devices (1992, Chapter 7) and Tektronix (1986).

In testing ADCs, the SNR usually is calculated using DSP techniques while applying a pure sine wave signal to the input of ADC. A typical test system is shown in Figure 3-8(a). The fast Fourier transform (FFT) processes a finite number of time samples and converts them into a frequency spectrum such as the one shown in Figure 3-8(b) for an AD676-type 16-bit 100 kSPS sampling ADC. The frequency spectrum then is used to calculate the SNR as well as the harmonics of the fundamental input signal.

The rms value of the signal is first computed. The rms value of all other frequency components over the Nyquist bandwidth (this includes not only noise but also distortion products) is computed. The ratio of these two quantities, expressed in decibels is the SNR. Various error sources in the ADC cause the measured SNR to be less than the theoretical value, $6.02N + 1.76$ dB.

3.2.3.1 Spurious Components and Harmonics

The peak spurious or peak harmonic component is the largest spectral component excluding the input signal and DC. This value is expressed in decibels relative to the rms value of a full-scale input signal as was shown in Figure 3-8(a). The peak spurious specification also occasionally is referred to as the *spurious free dynamic range* (SFDR). SFDR usually is measured over a wide range of input frequencies and at various amplitudes. It is important to note that the harmonic distortion or SFDR of an ADC is not limited by its theoretical SNR value. The SFDR of a 12-bit ADC may exceed 85 dB, while the theoretical SNR is only 74 dB. On the other hand, the SINAD of the ADC may be limited by poor

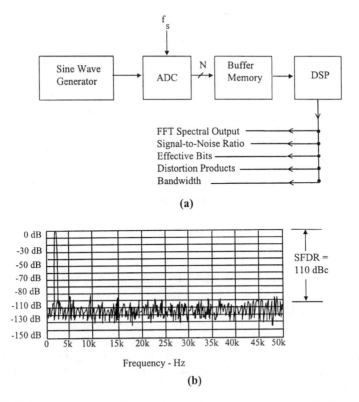

FIGURE 3-8 Testing an ADC for its performance parameters: (a) Test system, (b) Typical FFT output for an AD676 ADC. (Reproduced by permission of Analog Devices Inc.)

harmonic distortion performance, since the harmonic components are included with the quantization noise when computing the rms noise level. The SFDR of an ADC is defined as the ratio of the rms signal amplitude to the rms value peak spurious spectral content (measured over the entire first Nyquist zone, DC to $f_s/2$). The SFDR generally is plotted as a function of signal amplitude and may be expressed relative to the signal amplitude (dBc) or the ADC full scale (dBFS).

For a signal near full scale, the peak spectral spur generally is determined by one of the first few harmonics of the fundamental input signal. However, as the signal falls several decibels below full scale, other spurs generally occur that are not direct harmonics of the input signal due to the differential nonlinearity of the ADC transfer function. Therefore, the SFDR considers all sources of distortion, regardless of their origin.

The total harmonic distortion (THD) is the ratio of the rms sum of the harmonic components to the rms value of an input signal, expressed in a percentage or decibels. For input signals or harmonics above the Nyquist frequency, the aliased components are used in making the calculation. The THD usually is measured at several input signal frequencies and amplitudes. Figure 3-9 shows the

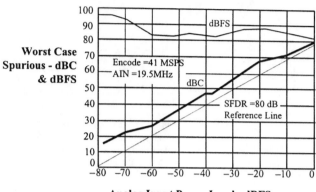

Analog Input Power Level - dBFS

FIGURE 3-9   SFDR vs. input power level for an AD9042. (Reproduced by permission of Analog Devices, Inc.)

SFDR performance of a 12-bit, 41 MSPS wideband ADC designed for communication applications (AD9042 from Analog Devices).

Note that a minimum of 80 dBc SFDR is obtained over the entire first Nyquist zone (DC to 20 MHz). The plot also shows SFDR expressed as dBFS. The SFDR generally is much greater than the ADCs theoretical N-bit SNR ($6.02N + 1.76$ dB). For example, the AD9042 is a 12-bit ADC with an SFDR of 80 dBc and a typical SNR of 65 dBc (the theoretical SNR is 74 dB). This is due to the fundamental distinction between noise and distortion measurements.

3.3 A/D Converter Errors

First, let us look at how bits are assigned to the corresponding analog values in a typical analog-to-digital converter. The method of assigning bits to the corresponding analog value of the sampled point often is referred to as *quantization* (see Figure 3-10(a)). As the analog voltage increases, it crosses transitions of "decision levels," which causes the ADC to change state. In an ideal ADC, the transitions are at half-unit levels, with Δ representing the distance between the decision levels. The Δ is often is referred to as the *bit size* or *quantization size*. The fact that Δ always has a finite size leads to uncertainty, since any analog value within the finite range can be represented. This quantization uncertainty is expressed as plus or minus half the least significant bit (LSB) as shown in Figure 3-10(b). As this plot shows, the output of an ADC may be thought of as the analog signal plus some quantizing noise. The more bits the ADC has, the less significant this noise becomes.

Certain parameters limit the rate at which an ADC can acquire a sample of the input waveform: the acquisition turn-on delay, acquisition time, sample or track time, and hold time. Figure 3-10(c) shows a graphic representation of the acquisition cycle of a typical ADC. The turn-on time (the time the device takes

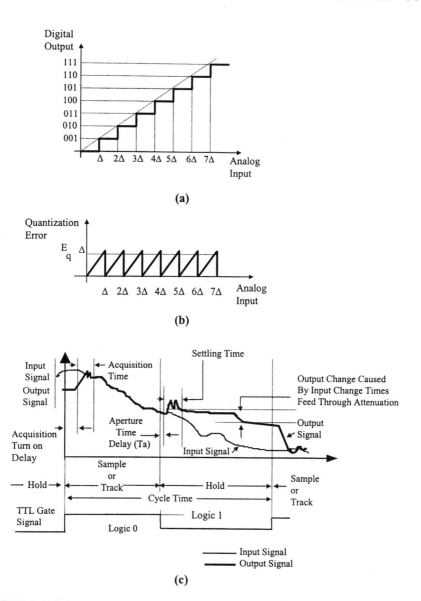

FIGURE 3-10 Quantization process and quantizing error: (a) Basic input/output characteristics, (b) Quantization error, (c) Basic timing of acquisition cycle of an ADC

to get ready to acquire a sample) is the first event. The acquisition time is next. This is the time the device takes to get to the point at which the output tracks the input sample, after the sample command or clock pulse. The aperture time delay is the time that elapses between the hold command and the point at which the sampling switch is completely open. The device then completes the hold cycle and the next acquisition is taken.

This process indicates that the real world of acquisition is not an ideal process at all, and the value sampled and converted could have some sources of error. Most of these errors increase with the sampling rate.

The approximation or "rounding" effect in A/D converters is called *quantization*, and the difference between the original input and the digitized output, the quantization error, is denoted here by ε_q. For the characteristic of Figure 3-10(a), ε_q varies as shown in Figure 3-10(b), with the maximum occurring before each code transition. This error decreases as the resolution increases, and its effect can be viewed as additive noise (quantization noise) appearing at the output. Thus, even an "ideal" *m*-bit ADC introduces nonzero noise in the converted signal simply due to quantization.

We can formulate the impact of quantization noise on the performance as follows. For simplicity, consider a slightly different input/output characteristic, shown in Figure 3-11(a), where code transitions occur at odd (rather than even)

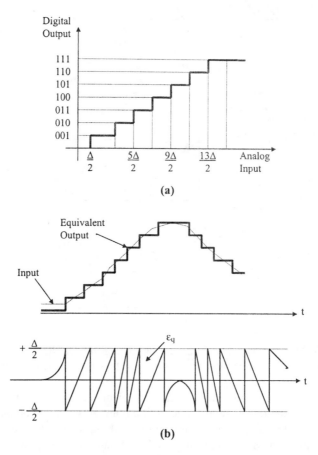

FIGURE 3-11 Modified ADC characteristics and effect of amplitude quantization on a time domain waveform: (a) Modified ADC characteristics, (b) Effect of quantization

multiples of $\Delta/2$. A time domain waveform therefore experiences both negative and positive quantization errors, as illustrated in Figure 3-11(b). To calculate the power of the resulting noise, we assume that ε_q is (i) a random variable uniformly distributed between $-\Delta/2$ and $+\Delta/2$, and (ii) independent of the analog input. While these assumptions are not strictly valid in the general case, they usually provide a reasonable approximation for resolutions above four bits. Razavi (1995) provides more details and the derivations of equations (3.2) and (3.3).

Full specification of the performance of ADCs requires a large number of parameters, some of which are defined differently by different manufacturers. Some important parameters frequently used in component data sheets and the like are described here. Figure 3-12 could be used to illustrate parameters such as differential nonlinearity (DNL), integral nonlinearity (INL), and offset and gain errors—all static parameters of the ADC process.

3.3.1

3.3.1.1 Differential Nonlinearity

Differential nonlinearity is the maximum deviation in the difference between two consecutive code transition points on the input axis from the ideal value of 1 LSB. The DNL is a measure of the deviation code widths from the ideal value of 1 LSB.

3.3.1.2 Integral Nonlinearity

The INL is the maximum deviation of the input/output characteristic from a straight line passed through its end points (line AB in Figure 3-12). The overall difference plot is called the *INL profile*. The INL is the deviation of code centers from the converter's ideal transfer curve. The line used as the reference may be drawn through the end points or may be a best-fit line calculated from the data.

The DNL and INL degrade as the input frequency approaches the Nyquist rate. The DNL shows up as an increase in quantization noise, which tends to elevate the converter's overall noise floor. Theoretical quantization noise for an

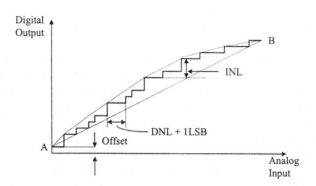

FIGURE 3-12 Static ADC metrics

ideal converter with the Nyquist bandwidth is

$$\text{rms quantization noise} = \frac{q}{\sqrt{12}} \tag{3.4}$$

where q is the weight of the LSB.

At the same time, because the INL appears as a bend in the converter's transfer curve, it generates spurious frequencies (spurs) not in the original signal information. The testing of ADC linearity parameters is discussed in Shill (1995).

3.3.1.3 Offset Error and Gain Error

The offset is a vertical intercept of the straight line through the end points. The gain error is the deviation of the slope of line AB from its ideal value (usually unity).

3.3.1.4 Testing of ADCs

A known periodic input is converted by an ADC under test at sampling times that are asynchronous relative to the input signal. The relative number of occurrences of the distinct digital output codes is termed the *code density*. For an ideal ADC, the code density is independent of the conversion rate and input frequency. These data are viewed in the form of a normalized histogram showing the frequency of occurrence of each code from zero to full scale. The code density data are used to compute all bit transition levels. Linearity, gain, and offset errors are readily calculated from a knowledge of the transition levels. This provides a complete characterization of the ADC in the amplitude domain.

The effect of some of these static errors in the frequency domain for high-speed ADCs is discussed in Louzon (1995). Doernberg, Lee, and Hodges (1984) provide ADC characterization methods based on code density test and spectral analysis using FFT.

3.4 Effects of Sample and Hold Circuits

The sample and hold amplifier (SHA) is a critical part of many data acquisition systems. It captures an analog signal and holds it during some operation (most commonly during analog-to-digital conversion). The circuitry involved is demanding, and unexpected properties of commonplace components such as capacitors and printed circuit boards may degrade SHA performance.

When a sample and hold amplifier is in the sample mode, the output follows the input with only a small voltage offset. In some SHAs, the output during the sample mode does not follow the input accurately and the output is accurate only during the hold period.

Today, high-density IC processes allow the manufacture of ADCs containing an integral SHA. Wherever possible, ADCs with an integral SHA (often known as *sampling ADCs*) should be used in preference to separate ADCs and SHAs.

The advantage of such a sampling ADC — apart from the obvious ones of smaller size, lower cost, and fewer external components — is that the overall performance is specified. The designer need not spend time ensuring that no specification, interface, or timing issues are involved in combining a discrete ADC and a discrete SHA.

3.4.1 Basic SHA Operation

Regardless of the circuit details or type of SHA in question, all such devices have four major components. The input amplifier, energy storage device (capacitor), output buffer, and switching circuits are common to all SHAs, as shown in the typical configuration of Figure 3-13(a).

The energy storage device, the heart of the SHA, almost always is a capacitor. The input amplifier buffers the input by presenting a high impedance to the signal source and providing current gain to charge the hold capacitor. In the track mode, the voltage on the hold capacitor follows (or tracks) the input signal (with some delay and bandwidth limiting). Figure 3-10 depicts this process. In the hold mode, the switch is opened and the capacitor retains the voltage present before it was disconnected from the input buffer. The output buffer offers a high impedance to the hold capacitor to keep the held voltage from discharging prematurely. The switching circuit and its driver form the mechanism by which the SHA is alternately switched between track and hold.

Four groups of specifications describe basic SHA operation: track mode, track-to-hold transition, hold mode, and hold-to-track transition. These specifications are summarized in Table 3-2, and some of the SHA error sources are shown in Figure 3-13(b). Because of both DC and AC performance implications for each of the four modes, properly specifying an SHA and understanding its operation in a system are complex matters.

3.4.1.1 Track Mode Specifications

Since an SHA in the sample (or track) mode is simply an amplifier, both the static and dynamic specifications in this mode are similar to those of any

TABLE 3-2 Sample and Hold Specifications

	Track Mode	Track-to-Hold Transition	Hold Mode	Hold-to-Sample Transition
Static	Offset	Pedestal	Droop	
	Gain error	Pedestal nonlinearity	Dielectric	
	Nonlinearity		absorption	
Dynamic	Settling time	Aperture delay	Feed through	Acquisition time
	Bandwidth	Time	Distortion	Switching transient
	Slew rate	Aperture jitter	Noise	
	Distortion	Switching transient		
	Noise	Settling time		

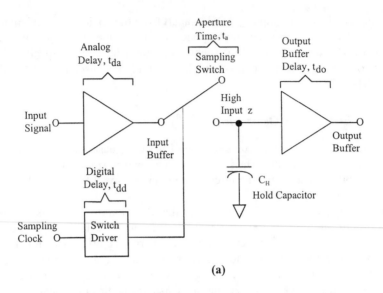

(a)

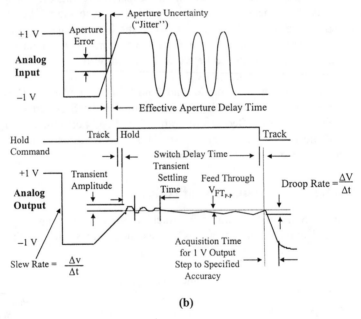

(b)

FIGURE 3-13 Sample and hold circuit and error sources: (a) Basic sample and hold circuit, (b) Some sources of error

amplifier. The principal track mode specifications are offset, gain, nonlinearity, bandwidth, slew rate, settling time, distortion, and noise; however, distortion and noise in the track mode often are of less interest than in the hold mode. Fundamental amplifier specifications are discussed in Chapter 2.

3.4.1.2 Track-to-Hold Mode Specifications

When the SHA switches from track to hold, generally a small amount of charge is dumped on the hold capacitor because of nonideal switches. This results in a hold mode DC offset voltage called *pedestal error*. If the SHA is driving an ADC, the pedestal error appears as a DC offset voltage that may be removed by performing a system calibration. If the pedestal error is a function of the input signal level, the resulting nonlinearity contributes to hold mode distortion. Pedestal errors may be reduced by increasing the value of the hold capacitor with a corresponding increase in acquisition time and a reduction in bandwidth and slew rate.

Switching from track to hold produces a transient, and the time required for the SHA output to settle to within a specified error band is called the *hold mode settling time*. Occasionally, the peak amplitude of the switching transient also is specified (see Figure 3-14).

3.4.1.3 Aperture and Aperture Time

Perhaps the most misunderstood and misused SHA specifications are those that include the word *aperture*. The most essential dynamic property of an SHA is its ability to disconnect quickly the hold capacitor from the input buffer amplifier (see Figure 3-13(a)).

The short (but nonzero) interval required for this action is called the *aperture time* (t_a). The actual value of the voltage held at the end of this interval is a function of both the input signal and the errors introduced by the switching operation itself. Figure 3-15 shows what happens when the hold command is applied with an input signal of arbitrary slope (for clarity, the sample-to-hold pedestal and switching transients are ignored). The value finally held is a delayed version of the input signal, averaged over the aperture time of the switch, as shown in Figure 3-15. The first-order model assumes that the final value of voltage on the hold capacitor is approximately equal to the average value of the signal applied to the switch over the interval during which the switch changes from a low to a high impedance (t_a).

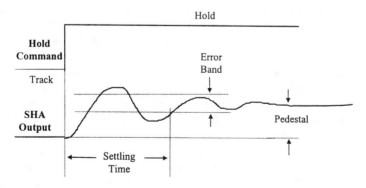

FIGURE 3-14 Hold mode settling time

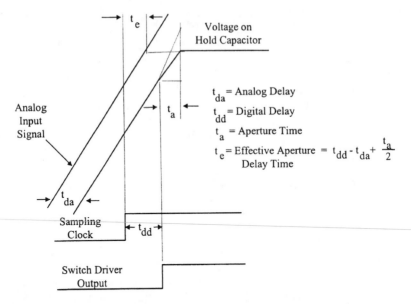

FIGURE 3-15 SHA waveforms

The model shows that the finite time required for the switch to open (t_a) is equivalent to introducing a small delay in the sampling clock driving the SHA. This delay is constant and may be either positive or negative. Called effective *aperture delay time* or simply *aperture delay* (t_e), it is defined as the time difference between the analog propagation delay of the front-end buffer (t_{da}) and the switch digital delay (t_{dd}) plus half the aperture time ($t_a/2$). The effective aperture delay time usually is positive but may be negative if the sum of half the aperture time ($t_a/2$) and the switch digital delay (t_{dd}) is less than the propagation delay through the input buffer (t_{da}). The aperture delay specification thus establishes when the input signal actually is sampled with respect to the sampling clock edge.

The aperture delay time can be measured by applying a bipolar sine wave signal to the SHA and adjusting the synchronous sampling clock delay such that the output of the SHA is 0 during the hold time. The relative delay between the input sampling clock edge and the actual zero crossing of the input sine wave is the aperture delay time (see Figure 3-16).

Aperture delay produces no errors but acts as a fixed delay in either the sampling clock input or the analog input (depending on its sign). If there is sample-to-sample variation in aperture delay (aperture jitter), then a corresponding voltage error is produced, as shown in Figure 3-17. This sample-to-sample variation in the instant that the switch opens, called *aperture uncertainty* or *aperture jitter*, usually is measured in rms picoseconds. The amplitude of the associated output error is related to the rate of change of the analog input. For any given value of aperture jitter, the aperture jitter error increases as the input dv/dt increases.

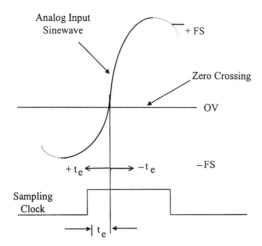

FIGURE 3-16 Measuring the effective aperture delay time

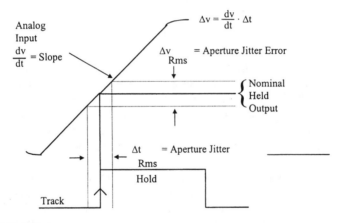

FIGURE 3-17 Effects of aperture jitter on SHA output

Measuring aperture jitter error in an SHA requires a jitter-free sampling clock and analog input signal source, because jitter (or phase noise) on either signal cannot be distinguished from the SHA aperture jitter itself—the effects are the same. In fact, the largest source of timing jitter errors in a system most often is external to the SHA (or the ADC if it is a sampling one), caused by noisy or unstable clocks, improper signal routing, and lack of attention to good grounding and decoupling techniques. SHA aperture jitter generally is less than 50 ps rms and less than 5 ps rms in high-speed devices.

Figure 3-18 shows the effects of total sampling clock jitter on the signal-to-noise ratio of a sampled data system. The total rms jitter will be composed of a number of components, the actual SHA aperture jitter often being the least of them.

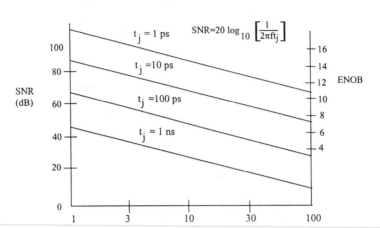

FIGURE 3-18 Effects of sampling clock jitter on the SNR. (Reproduced by permission of Analog Devices Inc.)

3.4.1.4 Hold Mode Droop

During the hold mode, there are errors due to imperfections in the hold capacitor, switch, and output amplifier. If a leakage current flows in or out of the hold capacitor, it will slowly charge or discharge and its voltage will change, an effect known as *droop* in the SHA output, expressed in V/μs. Droop can be caused by leakage across a dirty PCB if an external capacitor is used or by a leaky capacitor but most commonly is due to leakage current in semiconductor switches and the bias current of the output buffer amplifier. An acceptable value of droop is found when the output of an SHA does not change by more than 1/2 LSB during the conversion time of the ADC it is driving. See Figure 3-19.

Droop can be reduced by increasing the value of the hold capacitor, but this will increase acquisition time and reduce the bandwidth in the track mode. Even quite small leakage currents can cause troublesome droop when SHAs use small hold capacitors. Leakage currents in PCBs may be minimized by the intelligent use of guard rings. Details of planning a guard ring are discussed in Analog Devices (1995, Chapter 8).

3.4.1.5 Dielectric Absorption

Hold capacitors for SHAs must have low leakage, but another characteristic is equally important: low dielectric absorption. If a capacitor is charged, discharged, and then left on an open circuit, it will recover some of its charge. The phenomenon, known as *dielectric absorption*, can seriously degrade the performance of an SHA, since it causes the remains of a previous sample to contaminate a new one and may introduce random errors of tens or even hundreds of millivolts (see Figure 3-20). After discharge, C_D and R_S in circuit could cause the residual charge.

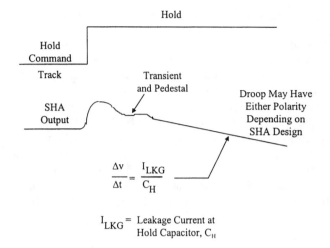

$$\frac{\Delta v}{\Delta t} = \frac{I_{LKG}}{C_H}$$

I_{LKG} = Leakage Current at Hold Capacitor, C_H

FIGURE 3-19 Hold mode droop

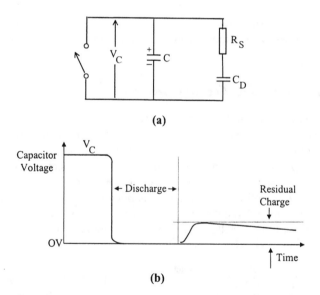

FIGURE 3-20 Dielectric absorption: (a) Model, (b) Waveform

Different capacitor materials have differing amounts of dielectric absorption: electrolytic capacitors are dreadful (and their leakage is high) and some high-K ceramic types are bad, while mica, polystyrene, and polypropylene generally are good. Unfortunately, dielectric absorption varies from batch to batch, and even occasional batches of polystyrene and polypropylene capacitors may be affected. Measuring hold mode distortion is discussed in Analog Devices (1995, Chapter 8).

3.4.1.6 Hold-to-Track Transition Specification

When the SHA switches from hold to track, it must reacquire the input signal (which may have made a full-scale transition during the hold mode). Acquisition time is the interval of time required for the SHA to reacquire the signal to the desired accuracy when switching from hold to track. The interval starts at the 50% point of the sampling clock edge and ends when the SHA output voltage falls within the specified error band (usually 0.1% and 0.01% times are given). Some SHAs also specify acquisition time with respect to the voltage on the hold capacitor, neglecting the delay and settling time of the output buffer. The hold capacitor acquisition time specification is applicable in high-speed applications, where the maximum possible time must be allocated for the hold mode. The output buffer settling time, of course, must be significantly smaller than the hold time.

3.5 SHA Architectures

There are numerous SHA architectures and we will examine a few of the most popular ones. For a more detailed discussion on SHA architectures, see Razavi (1995).

3.5.1 Open-Loop Architecture

The simplest SHA architecture is shown in Figure 3-21. The input signal is buffered by an amplifier and applied to the switch. The input buffer may either be open or closed loop and may or may not provide gain. The switch can be CMOS, FET, or bipolar (using diodes or transistors), controlled by the switch driver circuit. The signal on the hold capacitor is buffered by an output amplifier. This architecture sometimes is referred to as *open loop* because the switch is not inside a feedback loop. Note that the entire signal voltage is applied to the switch; therefore, it must have excellent common mode characteristics.

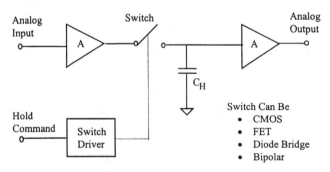

FIGURE 3-21 Open-loop SHA architecture

3.5.2 Open-Loop Diode Bridge SHA

Semiconductor diodes exhibit small on-resistance, large off-resistance, high-speed switching, and thus potential for the switching function in sampling circuits. A simplified diagram of a typical diode switch is shown in Figure 3-22(a). Here, four diodes form a bridge that provides a low-impedance path from V_{in} to V_{out} when current sources I_1 and I_2 are on and (in the ideal case) isolates V_{out} from V_{in} when I_1 and I_2 are off. Nominally, $I_1 = I_2 = I$. Implementation is shown in Figure 3-22(b).

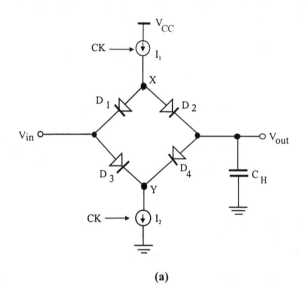

(a)

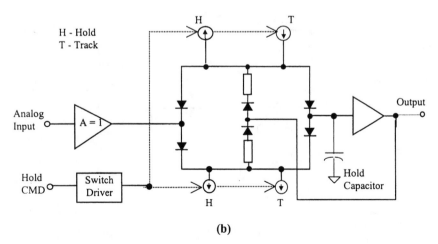

(b)

FIGURE 3-22 Diode bridge SHA: (a) Basic diode bridge, (b) Implementation. (Reproduced by permission of Analog Devices Inc.)

3.5.3 Closed-Loop Architecture

The SHA circuit shown in Figure 3-23 represents a classical closed-loop design and is used in many CMOS sampling ADCs. Since the switches always operate at virtual ground, there is no common mode signal across them. Switch S2 is required to maintain a constant input impedance and prevent the input signal from coupling to the output during the hold time. In the track mode, the transfer characteristic of the SHA is determined by the op amp and the switches introduce no DC errors because they are within the feedback loop. The effects of charge injection can be minimized by using the differential switching techniques shown in Figure 3-24.

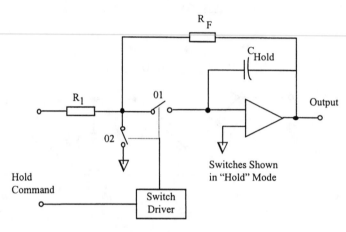

FIGURE 3-23 A closed-loop SHA

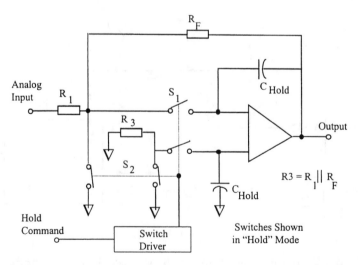

FIGURE 3-24 Differential switching for charge injection

3.6 ADC Architectures

Within the 1990s and the latter part of the 1980s, many architectures for A/D conversion have been implemented in monolithic form. Manufacturing process improvements achieved by mixed-signal product manufacturers have led to this unprecedented development, which was fueled by the demand from the product and system designers.

The most common ADC architectures in monolithic form are successive approximation, flash, integrating, pipeline, half flash (or subranging), two step, interpolative and folding, and sigma-delta (Σ-Δ). The following sections provide the basic operational and design details of these techniques.

While the Σ-Δ, successive approximation, and integrating types could give very high resolution at lower speeds, flash architecture is the fastest but with high power consumption. However, recent architecture breakthroughs have allowed designers to achieve a higher conversion rate at low power consumption with integral track and hold circuitry on a chip (McGoldrick, 1997). The AD9054 from Analog Devices is an example.

3.6.1 Successive Approximation ADCs

The successive approximation register (SAR) ADC architecture has been used for decades and still is a popular and cost effective form of converter for sampling frequencies up to few MSPS. A simplified block diagram of an SAR ADC is shown in Figure 3-25. On the start conversion command, all the bits of the successive approximation register are reset to 0 except the most significant bit (MSB), which is set to 1. Bit 1 is tested in the following manner. If the ADC output is greater than the analog input, the MSB is reset; otherwise, it is left set. The next most significant bit then is tested by setting it to 1. If the

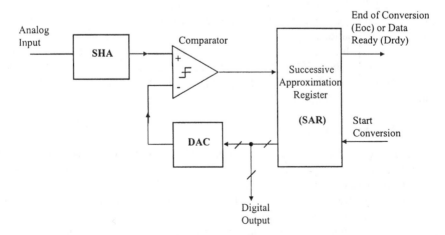

FIGURE 3-25 Block diagram of successive approximation ADCs

digital/analog converter (DAC) output is greater than the analog input, this bit is reset; otherwise it is left set. The process is repeated with each bit in turn. When all the bits have been set, tested, and reset or not as appropriate, the contents of the SAR correspond to the digital value of the analog input, and the conversion is complete.

An N-bit conversion takes N steps. On superficial examination, a 16-bit converter would seem to have a conversion time twice as long as an 8-bit one, but this is not the case. In an 8-bit converter, the DAC must settle to 8-bit accuracy before the bit decision is made, whereas in a 16-bit converter, it must settle to 16-bit accuracy, which takes a lot longer. In practice, 8-bit successive approximation ADCs can convert in a few hundred nanoseconds, while 16-bit ones generally take several microseconds.

The classic SAR ADC is only a quantizer — no sampling takes place — and for an accurate conversion, the input must remain constant for the entire conversion period. Most modern SAR ADCs are sampling types and have internal sample and hold so that they can process AC signals. They are specified for both AC and DC applications. An SHA is required in an SAR ADC because the signal must remain constant during the entire N-bit conversion cycle.

The accuracy of an SAR ADC depends primarily on the accuracy (differential and integral linearity, gain, and offset) of the internal DAC. Until recently, this accuracy was achieved using laser-trimmed thin-film resistors. Modern SAR ADCs utilize CMOS switched capacitor charge redistribution DACs. This type of DAC depends on the accurate ratio matching and the stability of on-chip capacitors rather than thin-film resistors. For resolutions greater than 12 bits, on-chip autocalibration techniques, using an additional calibration DAC and the accompanying logic, can accomplish the same thing as thin-film laser-trimmed resistors, at much less cost. Therefore, the entire ADC can be made on a standard submicron CMOS process.

The successive approximation ADC has a very simple structure, low power, and reasonably fast conversion times (<1 MSPS). It is probably most widely used ADC architecture and will continue to be used for medium-speed, medium-resolution applications.

Current 12-bit SAR ADCs achieve sampling rates up to about 1 MSPS, and 16-bit ones up to about 300 kSPS. Examples of typical state of the art SAR ADCs are the AD7892 (12 bits at 600 kSPS), the AD976/977 (16 bits at 100 kSPS), and the AD7882 (16 bits at 300 kSPS).

3.6.2 Flash Converter

Flash ADCs (sometimes called *parallel ADCs*) are the fastest ADCs and use large numbers of comparators. An N-bit flash ADC consists of 2^N resistors and $2^N - 1$ comparators arranged as in Figure 3-26. Each comparator has a reference voltage 1 least significant bit (LSB) higher than that of the one below it in the chain. For a given input voltage, all the comparators below a certain point will have their input voltage larger than their reference voltage and a 1 logic output, and all the comparators above that point will have a reference voltage larger than

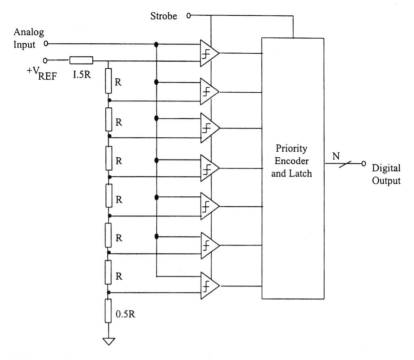

FIGURE 3-26 Flash or parallel ADC block diagram

the input voltage and a 0 logic output. The $2^N - 1$ comparator output therefore behaves like a mercury thermometer, and the output code at this point is sometimes called a *thermometer code*. Since $2^N - 1$ data output is not really practical, these are processed by a decoder to an N-bit binary output.

The input signal is applied to all the comparators at once, so the thermometer output is delayed by only one comparator delay from the input and the encoder N-bit output by only a few gate delays on top of that, so the process is very fast. However, the architecture uses large numbers of resistors and comparators and is limited to low resolutions; if it is to be fast, each comparator must run at relatively high power levels. Hence, the problems of flash ADCs include limited resolution, high power dissipation because of the large number of high-speed comparators (especially at sampling rates greater than 50 MSPS), and relatively large (and therefore expensive) chip sizes. In addition, the resistance of the reference resistor chain must be kept low to supply adequate bias current to the fast comparators, so the voltage reference has to source quite large currents (>10 mA).

In practice, flash converters are available up to 10 bits of resolution, but more commonly they have 8 bits of resolution. Their maximum sampling rate can be as high as 500 MSPS, and input full-power bandwidths are in excess of 300 MHz.

But, as mentioned earlier, full-power bandwidths are not necessarily full-resolution bandwidths. Ideally, the comparators in a flash converter are well matched both for DC and AC characteristics. Because the strobe is applied to all the comparators simultaneously, the flash converter is inherently a sampling converter. In practice, delay variations between the comparators and other AC mismatches cause a degradation in ENOB at high input frequencies. This is because the inputs are slewing at a rate comparable to the comparator conversion time.

The input to a flash ADC is applied in parallel to a large number of comparators. Each has a voltage-variable junction capacitance, and this signal-dependent capacitance results in all flash ADCs having reduced ENOB and higher distortion at high input frequencies. For more details, see Analog Devices (1996).

3.6.3 Integrating ADCs

The integrating ADC is a very popular architecture in applications where a very slow conversion rate is acceptable. A classic example is the digital multimeter.

All the converters discussed so far can digitize analog inputs at speeds of at least 10 kSPS. A typical integrating converter is slow relative to these high-speed converters. Useful for precisely measuring slowly varying signals, the integrating converter finds applications in low-frequency and DC measurement applications.

Integrating converters are based on an indirect conversion method. Here the analog input voltage is converted to a time period and later to a digital number using a counter. The integration eliminates the need for a sample/hold (S/H) circuit to "capture" the input signal during the measurement period. The two common variations of the integrating converter are the dual slope type and the charge balance or multislope type. The dual slope technique is very popular among instrument manufacturers because of its simplicity, low price, and better noise rejection. The multislope technique is an improvement on the dual slope method.

Figure 3-27(a) shows a typical integrating converter. It consists of an analog integrator, a comparator, a counter, a clock, and control logic. Figure 3-27(b) shows the circuit's charge (T_1) and discharge (T_2) waveforms. The conversion is started by closing the switch and thereby connecting the capacitor, C_{int}, to the unknown input voltage, V_{in}, through the resistor, R. This results in a linear ramp at the integrator output for a fixed period, T_1, controlled by the counter. The control circuit then switches the integrator input to the known reference voltage, V_{ref}, and the capacitor discharges until the comparator detects that the integrator has reached the original starting point. The counter measures the amount of time taken for the capacitor to discharge.

Because the values of the resistor, the integrating capacitor, and the frequency of the clock remain the same for both the charge and discharge cycles, the ratio of the charge time to the discharge time is equal to the ratio of the reference voltage to the unknown input voltage. The absolute values of the resistor, capacitor, and the clock frequency therefore do not affect the conversion accuracy. Furthermore, any noise on the input signal is integrated over the entire sampling period, which imparts a high level of noise rejection to the converter.

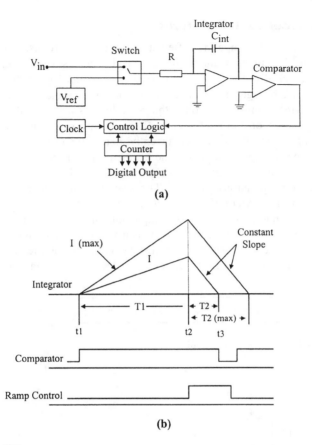

FIGURE 3-27 Diagrams of an integrating ADC: (a) Block diagram, (b) Timing diagram

By making the signal integration period an integral multiple of the line frequency period, the user can obtain excellent line frequency noise rejection.

A charge balance integrating converter incorporates many of the elements of the dual slope converter but uses a free-running integrator in a feedback loop. The converter continuously attempts to null its input by subtracting precise charge packets when the accumulated charge exceeds a reference value. The frequency of the charge packets (the number of packets per second) the converter needs to balance the input is proportional to that input. Clock-controlled synchronous logic delivers a serial output that a counter converts to a digital word in the circuit. Integrating converters in monolithic form typically are used in digital voltmeters due to their high resolution properties. Hybrid integrating converters with 22-bit resolutions were introduced to the market in the late 1980s. It therefore is possible to expect higher resolutions in the monolithic market as well. There could be many variations of this technique as applied to digital multimeters, and Kularatna (1996) is suggested for details.

3.6.4 Pipeline Architectures

The concept of a pipeline, often used in digital circuits, can be applied in the analog domain to achieve higher speed where several operations must be performed serially. Figure 3-28 shows a general (analog or digital) pipeline system. Here, each stage carries out an operation on a sample, provides the output for the following sampler, and once that sampler has acquired the data, begins the same operation on the next sample. Thus, at any given time, all the stages are processing different samples concurrently; hence, the throughput rate depends only on the speed of each stage and the acquisition time of the next sampler.

To arrive at a simple example of an analog pipeline, consider a two-step ADC, where four operations (coarse A/D conversion, interstage D/A conversion, subtraction, and fine A/D conversion) must be performed serially. As such, the ADC cannot begin to process the next sample until all four operations are finished. Now, suppose an SHA is interposed between the subtractor and the fine stage, as shown in Figure 3-29, so that the residue is stored before fine conversion begins. Thus, the front-end SHA, the coarse ADC, the interstage DAC, and the

FIGURE 3-28 A pipeline system

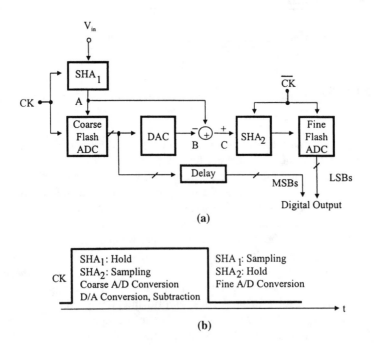

(a)

(b)

FIGURE 3-29 A two-step ADC pipeline: (a) Block diagram, (b) Clock waveform and related activities

subtractor can start processing the next sample while the fine ADC operates on the previous one, allowing potentially faster conversion. More details on pipeline architectures can be found in Louzon (1995).

3.6.5 Half-Flash ADCs

Although it is not practical to make them with high resolution, flash ADCs often are used as subsystems in "subranging" ADCs (sometimes known as *half-flash ADCs*), which are capable of much higher resolutions (up to 16 bits).

A block diagram of an 8-bit subranging ADC based on two 4-bit flash converters is shown in Figure 3-30. Although 8-bit flash converters are readily available at high sampling rates, this sample will be used to illustrate the theory. The conversion process is done in two steps. The four most significant bits are digitized by the first flash (to better than 8-bit accuracy) and the 4-bit binary output is applied to 4-bit DAC (again, better than 8-bit accuracy). The DAC output is subtracted from the held analog input, and the resulting residue signal is amplified and applied to the second 4-bit flash. The output of the two flash converters are combined into a single 8-bit binary output word. If the residue signal range does not exactly fill the range of the second flash converter, nonlinearities and perhaps missing codes[1] will result.

Modern subranging ADCs use a technique called *digital correction* to eliminate problems associated with the architecture of Figure 3-30. A simplified block diagram of a 12-bit digitally corrected subranging ADC is shown in Figure 3-31. An example of such a practical ADC is the AD9042 from Analog Devices, a 12-bit, 41 MSPS device. Key specifications of the AD9042 are given in Table 3-3.

Note that a 6-bit and 7-bit ADC have been used to achieve an overall 12-bit output. These are not flash ADCs but utilize a magnitude-amplifier (MagAmp™) architecture. (See Chapter 4 in Analog Devices (1996) for MagAmp™ basics.)

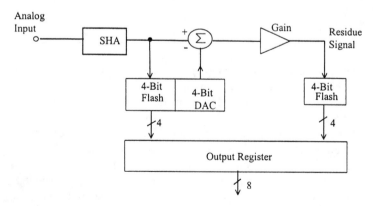

FIGURE 3-30 8-bit subranging ADC

---

[1] A converter must be able to correspond all possible digital outputs to an analog input. If it is unable to do so (due to excessive DNL) it is said to have missing codes.

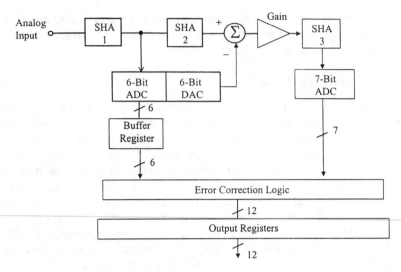

FIGURE 3-31 Pipeline subranging ADC with a digital error correction. (Reproduced by permission of Analog Devices Inc.)

TABLE 3-3 Key Specifications of the AD9042. (Reproduced by permission of Analog Devices Inc.)

Parameter	Value
Input range	1 V peak-to-peak, $V_{cm} = +2.4$ V
Input impedance	250 Ω to V_{cm}
Effective input noise	0.33 LSBs rms
SFDR at 20 MHz input	80 dB minimum
SINAD (S/N +D) at 20 MHz input	67 dB
Digital outputs	TTL compatible
Power supply	Single +5 V
Power dissipation	595 mW
Fabrication	High-speed dielectrically isolated complementary bipolar process

If there were no errors in the first-stage conversion, the 6-bit "residue" signal applied to the 7-bit ADC by the summing amplifier would never exceed one-half the range of the 7-bit ADC. The extra range in the second ADC is used in conjunction with the error correction logic (usually just a full adder) to correct the output data for most of the errors inherent in the traditional uncorrected subranging converter architecture. It is important to note that the 6-bit DAC must be better than 12-bit accurate, because the digital error correction does not correct for DAC errors. In practice, "thermometer" or "fully decoded" DACs using one current switch per level (63 switches in the case of a 6-bit DAC) often are used instead of a "binary" DAC to ensure excellent differential and integral linearity and minimum switching transients (Analog Devices, 1996).

The second SHA delays the held output of the first SHA while the first-stage conversion occurs, thereby maximizing throughput. The third SHA "deglitches" the residue output signal, allowing a full conversion cycle for the 7-bit ADC to make its decision (the 6- and 7-bit ADCs in the AD9042 are bit-serial MagAmp ADCs, which require more settling time than a flash converter). Additional shift registers in series with the digital output of the first-stage ADC ensure that its output ultimately is time-aligned with the last 7 bits from the second ADC when their outputs are combined in the error correction logic. A pipeline ADC therefore has a specified number of clock cycles of latency—pipeline delay—associated with the output data. The leading edge of the sampling clock (for sample, N) is used to clock the output register, but the data that appears as a result of that clock edge corresponds to sample $N - L$, where L is the number of clock cycles of latency; in the case of the AD9042, two clock cycles of latency.

The error correction scheme described previously is designed to correct for errors made in the first conversion. Internal ADC gain, offset, and linearity errors are corrected as long as the residue signal falls within the range of the second-stage ADC. These errors will not affect the linearity of the overall ADC transfer characteristic. Errors made in the final conversion, however, translate directly as errors in the overall transfer function. Also, linearity errors or gain errors either in the DAC or the residue amplifier will not be corrected and will show up as nonlinearities or nonmonotonic behavior in the overall ADC transfer function.

So far, we have considered only two-stage subranging ADCs, as these are easiest to analyze. There is no reason to stop at two stages, however. Three-pass and four-pass subranging pipeline ADCs are quite common and can be made in many different ways, usually with digital error correction. For details, see Analog Devices (1996).

3.6.6 Two-Step Architectures

The exponential growth of power, die area, and input capacitance of flash converters as a function of resolution makes them impractical for resolutions above 8 bits in general. These resolutions call for topologies that provide a more relaxed trade-off among the parameters. Two-step architectures trade speed for power, area, and input capacitance.

In a two-step ADC, first a coarse analog estimate of the input is obtained to yield a small voltage range around the input level. Subsequently, the input level is determined with higher precision within this range. Figure 3-32(a) illustrates a two-step architecture consisting of a front-end SHA, a coarse flash ADC stage, a DAC, a subtractor, and a fine flash ADC stage. We describe its operation using the timing diagram shown in the Figure 3-32(b).

For $t < t_1$, the SHA tracks the analog input. At $t = t_1$, the SHA enters a hold mode and the first flash stage is strobed to perform the coarse conversion. The first stage then provides a digital estimate of the signal held by the SHA (V_A), and the DAC converts this estimate to an analog signal (V_B), which is a coarse approximation of the SHA output. Next, the subtractor generates an output equal to the difference between V_A and V_B (V_C, called the *residue*), which is subsequently digitized by the fine ADC. Comparison of timing in flash and two-step architectures is shown in Figure 3-32(c). For more details, see Razavi (1996).

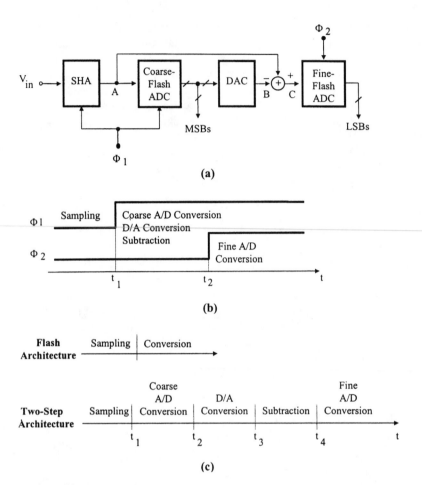

FIGURE 3-32 Two-step architecture: (a) Block diagram, (b) Timing, (c) Comparison of timing in flash and two-step architectures

A two-step ADC need not employ two separate flash stages to perform the coarse and fine conversions. One stage can be used for both; and such an architecture, shown in Figure 3-33, is called *recycling architecture*.

Here, during the coarse conversion, the flash stage senses the front-end SHA output, V_A, and generates the coarse digital output. This digital output then is converted to analog by the DAC and subtracted from V_A by the subtractor. During fine conversion, the subtractor output is digitized by the flash stage. Note that, in this phase, the ADC full-scale voltage must be equal to that of the subtractor output. Therefore, for proper fine conversion, either the ADC reference voltage must be reduced or the residue must be amplified.

While reducing area and power dissipation by roughly a factor of 2 relative to two-stage ADCs, recycling converters suffer from other limitations. The

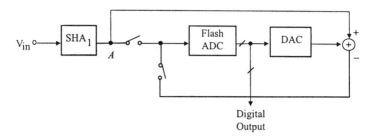

FIGURE 3-33 Recycling ADC architecture

converter must now employ either low-offset comparators (if the subtractor has a gain of 1), inevitably slowing down the coarse conversion, or a high-gain subtractor, increasing the interstage delay. This is in contrast with two-stage ADCs, where the coarse stage comparators need not have a high resolution and hence can operate faster.

3.6.7 Interpolating and Folding Architectures

To maintain the one-step nature of the flash-type architectures, without adding sample-and-hold circuits to the ADC, several other architectures are available. Among these techniques, interpolation and folding have proven quite beneficial. Earlier, these techniques had been applied predominantly to bipolar circuits; recently CMOS devices have entered the market.

As a comprehensive discussion on these techniques is beyond the scope of this chapter, only basic approach in the design is discussed here.

3.6.7.1 Interpolating Architectures

To reduce the number of preamplifiers at the input of a flash ADC, the difference between the analog input and each reference voltage can be quantized at the output of each preamplifier. This is possible because of the finite gain — hence, nonzero linear input range — of typical preamplifiers used as the front end of comparators.

We illustrate this concept in Figure 3-34(a). In Figure 3-34(a), preamplifiers A_1 and A_2 compare the analog input with V_{r1} and V_{r2}, respectively. In Figure 3-34(b), the input/output characteristics of A_1 and A_2 are shown. Assuming zero offset for both preamplifiers, we note that $V_{X1} = V_{Y1}$ if $V_{in} = V_{r1}$, and $V_{X2} = V_{Y2}$ if $V_{in} = V_{r2}$. More important, $V_{X2} = V_{Y1}$ if $V_{in} = V_m = (V_{r1} + V_{r2})/2$; that is, the polarity of the difference between V_{X2} and V_{Y1} is the same as that of the difference between V_{in} and V_m.

The preceding observation indicates that the equivalent resolution of a flash stage can be increased by "interpolating" between the output of preamplifiers. For example, Figure 3-34(c) shows how an additional latch detects the polarity of the difference between the single-ended output of two adjacent preamplifiers. Note that in contrast with a simple flash stage, this approach halves the number of preamplifiers but maintains the same number of latches.

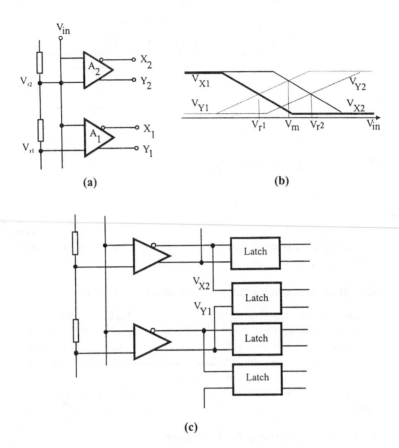

FIGURE 3-34 Interpolating architecture: (a) Basic block, (b) Interpolation between output of two amplifiers, (c) Interpolation in a flash ADC

The interpolation technique of Figure 3-34(c) substantially reduces the input capacitance, power dissipation, and area of flash converters, while preserving the one-step nature of the architecture. This is possible because all the signals arrive at the input of the latches simultaneously and hence can be captured on one clock edge. Since this configuration doubles the effective resolution, we say it has an interpolation factor of 2. For further details on this architecture, see Razavi (1996).

3.6.7.2 Folding Architectures

Folding architectures have evolved from flash and two-step topologies. Folding architectures perform analog preprocessing to reduce hardware while maintaining the one-step nature of flash architectures.

The basic principle in folding is to generate a residue voltage through analog preprocessing and subsequently digitize that residue to obtain the least significant bits. The most significant bits can be resolved using a coarse flash stage that operates in parallel with the folding circuit and, hence, samples the signal at approximately the same time that the residue is sampled. Figure 3-35 depicts the generation of residue in two-step and folding architectures. In a two-step architecture, coarse A/D conversion, interstage D/A conversion, and subtraction must be completed before the proper residue becomes available. In contrast, folding architectures generate the residue "on the fly" using simple wideband stages.

To illustrate this principle, we first describe a simple, ideal approach to folding. Consider two amplifiers, A_1 and A_2, with the input/output characteristics depicted in Figure 3-36(a). The active region of one amplifier is centered around $(V_{r2} + V_{r1})/2$ and that of the other around $(V_{r3} + V_{r2})/2$, and $V_{r3} - V_{r2} = V_{r2} - V_{r1}$. Each amplifier has a gain of 1 in the active region and 0 in the saturation region. If the outputs of the two amplifiers are summed, the "folding" characteristic of Figure 3-36(b) results, yielding an output equal to $V_{in} - V_{r1}$ for $V_{r1} < V_{in} < V_{r2}$ and $(-V_{in} + V_{r2} + \Delta)$ for $V_{r2} < V_{in} < V_{r3}$, where Δ is the value of the summed characteristics at $V_{in} = V_{r2}$. If V_{r1}, V_{r2}, and V_{r3} are the reference voltages in an ADC, then these two regions can be viewed as the residue characteristics of the ADC for $V_{r1} < V_{in} < V_{r3}$. To understand why, we compare this characteristic with that of a two-step architecture, as shown in Figure 3-36(c). The two characteristics are similar except for a negative sign and a vertical shift in the folding output for $V_{r2} < V_{in} < V_{r3}$. Therefore, if the system accounts for the sign reversal and level shift, the folding output can be used as the residue for fine digitization.

Figure 3-37(a) shows an implementation of folding. Here, four differential pairs process the difference between V_{in} and $V_{r1}, \ldots, V_{r4}$, and their output currents are summed at nodes X and Y. Note that the outputs of adjacent stages

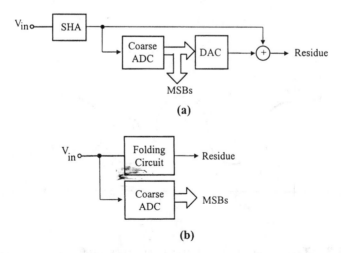

(a)

(b)

FIGURE 3-35 Generation of residue: (a) Two-step architecture, (b) Folding architecture

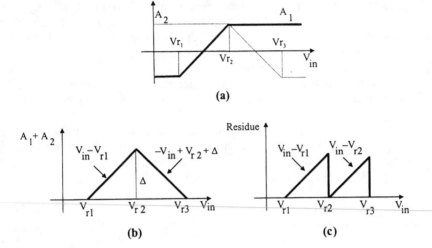

FIGURE 3-36 The concept of folding: (a) Input/output characteristics of two amplifiers, (b) Sum of characteristics in (a), (c) Residue in two-step ADC

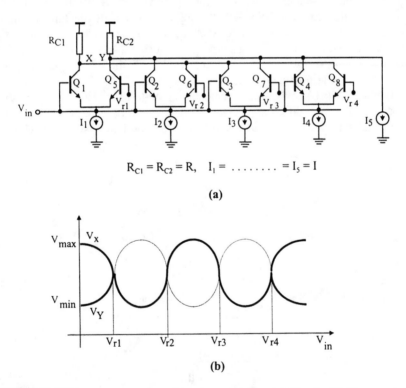

FIGURE 3-37 Folding circuit and its characteristics: (a) Circuit, (b) Characteristics

are added with opposite polarity; for example, as V_{in} increases, Q_1 pulls node X low while Q_2 pulls node Y low. Current source I_5 shifts V_Y down by IR. To explain the operation of the circuit, we consider its input/output characteristics, plotted in Figure 3-37(b). For V_{in} well below V_{r1}, $Q_1 - Q_4$ are off, $Q_5 - Q_8$ are on, I_2 and I_4 flow through R_{C1}, and I_1, I_3, and I_5 flow through R_{C2}. As V_{in} exceeds V_{r1} by several V_T, Q_5 turns off, allowing V_X and V_Y to reach V_{min} and V_{max}, respectively. As V_{in} approaches V_{r2}, Q_2 begins to turn on and the circuit behaves as before. Considering the differential output, $V_X - V_Y$, we note that the resulting characteristic exhibits folding points at $(V_{r1} + V_{r2})/2$, $(V_{r2} + V_{r3})/2$, and so forth. As V_{in} goes from below V_{r1} to above V_{r4}, the slope of $V_X - V_Y$ changes sign four times; hence, we say the circuit has a folding factor of 4.

The simplicity and speed of folding circuits have made them quite popular in A/D converters, particularly because they eliminate the need for sample-and-hold amplifiers, D/A converters, and subtractors. Nevertheless, several drawbacks limit their use at higher resolutions (Razavi, 1995).

3.6.8 Sigma-Delta Converters

Sigma-Delta analog/digital converters (Σ-Δ ADCs) have been known for nearly 30 years, but only recently has the technology (high-density digital very large-scale ICs) existed to manufacture them as inexpensive monolithic integrated circuits. They are used in many applications where a low-cost, low-bandwidth, high-resolution ADC is required.

The literature contains innumerable descriptions of the architecture and theory of Σ-Δ ADCs (Candy and Temes, 1992). As a text of this nature is not appropriate for describing their mathematical analysis and background, this section has been written to classify the subject. A practical monolithic Σ-Δ ADC contains very simple analog circuit blocks (a comparator, a switch, and one or more integrators and analog summing circuits) and quite complex digital computational circuitry. The circuitry consists of a digital signal processor that acts as a filter (generally, but not invariably, a low-pass filter). It is not necessary to know how the filter works to appreciate what it does. To understand how a Σ-Δ ADC works, one should be familiar with the concepts of oversampling, noise shaping, digital filtering, and decimation. We briefly discuss these concepts in Σ-Δ converters.

3.6.8.1 Key Concepts Behind the Σ-Δ ADC

Figure 3-38 shows the transfer characteristic of a 3-bit unipolar Σ-Δ ADC. The input to an ADC is analog and is not quantized, but its output is quantized. The transfer characteristic therefore consists of eight horizontal steps (when considering the offset, gain, and linearity of an ADC, we consider the line joining the midpoints of these steps). Digital full scale (all ones) corresponds to 1 LSB below the analog full scale (the reference or some multiple of it). This is because, as mentioned previously, the digital code represents the normalized ratio of the

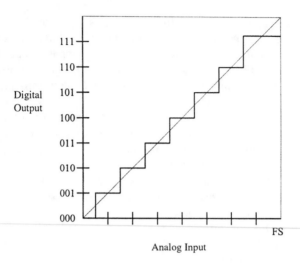

FIGURE 3-38 Transfer characteristics of a 3-bit unipolar ADC

analog signal to the reference, and if this were unity, the digital code would be all zeros and 1 in the bit above the MSB.

The (ideal) ADC transitions take place at 0.5 LSB above 0 and thereafter every LSB, up to 1.5 LSB below analog full scale. Since the analog input to an ADC can take any value but the digital output is quantized, there may be a difference of up to 0.5 LSB between the actual analog input and the exact value of the digital output. This is known as the *quantization error* or *quantization uncertainty*. In AC (sampling) applications, this quantization error gives rise to quantization noise. If we apply a fixed input to an ideal ADC, we always will obtain the same output and the resolution will be limited by the quantization error.

Suppose, however, that we add some AC (dither) to the fixed signal, take a large number of samples, and prepare a histogram of the results. We will obtain something like the result in Figure 3-39. If we calculate the mean value of a large number of samples, we will find that we can measure the fixed signal with greater resolution than that of the ADC we are using. This procedure is known as *oversampling*.

The AC (dither) that we add may be a sine wave, a triangular wave, or Gaussian noise (but not a square wave); and with some types of sampling ADCs (including Σ-Δ ADCs), an external dither signal is unnecessary, since the ADC generates its own. Analysis of the effects of differing dither waveforms and amplitudes is complex and, for the purposes of this section, unnecessary. What we need to know is that with the simple oversampling described here, the number of samples must be doubled for each bit of increase in resolution.

If, instead of a fixed DC signal, the signal that we are oversampling is AC, then it is not necessary to add a dither signal to it to oversample, since the signal

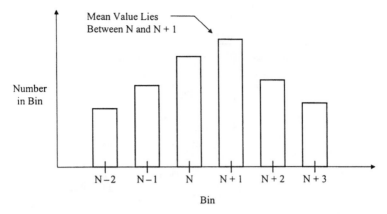

FIGURE 3-39 Oversampling

is moving anyway. (If the AC signal is a single-tone, harmonically related to the sampling frequency, dither may be necessary, but this is a special case.)

Consider the technique of oversampling with an analysis in the frequency domain. Where a DC conversion has a quantization error of up to 1/2 LSB, a sampled data system has quantization noise. As we already have seen, a perfect classical N-bit sampling ADC has an rms quantization noise of $q/\sqrt{12}$ uniformly distributed within the Nyquist band of DC to $f_s/2$ (where q is the value of an LSB and f_s is the sampling rate), giving us an SNR of $(6.02N + 1.76)$ dB with full-scale sine wave inputs (e.g., equation (3.1); see Figure 3-40).

If the ADC is less than perfect and its noise is greater than its theoretical minimum quantization noise, then its effective resolution will be less than N bits. Its actual resolution (often known as its *effective number of bits*, ENOB) will be defined by equation (3.2).

If we choose a much higher sampling rate (K times f_s, as in Figure 3-41(a)), the quantization noise is distributed over a wider bandwidth as shown in

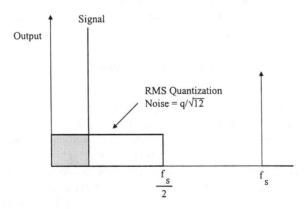

FIGURE 3-40 Sampling ADC quantization noise

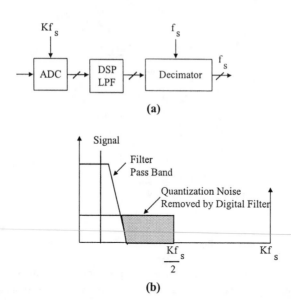

(a)

(b)

FIGURE 3-41 Oversampling and digital filtering: (a) Block diagram, (b) Output vs. frequency

Figure 3-41(b). If we then apply a digital low-pass filter to the output, we remove much of the quantization noise but do not affect the wanted signal, so the ENOB is improved. We have performed a high-resolution A/D conversion with a low-resolution ADC.

Since the bandwidth is reduced by the digital output filter, the output data rate may be lower than the original sampling rate and still satisfy the Nyquist criteria. This may be achieved by passing every Mth result to the output and discarding the remainder, a process is known as *decimation* by a factor of M. Here, M can have any integer value, provided that the output data rate is more than twice the signal bandwidth. Decimation causes no loss of information (Figure 3-42). As shown in Figure 3-42, after sampling at f_s and filtering, the output data rate may be reduced to f_s/M with no loss of information.

3.6.8.2 Block Diagram of a Σ-Δ ADC

Oversampled Σ-Δ ADCs in recent years have become more prevalent for high-accuracy, 12 bit to beyond 22 bit, A/D conversion of DC through moderately high (hundreds of kHz) AC signals. Their greatest advantage is that they trade greatly reduced analog circuit accuracy requirements for increased digital circuit complexity. This is a distinct advantage for 1–2 micron and submicron very large-scale integrated (VLSI) digital circuit technologies. VLSI circuit techniques can achieve circuit densities of hundreds of thousands of gates, allowing complex digital filters to be integrated on the chip. The result is high precision at low cost.

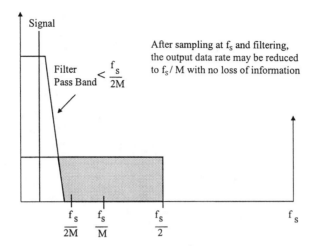

FIGURE 3-42 Decimation process

The basic oversampled Σ-Δ A/D converter (Figure 3-43) is an integrating A/D converter. The single-bit feedback D/A converter output is subtracted from the analog input signal, V_{in}, in the summing amplifier. The resulting error signal from the summing amplifier output is low-pass filtered by the integrator and the integrated error signal polarity is detected by the single comparator. This comparator effectively is a 1-bit A/D converter.

The output of the comparator drives the 1-bit DAC to a 1 or 0 (a 1, if during the previous sample time the integrator output was detected by the comparator as being too low, that is, below 0 V; a 0, if the difference detected during the previous sample was too high, that is, above 0 V reference of the comparator).

The 1-bit D/A converter, as in successive approximation A/D converters, provides the negative feedback. This negative feedback for a 1 in the D/A

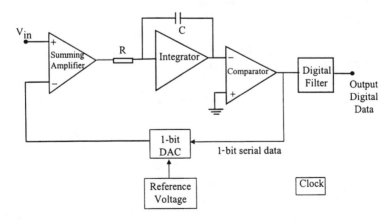

FIGURE 3-43 Oversampled Σ-Δ ADC

converter always is in a direction to drive the integrator output toward 0 V. The D/A converter output for a 1 input would be the reference voltage. The reference voltage would be equal to or exceed the expected full-scale analog input signal voltage. Then, for a small value of V_{in}, the integrator would take many clock pulses to cross 0 V, after a single 1 was generated, during which time the comparator is sending zeros to the digital filter. If V_{in} were at full scale, the integrator would cross 0 every clock time and the comparator output would be a string of alternate ones and zeros. The digital filter's function is to determine a digital number at its output that is proportional to the number of ones in the previous bit stream from the comparator. Various types of digital filters are used to perform this computational function, which is the most complex function in this type of D/A converter. However, complex digital computations can be performed readily in a VLSI circuit.

Oversampled converters sample at much higher rates than the Nyquist rate. The *oversampling ratio* is equal to the actual sampling rate divided by the Nyquist rate. The oversampling rate can be hundreds to thousands of times the analog input signal frequency bandwidth. Since each sample is in a 1-bit low accuracy conversion, sampling rates can be very high.

In cases where antialiasing filtering is required on the input analog signal, the filter does not require the sharp cutoff characteristics as would be required to limit broadband signals prior to a successive approximation-type A/D converter operating at or near a small multiple of the Nyquist sampling rate. The reason is that the oversampling rate is many times the Nyquist rate. Therefore, a simple RC filter is adequate to prevent aliasing. The input filter can pass frequencies many times higher than the frequencies of interests before filter cutoff is required. The bandwidth of signals converted can be increased significantly by a sigma-delta A/D converter at a given clock sampling rate by using a multibit A/D and D/A converter rather than a single-bit A/D and D/A converter. Digital filter design is another variable affecting the bandwidth of signals that can be accurately converted for a given oversampling rate.

However, the greatest advantage of an oversampled Σ-Δ converter is that it requires only a single-bit A/D and D/A converter with a relatively inaccurate differential summing amplifier and integrator (low-pass filter). These analog circuits are much easier to implement in a digital VLSI circuit than the accurate analog circuits required in parallel- and successive-approximation A/D converters that require precision resistors or capacitors. This is especially true when accuracies exceed 12 bits.

Based on the preceding description, we can show that the quantized signal bounces between two levels, keeping its mean equal to the input, when the input to the modulator is a DC signal. Figure 3-44 shows the quantized signal and the integrated output when the input signal is $3\Delta/7$ above 0 for a quantization level of Δ. The figure indicates that the oscillation may be repetitive (it returns to its starting condition after seven clock periods). The frequency of repetition depends on the input level.

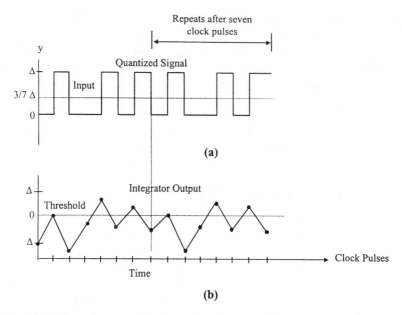

FIGURE 3-44 Waveforms in a Σ-Δ circuit for a constant input situated at $3\Delta/7$ above a quantization level: (a) Quantized signal and input, (b) Integrator output

By using more than one integration and summing stage in the Σ-Δ modulator, higher orders of quantization noise shaping and better ENOB for a given oversampling ratio can be achieved (Analog Devices, 1995). A block diagram of a second-order Σ-Δ ADC is shown in Figure 3-45.

3.6.9 Self-Calibration Techniques

Integral linearity of data converters usually depends on the matching and linearity of integrated resistors, capacitors, or current sources; and it is typically limited to approximately 10 bits with no calibration. For higher resolutions, means must be sought that can reliably correct nonlinearity errors. This often is accomplished by either improving the effective matching of individual

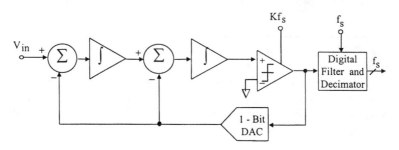

FIGURE 3-45 Second-order Σ-Δ ADC

devices or correcting the overall transfer characteristics. Since high-resolution A/D converters typically employ a multistep architecture, they often impose two stringent requirements: small integral nonlinearity in their interstage DACs and precise gain (usually a power of 2) in their interstage subtractors/amplifiers. These constraints in turn demand correction for device mismatches if resolutions above 10 bits are required.

ADC calibration techniques can be in two forms: use of analog processing techniques for correction of nonidealities and digital calibration techniques. A description of these techniques is beyond the scope of this chapter. For details, see Razavi (1995), Kularatna (1996), and O'Leary (1995).

3.6.10 Figure of Merit for ADCs

The demand for lower power-dissipating electronic systems has become a challenge to the IC designer, including designers of ADCs. As a result, a figure of merit was devised by the ISSCC[1] Program Committee to compare available and future sampling-type ADCs. The figure of merit (FOM) is based on an ADC's power dissipation, its resolution, and its sampling rate. The FOM is derived by dividing the device's power dissipation (in watts) by the product of its resolution (in 2^n bits) and its sampling rate (in hertz). The result is multiplied by 10^{12}. This is expressed by the equation

$$\text{FOM} = \frac{PD}{R \times SR} 10^{12} \tag{3.5}$$

where

$$PD = \text{Power dissipation (in watts)};$$

$$R = \text{Resolution (in } 2^n \text{ bits)};$$

$$SR = \text{Sampling rate (in hertz)}.$$

Therefore, a 12-bit ADC sampling at 1 MHz and dissipating 10 mW has a figure of merit rounded off to 2.5. This figure of merit is expressed in the units of picojoules of energy per unit conversion (pj)/conversion. For details and a comparison of performance of some monolithic ICs, see Goodenough (1995).

3.7 D/A Converters

Digital-to-analog conversion is an essential function in data processing systems. D/A converters provide an interface between the digital output of signal processes and the analog world. Moreover, as discussed previously, multistep ADCs employ interstage DACs to reconstruct analog estimates of the input signal.

---

[1] ISSCC, International Solid State Circuits Conference.

Each of these applications imposes certain speed, precision, and power dissipation requirements on the DAC, mandating a good understanding of various D/A conversion techniques and their trade-offs.

3.7.1 General Considerations

A digital-to-analog converter produces an analog output, A, proportional to the digital input D:

$$A = \alpha D \tag{3.6}$$

where α is a proportionality factor. Since D is a dimensionless quantity, α sets both the dimension and the full-scale range of A. For example, if α is a current quantity, I_{REF}, then the output can be expressed as

$$A = I_{\text{REF}}D \tag{3.7}$$

In some cases, it is more practical to normalize D with respect to its full-scale value, 2^m, where m is the resolution. For example, if α is a voltage quantity, V_{REF},

$$A = V_{\text{REF}}\frac{D}{2^m} \tag{3.8}$$

From (3.7) and (3.8), we can see that, in a D/A converter, each code at the digital input generates a certain multiple or fraction of a reference at the analog output. In practical monolithic DACs, conversion can be viewed as a reference multiplication or division function, where the reference may be one of the three electrical quantities: voltage, current, or charge.

The accuracy of this function determines the linearity of the DAC, while the speed at which each multiple or fraction of the reference can be selected and established at the output gives the conversion rate of the DAC. Figure 3-46 shows the input/output characteristic of an ideal 3-bit D/A converter. The analog levels generated at the output follow a straight line passed through the origin and the full-scale point.

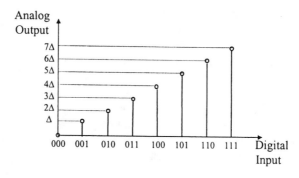

FIGURE 3-46 Input/output characteristic of an ideal 3-bit DAC

TABLE 3-4 Binary, Thermometer, and One-of-n Codes

Decimal	0	1	2	3
Binary	00	01	10	11
Thermometer	0	0	0	0
	0	0	0	1
	0	0	1	1
	0	1	1	1
One of n	0	0	0	0
	0	0	0	1
	0	0	1	0
	0	1	0	0

We should mention that, in some applications such as "companding" (compressing and expanding) DACs, the desired relationship between D and A is nonlinear (Sheer, 1988), but in this chapter we discuss only "linear" or "uniform" DACs; that is, those that ideally behave according to equation (3.7) or (3.8).

The digital input to a DAC can assume any predefined format but eventually must be of a form easily convertible to analog. Table 3–4 shows three formats often used in DACs: binary, thermometer, and one-of-n codes. The latter two are shown in column form to make visualization easier.

3.7.2 Performance Parameters and Data Sheet Terminology

In manufacturers, data books, many terms are used to characterize DACs. The following is a basic guideline only, and the reader is referred to manufacturers, data sheet guidelines for a more application-oriented description. Figure 3-47 illustrates some of these metrics that are listed in Table 3-5.

Among these parameters, DNL and INL usually are determined by the accuracy of reference multiplication or division, settling time and delay are functions of output loading and switching speed, and glitch impulse depends on the D/A converter architecture and design.

3.7.3 Voltage Division

A given reference voltage, V_{REF}, can be divided into N equal segments using a ladder composed of N identical resistors $R_1 = R_2 = \cdots = R_N$ (N typically is a power of 2; Figure 3-48(a)). An m-bit DAC requires a ladder with 2^m resistors, manifesting the exponential growth of the number of resistors as a function of resolution.

An important aspect of resistor ladders is the differential and integral nonlinearity they introduce when used in D/A converters. These errors result from mismatches in the resistors composing the ladder.

The DACs most commonly used as examples of simple DAC structures are binary weighted DACs or ladder networks, but although simple in structure, these require quite complex analysis. The simplest structure of all is the Kelvin divider shown in Figure 3-48(b). An N-bit version of this DAC simply consists

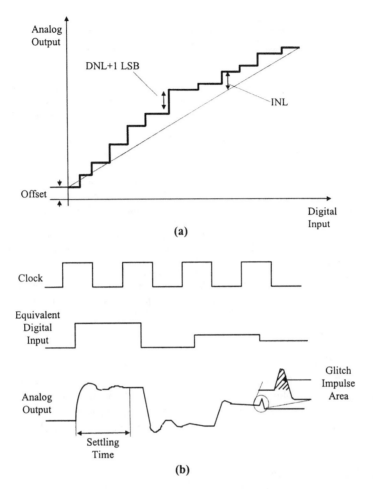

FIGURE 3-47 Parameters of DACs: (a) Static parameters, (b) Dynamic parameters

of 2^N equal resistors in series. The output is taken from the appropriate tap by closing one of the 2^N switches.

3.7.4 Current Division

Instead of using voltage division, current division techniques can be used in DACs. Figure 3-49(a) shows how a reference current I_{REF} can be divided into N equal currents using N identical (bipolar or MOS) transistors. These currents can be combined to provide binary weighting as depicted in Figure 3-49(b), using a 3-bit case as the example. In this simple implementation, an m-bit DAC requires $2^m - 1$ transistors, resulting in a large number of devices for $m > 7$.

While conceptually simple, the implementation of Figure 3-49(a) has two drawbacks: the stack of current division transistors on top of I_{REF} limits output

TABLE 3-5 DAC Performance Parameters

Parameter	Description
Differential nonlinearity (DNL)	Maximum deviation in the output step size from the ideal value of one least significant bit (LSB)
Integral nonlinearity (INL)	Maximum deviation of the input/output characteristic from a straight line passed through its end points. The difference between the ideal and actual characteristics is called the *INL profile*.
Offset	Vertical intercept of the straight line passed through the end points.
Gain error	Deviation of the slope of the line passed through the end points from its ideal value (usually unity).
Settling time	Time required for the output to experience a full-scale transition and settle within a specified error band around its final value.
Glitch impulse area	Maximum area under any extraneous glitch that appears at the output after the input code changes, also called *glitch energy* in the literature, even though it has no energy dimension.
Latency	Total delay from the time the digital input changes to the time the analog output has settled within a specified error band around its final value. Latency may include multiples of the clock period if the digital logic in the DAC is in a pipeline.
Signal-to-noise (+ distortion) ratio (SNDR or SINAD)	Ratio of the signal power to the total noise and harmonic distortion at the output when the input is a (digital) sinusoid.

voltage range, and I_{REF} must be N times each of the output currents. This requires a high-current device for the I_{REF} source transistor. There are techniques for alleviating these problems (Razavi, 1995). DACs that employ current division suffer from three sources of nonlinearity: current source mismatch, finite output impedance of current sources, and voltage dependence of the load resistor that converts the output current to voltage.

3.7.5 Charge Division

A reference charge, Q_{REF}, can be divided into N equal packets using N identical capacitors configured as in Figure 3-50. In this circuit, before S_1 turns on, C_1 has a charge equal to Q_{REF}, while $C_2, \ldots, C_N$ have no charge. When S_1 turns on, Q_{REF} is distributed equally among $C_1, \ldots, C_N$ yielding a charge of Q_{REF}/N on each. Further subdivision can be accomplished by disconnecting one of the capacitors from the array and redistributing its charge among some other capacitors.

While the circuit of Figure 3-50 can operate as a D/A converter if a separate array is employed for each bit of the digital input, the resulting complexity prohibits its use for resolutions above 6 bits. A modified version of this circuit is shown in Figure 3-51(a). Here, identical capacitors $C_1 = \cdots = C_N = C$ share the same top plate, and their bottom plates can be switched from ground to a reference voltage, V_{REF}, according to the input thermometer code. In other words, each capacitor can inject a charge equal to CV_{REF} onto the output node, producing an output voltage proportional to the height of the thermometer code.

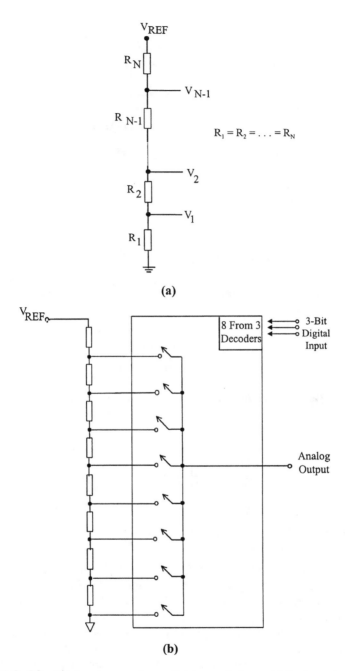

FIGURE 3-48 DAC using a voltage division technique: (a) Basic resistor ladder, (b) Kelvin divider (3-bit DAC example)

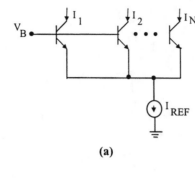

(a)

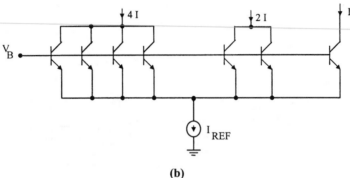

(b)

FIGURE 3-49 Current division: (a) Uniform division, (b) Binary division

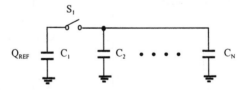

FIGURE 3-50 Simple charge division

The circuit operates as follows. First, S_P is on and the bottom plates of $C_1, \ldots, C_N$ are grounded, discharging the array to 0 (Figure 3-51(b)). Next, S_P turns off, and a thermometer code with height j is applied at $D_1, \ldots, D_N$, connecting the bottom plate of $C_1, \ldots, C_j$ to V_{REF} and generating an output equal to jV_{REF}/N (Figure 3-51(c)). This circuit, in a strict sense, is a voltage divider rather than a charge divider. In fact, the expression relating its output voltage to V_{REF} and the value of the capacitors is quite similar to that of resistor ladders. Nonetheless, in considering nonlinearity and loading effects, it is helpful to remember that the circuit's operation is based on charge injection and redistribution.

The nonlinearity of capacitor DACs arises from three sources: capacitor mismatch, capacitor nonlinearity, and the nonlinearity of the junction capacitance

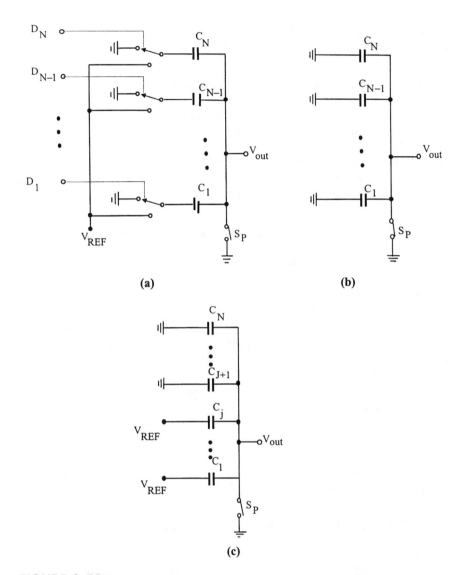

FIGURE 3-51 Modified charge division: (a) Configuration, (b) Circuit of (a) in discharge mode, (c) Circuit of (a) in evaluate mode

of any switches connected to the output code. For details and implementation of capacitor DACs, see Razavi (1995).

3.7.6 DAC Architectures

With the basic principles of D/A conversion explained, we can study this function from architectural perspective. This section describes D/A converter architecture based on resistor ladders and current steering arrays, with an emphasis on stand-alone applications. While capacitor DACs frequently are used in ADCs, they have not been popular as stand-alone circuits.

3.7.6.1 Resistor Ladder DAC Architectures

The simplicity of resistor ladder DACs using MOS switches makes these architectures attractive for many applications. Simple ladder networks with simple voltage division as per Section 3.7.3 have several drawbacks: they require a large number of resistors and switches (2^m, where m is the resolution) and exhibit a long delay at the output. Consequently, alternative ladder topologies have been devised to improve the speed and resolution.

3.7.6.1.1 Ladder Architecture with Switched Subdividers

In high-resolution applications, the number of devices in a DAC can be prohibitively large. It therefore is plausible to decompose the converter into a coarse section and a fine section so that the number of devices become proportional to approximately $2^m/2$ rather than 2^m, where m is the overall resolution. Such an architecture is shown in Figure 3-52(a). In this circuit, a primary ladder divides the main reference voltage, generating 2^j equal voltage segments. One of these segments is selected by the j most significant bits of $(k + j) = m$. If $k = j$, the number of devices in this architecture is proportional to $2^m/2$. It also is possible to utilize more than two ladders to further reduce the number of devices at high resolutions.

Figure 3-52(b) depicts a simple implementation of this architecture using MOS switches that are driven by one-of-n codes in both stages (Sheer, 1988). Depending on the environment, these codes are generated from binary or thermometer code inputs. The details and drawbacks of this implementation are discussed by Razavi (1995).

3.7.6.1.2 Intermeshed Ladder Architecture

Some of the drawbacks of ladder DACs can be alleviated through the use of intermeshed ladder architectures (Razavi, 1995). In these architectures, a primary ladder divides the main reference voltage into equal segments, each of which is subdivided by a separate, fixed secondary ladder. Figure 3-53 illustrates such an arrangement (Analog Devices, 1996), where all the switches are controlled by a one-of-n code.

The intermeshed ladder has several advantages over single-ladder or switched-ladder architectures. This configuration can have smaller equivalent resistance at each tap than a single-ladder DAC having the same resolution, allowing faster recovery. Also, since the secondary ladders do not switch, their loading on the primary ladder is constant and uniform. Furthermore, the DNL resulting from finite on-resistance of switches does not exist here.

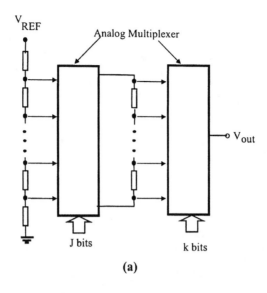

(a)

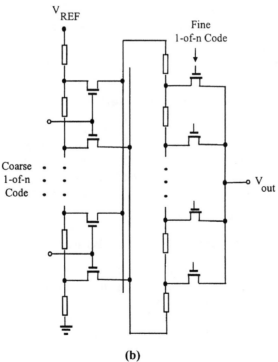

(b)

FIGURE 3-52 Resistor ladder DAC with a switched subdivider: (a) Block diagram, (b) Implementation

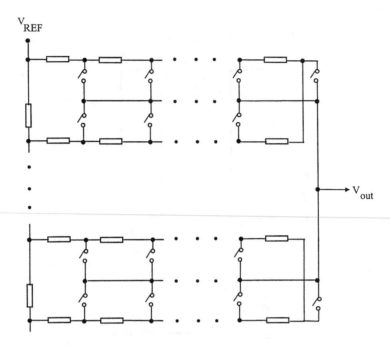

FIGURE 3-53 Intermeshed resistor ladder DAC with one-level multiplexing

3.7.6.2 Current Steering Architecture

Most high-speed D/A converters are based on a current steering architecture. Since these architectures can drive resistive loads directly, they require no high-speed amplifiers at the output and potentially are faster than other types of DACs. While the high-speed switching of bipolar transistors makes them the natural choice for current-steering DACs, many designs have been recently reported in CMOS technology as well.

3.7.6.2.1 R-2R Network-Based Architectures

To realize binary weighting in a current steering DAC, an R-2R ladder can be incorporated to relax device scaling requirements. Figure 3-54(a) illustrates an architecture that employs an R-2R ladder in the emitter network. A network with an R-2R ladder in collector networks is shown in Figure 3-54(b). For details, see Razavi (1995) and Hoeschele (1994).

3.7.6.3 Other Architectures

Other architectures for DACs include segmented current steering versions, multiplying DACs, and Σ-Δ types. This chapter does not permit discussing these, so see Analog Devices (1995), Razavi (1995), and Hoeschele (1994).

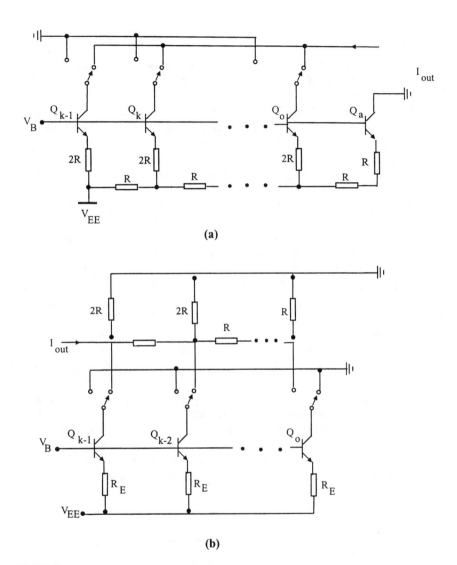

FIGURE 3-54 Current steering DAC with an R-2R ladder: (a) R-2R ladder in the emitter, (b) R-2R ladder in the collector

3.8 Data Acquisition System Interfaces

3.8.1 Signal Source and Acquisition Time

Continued demand for lower-power, lower-cost systems increases the likelihood that a mixed-signal design will operate from a single 3.3 or 5 V power supply. Doing away with traditional ±15 V analog power supplies can help you to meet your power and cost goals, but it also will eliminate some of your options.

Most low-voltage ADC and data acquisition system chips are designed for easy analog and digital interfaces. The ICs' digital interfaces generally are compatible with popular microcontrollers, and the devices almost always can accept analog input signals that range from ground to the positive supply voltage; the span is set by an internal or external bandgap voltage reference. Virtually all ADCs that operate from 5 V or less are CMOS devices that use arrays of switches and capacitors to perform their conversions. Although the architectural details vary from design to design, the input stage of this type of converter usually includes a switch and a capacitor that present a transient load to the input signal source. The simplified schematic of Figure 3-55 shows how these input stages affect the circuits that drive them.

R_{ON} is not a separate component; it is the on-resistance of the internal analog switch. Sampling capacitor C_s connects to an internal bias voltage, whose value depends on the ADC's architecture. In a sampling ADC, the switch closes once per conversion, during the acquisition (sampling) time. The on-resistance of the sampling switches ranges from about 5 to 10 kΩ in many low-resolution successive-approximation ADCs to 70 Ω in some multistep or half-flash converters. The capacitors can be as small as 10 pF in lower-resolution successive-approximation converters and 100 pF or more in higher-resolution devices.

When the sampling switch closes, the capacitor begins to charge through the switch and source resistance. After a time interval that usually is controlled by counters or timers within the ADC, the switch opens and the capacitor stops charging. The acquisition time described in Figure 3-10(c) is actually the time during which the switch is closed and the capacitor charges. As long as the source impedance is low enough, the capacitor has time to charge fully during the sampling period and no conversion errors occur. Most input stages are conservatively designed and can work properly at their rated speeds with a reasonable source resistance (1 kΩ is common). Larger source impedance slows the charging of the sampling capacitor and can cause significant errors unless you take steps

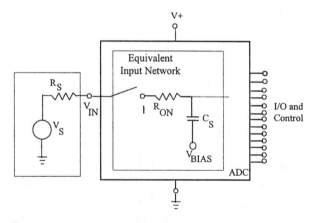

FIGURE 3-55 Simplified interface between a low-voltage ADC and a signal source

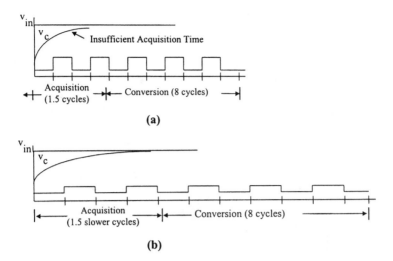

FIGURE 3-56 The effect of source resistance: (a) Insufficient acquisition time, (b) Slowing the clock to increase acquisition time

to avoid them. Figure 3-56 illustrates this. Figure 3-56(a) indicates the case of insufficient acquisition time. Figure 3-56(b) shows a case in which the problem could be solved by slowing the clock. For further details, see Lancanette (1994) and Swager (1994).

3.8.2 The Amplifier-ADC Interface

Operational amplifiers nearly always are present in data acquisition systems, performing basic signal conditioning ahead of the ADC. Their interactions with ADCs affect system performance. Although many amplifiers are good at driving a variety of static loads, the switched nature of the ADC input stage can introduce problems with some amplifiers, especially the low-power, low-speed devices most likely to be used in 3 and 5 V systems. Using the simple model in Figure 3-57(a), the load presented to the amplifier by the ADC input keeps switching abruptly between an open circuit and a series RC network connected to an internal voltage source. The op amp's response to the sudden load-current and impedance change depends on several parameters. Among them are the device's gain-bandwidth product, slew rate, and output impedance.

Selecting the appropriate drive amplifier for an ADC involves many considerations. Because the ADC drive amplifier is in the signal path, its error sources (both DC and AC) must be considered in calculating the total error budget. Ideally, the AC and DC performance of the amplifier should be such that there is no degradation of the ADC performance. Achieving this rarely is possible, however; therefore, the effects of each amplifier error source on system performance should be evaluated individually.

Evaluating and selecting op amps based on the DC requirements of the system is a relatively straightforward matter. For many applications, however,

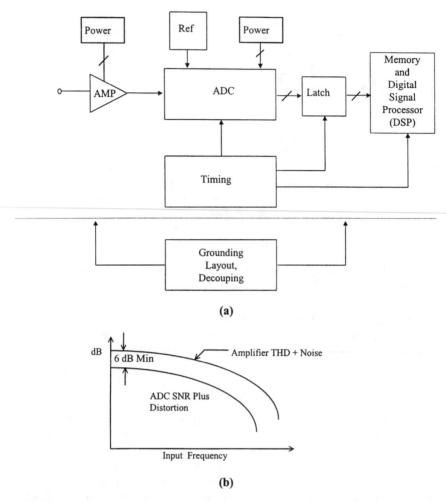

FIGURE 3-57 ADC-amplifier interface: (a) Basic elements, (b) Performance expected from ADC and amplifier. (Source: Analog Devices.)

it is more desirable first to select an amplifier on the basis of AC performance (bandwidth, THD, noise, etc.). The AC characteristics of ADCs are specified in terms of SNR, ENOBs, and distortion. The drive amplifier should have performance better than that of the ADC so that maximum dynamic performance is obtained (see Figure 3-57(b)). If the amplifier AC performance is adequate, the DC specifications should be examined in terms of system performance. Table 3-6 summarizes the ADC drive amplifier considerations. Further details can be found in Analog Devices (1995).

Other considerations in interfacing are the input clamping and protection, drive amplifier noise configurations, ADC reference voltage considerations,

TABLE 3-6 ADC Drive Amplifier Considerations

Performance Requirements	*Parameter*
AC performance	Bandwidth, settling time, harmonic distortion, total harmonic distortion, noise, THD + noise
DC performance	Gain, offset, drift, gain nonlinearity
General	As a general principle, select first for AC performance, then evaluate DC performance. Always consult the data sheet for recommendations

settling time considerations, and the like. These are beyond the scope of this chapter, and the reader is referred to Analog Devices (1992, Chapter 7; 1995; 1996), Lacanette (1994), and Swager (1994).

Interfaces between data converters and digital signal processors are discussed in Chapter 5.

References

Analog Devices Inc. 1992 Amplifier Applications Guide. Analog Devices Inc., USA, 1992.

Analog Devices Inc. Linear Design Seminar. Analog Devices Inc., USA, 1995.

Analog Devices Inc. High Speed Design Techniques. Analog Devices Inc., USA, 1996.

Candy, James C., and Gabor C. Temes. *Over-Sampling Delta-Sigma Data Converters—Theory, Design and Simulation.* Piscataway, NJ: IEEE Press, 1992.

Doernberg, Joe, Hae-Seung Lee, and David Hodges. "Full-Speed Testing of A/D Converters." *IEEE Journal of Solid-State Circuits* SC-19, no. 6 (December 1984), pp. 820–827.

Goodenough, Frank. "ADCs Move to Cut Power Dissipation." *Electronic Design* (January 9, 1995), pp. 69–74.

Hoeschele, David F. *Analog-to-Digital and Digital-to-Analog Conversion Techniques*, 2d ed. New York: John Wiley Interscience, 1994.

Kularatna, N. "Modern Electronic Test & Measuring Instruments"; IEE Electrical Measurement Series, Vol 10, IEE (London), 1996.

Lacanette, Kerry. "To Build Data Acquisition Systems That Run from 5 or 3.3 V, Know Your ICs." *EDN* (September 29, 1994), pp. 89–98.

McGoldrick, Paul. "Architectural Breakthrough Moves Conversion into Mainstream." *Electronic Design* (January 20, 1997), pp. 67–72.

O'Leary, Sean. "Self-Calibrating ADCs Offer Accuracy, Flexibility." *EDN* (June 22, 1995), pp. 77–85.

Phillip Louzon: "Decipher High-Sample Rate ADC Specs"; *Electronic Design*, March 20, 1995, pp. 91–100.

Razavi, Behad. *Data Conversion System Design.* Piscataway, NJ: IEEE Press, 1995.

Sheer, David. "Monolithic High-Resolution ADCs." *EDN* (May 12, 1988), pp. 116–130.

Shill, Mark A. "Servo Loop Speeds Tests of 16-Bit ADCs." *Electronic Design* (February 6, 1995), pp. 93–109.

Swager, Anne Watson. "Evolving ADCs Demand More from Drive Amplifiers." *EDN* (September 29, 1994), pp. 53–62.

Tektronix Inc. "Digital Oscilloscope Concepts." Engineering note 37W-6136 (April 1986).

CHAPTER 4

Microprocessors and Microcontrollers

Coauthor: Kithsiri Samarasinghe

4.1 Introduction

The human race accumulates knowledge and develops intelligence from generation to generation. The engineered systems around us have increased the amount of intelligence embedded in the microelectronic components used in the industrial designs as well as consumer electronics such as personal phones, watches, television, entertainment systems, and cameras. The components became more sophisticated as engineers and software designers acquired more knowledge and experience on processor families since the commercial single-chip microprocessor was developed in 1971. As system designers strive to combine human intelligence with nanosecond- and picosecond-order microelectronics, new generations of microprocessor-based systems continue to evolve. As the technology advances, the microelectronic systems not only gain in speed but become small, more compact, lightweight, and affordable, making the way to invade every corner of the industry and society. Microprocessors and microcontrollers have become indispensable in system design. This chapter provides an overview of the fundamentals and applications of microprocessors and microcontrollers.

4.2 What Is a System?

The word *system* can be used to describe and understand a complex whole. It is a way of limiting our thoughts to those aspects important for us to understand in any complex whole at the depth required. Therefore, a system is "an adequate way of describing a complex whole." Any system can be viewed as a single unit or a connected group of subsystems or subunits.

FIGURE 4-1 The three components of a system

4.2.1 A System as a Single Unit — The External View

When searching for the external view of a system, we ignore its internal details and try to identify its inputs, process, and output (Figure 4-1). The process takes place within the system "box." The process acts on the inputs to change the internal condition of the system or produce an output.

4.2.2 A System as a Connected Group of Subsystems — The Internal View

When a system is viewed internally as a connected group of subunits, it can be drawn as a block diagram. Each block in the system can be treated as a separate system for further analysis, and each block's external and internal views can be separately drawn. In this manner, the most complex whole can be decomposed into manageable, understandable parts.

4.2.3 Combinational and Sequential Systems

In a combinational system, the results of the process are available only when all the inputs are applied and the output does not depend on the sequence in which the inputs are applied but on the combination of the inputs used.

In a sequential system, the output of the process depends completely on the sequence in which the inputs are applied. The result is different when the sequence of inputs is changed. Normally, all inputs are not required to produce an output. The process is sequential and the subunits do not work simultaneously.

4.3 Central Control

Consider a simple machine, such as a simple domestic washing machine (Figure 4-2). Its internal components easily can be identified.

The subsystems must communicate with each other and information in the form of signals must pass between them to carry out the process. A sequential system operates in a sequence of steps. In between steps can be a pause, while a signal is passed. During the pause, all internal signals are maintained by the system. Such a pattern of internal signals is known as an *internal condition* (or *internal state*) of the system. During the process, the machine proceeds from one state to another, in a sequential manner. These changes are called *state transitions*.

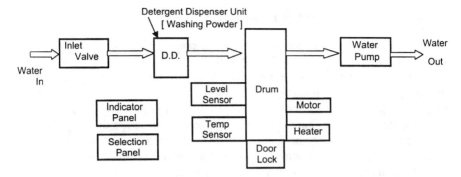

FIGURE 4-2 Internal view of a washing machine

The process acts on the inputs of the system to produce an output or transfer the system from one state to another.

In a sequential system, the subunits are linked during the process. Subunits must communicate with each other to carry out their jobs. At its maximum, central control would completely ban direct communication between subsystems; all communication would have to be channeled through the central controller.

Washing is a sequential process; therefore, it conveniently can be performed through a central controller. Figure 4-3 shows the links when the system is put under the central control.

All subsystems attached to the central controller are called its *peripherals*. In the next level of analysis we treat the central controller as a system having its own inputs, output, and a process.

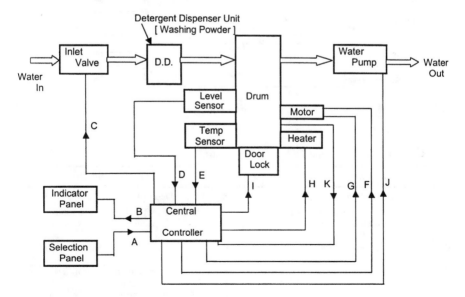

FIGURE 4-3 The washing process under central control

4.3.1 The Central Controller as a System

Figure 4-4 shows an external view of the controller and Table 4-1 summarizes the input to and output of the controller.

Each input conveys some information to the controller, keeping it well informed of what is happening. The controller has complete control over its peripherals. It makes decisions based on the information conveyed by the sensors and generates a set of commands to its peripherals. These are the output of the central controller. Thus, the behavior of the overall system is fully determined and controlled by the central controller. Depending on the system being controlled the central controller could be any of the following:

- Mechanical (with rods and levers and so forth).
- Electromechanical (with relays and breakers and the like).
- Pneumatic (with valves, compressors, and so on).
- Microelectronic (with microprocessors, memory, and I/O interface ICs).
- Biological (with brain cells).

Microelectronic central controllers have the advantages of high reliability, low maintenance, low cost, compactness, and high flexibility.

Next the process of the system should be designed. The process is the *sequence of events* that takes place during the operation of a system. A flowchart is the tool for designing a process.

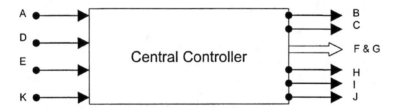

FIGURE 4-4 External view of the central controller

TABLE 4-1 Inputs to and Output of the Controller

Inputs to the Controller		Output of the Controller	
Signal	Description	Signal	Description
A	The values selected by the user, such as the temperature or start button	B	Output to LEDs on the indicator panel
D	Water level ("drum full")	C	Opening and closing of the inlet valve
E	Temperature	F–G	Agitator motor control
		H	Heater control
		I	Door control
K	The door lock sensor	J	Water pump control

4.3.2 The Washing Process

When planning the process, we first write down *macro* instructions, ignoring small details. Then each macro is decomposed into smaller (micro) steps. The macro steps of the washing process are

1. Receive a load of clothes.
2. Receive the detergent.
3. Receive the start signal.
4. Read the panel selection.
5. Fill the drum with water and heat up to set temperature.
6. Agitate for 30 minutes.
7. Empty water from the drum.
8. Refill the drum with fresh water.
9. Agitate for 5 minutes.
10. Drain the water while spinning the drum.
11. Indicate completion of washing process.

4.3.3 Flowchart of the Process

The microelectronic controller can do only one small job at a time and so needs a more detailed sequence of instructions. A flowchart specifies those details: the activities, events, and sequence (flow). Figure 4-5 depicts the flowchart of the washing process.

4.3.4 Writing the Program

The flowchart carries instructions and relationships but these are not understood by the controller. Therefore, the steps in the flowchart must be converted into instructions to the microcontroller, called *program coding*. Coding is a translation. The flowchart does not carry the syntax of a computer language. After coding, it carries the syntax and the program can be stored in the semiconductor memory of the controller.

The programming process has several basic steps to be planned and carried out:

1. Study the block diagram.
2. Identify the sequence of macro steps.
3. Break down the macro step into smaller steps.
4. Draw a flowchart.
5. Code the program.
6. Store the coded program in the memory of the controller.

4.3.5 Centralized vs. Decentralized Control

Is central control applicable to all situations? Certainly not. It may be the best approach for small, dedicated systems in which the system is not very complex and its parts are close together. Think of yourself. You are a *biological system* under central control. All your sensors (senses) are wired to the human brain,

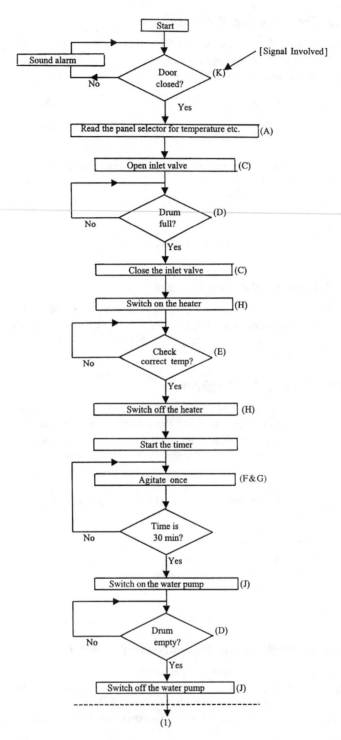

FIGURE 4-5 Continuation of flowchart for the washing process

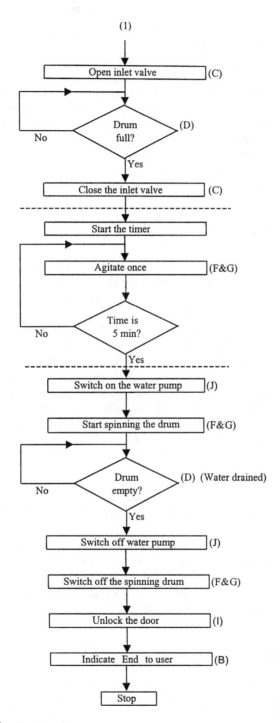

FIGURE 4-5 Continued

the great central controller. The brain collects information from the sensors, processes them, makes decisions based on them, and issues commands to the various actuators (muscles) of the body. It works.

But think of large *social systems* like organizations and governments, which have to manage diversity in a wide geographical spread. They attempt to decentralize control to achieve greater efficiency and effectiveness. A computer network spreads its computing power to a set of scattered PCs to decentralize control. For large systems having widespread entry/exit points, decentralization may be a better technique. However, at the periphery of such a system, the nodes may be under a central control. Therefore, the social system may be treated as a collection of subsystems, each under its own central control.

4.4 Stored Program Control

Inside the central controller is a program stored in the memory subsystem that will determine the sequence of steps the controller has to carry out. A program is a sequence of instructions arranged in a particular, meaningful order to produce a useful result from the system. The program controls the process. Therefore, the control is centralized at two levels: at the central controller, and at a program device (memory) within the central controller.

For example, a read-only memory (ROM) may hold the sequences of instructions needed to perform the job. The system is under the control of a stored program that carries the know-how and expertise of the team that developed the program. The team's intelligence is at work even in its absence. This is the advantage of the stored program. It creates portable human intelligence. If you write a program, in your absence, your intelligence will be at work, controlling all the systems that carry your program. You can package and market your intelligence, and it will remain in the world.

4.5 Inside the Microelectronic Central Controller

Figure 4-6 shows the internal view of a central controller that consists of three subsystems: processing, memory, and input/output.

The three subsystems are linked to each other using *buses*, a group of conductors running in parallel. They are used to interconnect subsystems. The subsystems communicate through the buses, sending digital data and messages.

4.5.1 The Processing Subsystem

The processing subsystem is the programmable VLSI device called the *microprocessor*. It should be provided with a power supply, a clock, and instructions (operational codes). When these are provided, it takes overall control of the system.

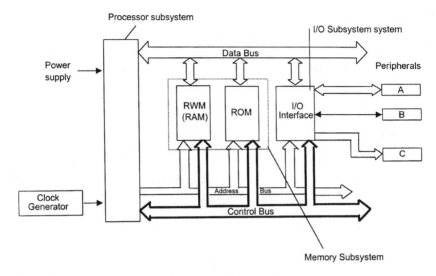

FIGURE 4-6 Internal view of a microelectronic central controller

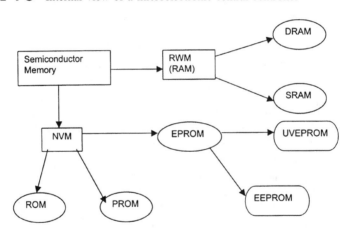

Rwm – Read/write memory
NVM – Nonvolatile memory

FIGURE 4-7 Classification of semiconductor memory

4.5.2 The Memory Subsystem

The memory system contains programs and data. It consists of two main types of memory, which have several subcategories, as shown in Figure 4-7.

4.5.2.1 Read/Write Memory

Read/write memory is also known as *random access memory* (RAM). RAM is fast compared to other forms of memory. RAM is temporary storage. When electrical power is switched off, RAM looses all its contents (volatility). Therefore, it is used to store programs and data temporarily, while the system is running. Memory speed is measured by the access time in nanoseconds. The smaller the access time, the faster the memory. The parity technique is used to check the integrity of the data bytes. Therefore, a parity bit is saved along with each byte in a memory location. But parity can detect only single-bit errors. A normal RAM can be accessed from only one set of bus lines, but dual-ported RAM supports two separate buses. This feature is useful in applications like video RAM, where the microprocessor and the cathode ray tube controller demand access to the same RAM. Another version, nonvolatile RAM, keeps the memory contents even in the absence of an external power supply.

Depending on the operating principle, there are two main categories of RAM: static RAM (SRAM) and dynamic RAM (DRAM). DRAM remembers data for only a short period of time. Therefore, it has its own circuitry to refresh itself (i.e., to remind itself of what it has remembered). It uses capacitors to remember the logic's ones and zeros. When the capacitor is charged it is in the logic 1 state. But capacitors remain charged for only a short period of time, and therefore, the refreshing circuit needs to keep recharging them. This circuitry has to look at each location and recharge the capacitors to their previous levels. DRAM has a very high packing density and therefore offers lower cost/bit than SRAM.

Static RAM, on the other hand, uses flip-flop circuits to store bits. But one flip-flop takes much more space than the capacitors used in DRAM. Therefore, packing density is low. But they are faster than capacitors. Access times often are as low as 8–14 ns. Also, they need no refreshing. The design variations in RAM ICs are identified as FPM RAM, EDO RAM, and SDRAM. In FPM (fast page model) RAM, a complete address is sent only once for a column, in its internal matrix. Then, for all other memory locations in that column, only the row address is supplied. Within that memory page, the access is faster; therefore, it is faster in accessing bytes sequentially. EDO (extended data out) RAM starts getting the next address before it finishes reading the last address. It holds old data for a longer period of time for the microprocessor to read, while it is getting the next address from the microprocessor. Therefore, the time gap between two successive readings can be reduced, increasing performance. Synchronous dynamic RAM (SDRAM) is a faster version of DRAM. Average access time can be as low as 8–10 ns. SDRAM employs an addressing modification. After one address operation, for each clock cycle, it gives the next sequential address. Therefore, it offers other advantages of DRAM with low access time.

4.5.2.2 Nonvolatile Memory

The contents of nonvolatile memory will not be erased even when the power is switched off. In this category of ICs, the ratio of read operations to write operations is very high. Nonvolatile memory includes ICs such as ROM, PROM, UVEPROM, and EEPROM.

Read-only memory is a form of permanent storage. ROM is written during the manufacturing process with user-supplied data. Once written, it can never be changed. Because information on a ROM cannot be changed, it is useful to designers for storing parts that have no need to change (such as the primary loader for starting). RAM often is faster than ROM. Therefore, some designs "shadow" the ROM (copy it into RAM and then always access the RAM version). Although this wastes RAM, a designer may trade off performance for space.

Programmable read-only memory (PROM) is a form of permanent storage that you can write to, but just once for its lifetime. A blank PROM has all ones. After the user writes data to a PROM programmer unit, it can only be read. Pin-compatible PROMs can be used in place of ROMs during software development.

Erasable programmable read-only memory (EPROM) can be programmed and reprogrammed by the user with the assistance of a PROM programming unit and an ultraviolet eraser. There are two types of EPROMs: ultraviolet EPROM (UVEPROM) and electrically erasable PROM (EEPROM). UVEPROM can be erased only by exposing it to ultraviolet light of specified intensity over a specified duration. But no erasing of selective locations is possible, because the UV light causes bulk erasing. One disadvantage is that, even to modify few locations, the whole IC needs to be erased and reprogrammed.

EEPROM is a form of EPROM that can be erased electrically. Erasing is achieved by sending special electrical voltage. Erasing of selective locations is possible. During reading, they work just like any other EPROM. EEPROMs can be rewritten a few million times.

One form of EEPROM, known as *flash ROM*, can be erased in a "flash." This form of EEPROM is manufactured so that it usually is erased in blocks of memory rather than character by character, and it is cheaper to manufacture. Other forms of nonvolatile memory include battery-backed RAMs and ferroelectric memory.

4.5.3 The Input/Output Interface Subsystem

Any external device connected to a microprocessor-based central controller is called a *peripheral* (e.g., the keyboard, VDU, mouse, printer, scanner, light pen, joystick, hard and floppy disk drives, CD drive, plotter, magnetic tape drive, and transducers, both sensors and actuators). The number of peripherals can be tailored to fit the requirements of the user. Therefore, computer systems have become very flexible and adaptable. Peripherals are connected through ports ready-made for the export and import of information. A port can be programmed to do either input operations, output operations, or both. Ports are either parallel ports and serial ports.

The input/output (I/O) interface subsystem allows a digital message to be passed between the central controller and the peripheral devices or I/O devices. The interface comes between a peripheral device and the microprocessor buses, providing communication between peripherals and the microprocessor. Peripherals usually are slower than the microprocessor, creating a timing problem. Furthermore, connecting them to the buses of the central controller presents additional problems, such as electrical buffering, code conversion, and analog-to-digital conversion. The interface subsystem is there to solve these problems. A few common examples of peripheral interfacing devices follow:

PIO — parallel input/output device.
PIA — programmable interface adapter (the same as PIO).
USRT — universal synchronous receiver-transmitter.
UART — universal asynchronous receiver-transmitter.
CTC — counter-timer circuit.
FDC — floppy disk controller.
HDC — hard disk controller.
CRTC — cathode ray tube controller.

4.5.4 Buses

The three subsystems are joined by the three conventional data buses. The width of the data bus (number of lines) is an important figure for the system (e.g., 8 bit, 16 bit, 32 bit, 64 bit). It carries both data bytes and instruction operational codes. Each memory location and each port have a unique address, given to them by the system designer. When the microprocessor wants to exchange information with a memory IC or a port, its address first is deposited on the address bus. An address bus broadcasts it to all subsystems. Therefore, it is unidirectional. With an n-bit address bus, the microprocessor can generate 2^n different addresses. Control lines can be further categorized into "system status" lines, which bring in information, and "system control" lines, which carry the commands generated by the microprocessor to the other ICs in the system.

4.6 Microprocessor Architecture

4.6.1 External View of a Microprocessor

To simplify the external view of a microprocessor, its pins can be categorized into five groups as shown in Table 4-2. A typical microprocessor, Z80, and its pin diagram and signal classification are shown in Figure 4-8.

Even though these lines appear distinct in the Z80 block diagram, in some microprocessors certain lines are made to perform different functions during different parts of their computing cycles. To reduce the pin count of the IC, certain pins are assigned with two nonoverlapping functions, a technique called *pin multiplexing* (e.g., in the Z80, address and data lines are multiplexed on the same set of pins). At one time, addresses appear on the lines and in the next moment data appear on the same lines. Additional control signals are sent to the system to identify them; for example, DS (data strobe, this pin goes high when data come out) or AS (address strobe, this pin goes high when an address

TABLE 4-2 Signal Groups of a Microprocessor

Group	*Description*
Datalines	These lines carry data bytes and operational codes between the microprocessor and the external ICs; they are bi-directional
Address lines	These carry memory addresses and addresses of input/output devices from the microprocessor to other subsystems; therefore they are unidirectional
CPU control (system status) lines	These lines carry signals that provide information to the CPU relating the present state of the system; the microprocessor reacts to these signals; therefore they also are called *CPU control signals* (e.g., the interrupt line, reset line, hold line)
System control/CPU status lines	These carry signals that provide information to the system relating to the current status of the microprocessor; the system reacts to these signals (e.g., the R/W line)
Service lines	These provide basic services needed for the microprocessor to operate (e.g., the power supply, external clock requirements)

comes out). The system is responsible for separating addresses and data using the information of the AS and DS pins. This is called *demultiplexing*.

4.6.2 Internal View of a Microprocessor

A typical microprocessor consists of three main components: the arithmetic and logic unit (ALU), the control unit, and a set of registers (Figure 4-9). The register set can be further categorized into general purpose and special purpose registers.

4.6.2.1 Arithmetic and Logic Unit (ALU)

The ALU performs arithmetic operations such as addition and subtraction and logic operations such as AND, OR, Exclusive OR, Complement, and Compare. Multiplication and division are not available as direct operations in most cases, but the control unit may use a program sequence (i.e., multiple addition and shifts) to generate a multiplication. Hence, these operations take much longer to execute than direct operations. In some operations like shifting both arithmetic and logical versions are available. Shift operations are typically 1-bit shifts, with multiple-bit shifts performed as a successive sequence of single-bit shifts. Shift operations can be

- Logical shifts that do not preserve the sign and fill empty bits with zero.
- Arithmetic shifts that preserve the sign and fill empty bits with zeros or an extended sign bit.
- Cyclical shifts that simply rotate the contents.

Bit operations are available to set or clear bits or test whether a specific bit is set or reset.

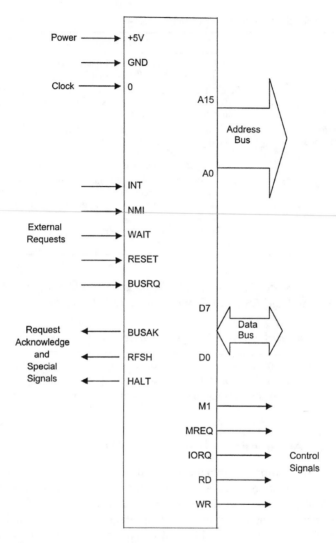

FIGURE 4-8 A typical microprocessor, Z80, pin diagram and classification. (Source: Zilog Z80 databook.)

4.6.2.2 The Control Unit (Micro Code Interpreter)

The control unit or sequencer decodes the instruction given to it and performs the appropriate sequence of actions. The timing and sequencing of all the minute steps to execute the instruction are done by the control unit.

4.6.2.3 The Registers

A microprocessor has several special purpose registers. Typical special purpose registers are the accumulator(s), program counter, stack pointer, flag register, and index register. The *accumulator* is used to feed one input number to the ALU and, immediately after the operation, collect and store the result of the operation. The *program counter* contains the address of the next program instruction to be executed. Fetching the next instruction starts when the program counter deposits this address on the address bus. *Index registers* are used to offset addresses in memory when addressing data tables and blocks.

The *status register* consists of a set of 1-bit indicators known as *flags*. Each flag may be set or reset to indicate a condition. Therefore, the status register also is known as the *condition code register*. The number and the designations of these indicators vary from processor to processor. However, carry (C) flags, zero (Z) flags, negative (N) flags, overflow (V) flags, and interrupt (I) flags are common to almost all microprocessors. The results of the arithmetic and logic operations typically affect the flags. For example, if the result is 0, a Z flag will be set (Z = 1); if it is negative, an N flag gets set. Using these flags, the programmer can program the microprocessor to make decisions during the actual operation.

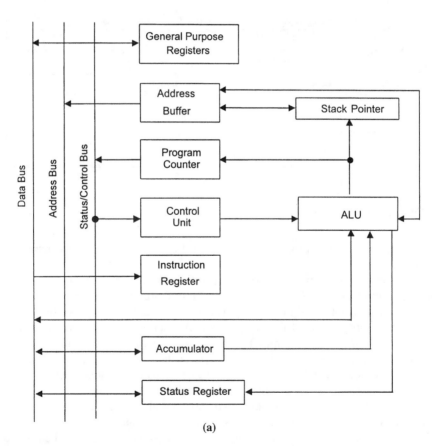

(a)

FIGURE 4-9 The basic architecture of a microprocessor: (a) Conceptual model, (b) Practical implementation in Z80 (Source: Zilog Z80 databook.)

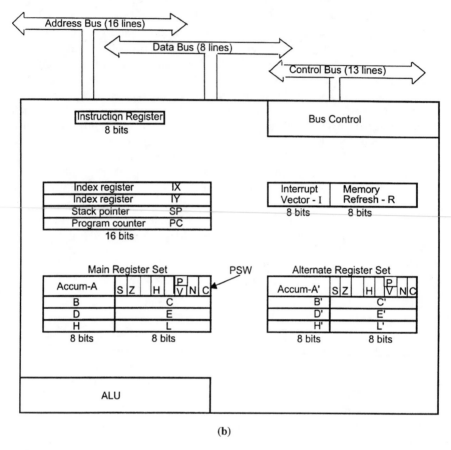

(b)

FIGURE 4-9 Continued

These are useful when programming backward or forward conditional jumps of a flowchart.

In applications, it becomes necessary to temporarily save registers and other data for future use, such as when a subprogram (subroutine) is executed or the program is interrupted and the interrupt is to be entertained. A stack is a sequential block of memory reserved for this purpose. A stack pointer is used to manage this memory block as a last-in-first-out (LIFO) data structure. The only byte that can be accessed on the stack is the last byte pushed into it, which is at the top of the stack. A *stack pointer* is the register that keeps this address. Furthermore, it employs destructive reading (i.e., the byte is erased when read).

Many ALU operations require two input numbers. In accumulator-based microprocessors, the first number usually comes from the accumulator and the other must be retrieved from memory into an internal data buffer and then fed to the ALU. In register-based microprocessors, a bank of general purpose registers may have direct access to the ALU, and both operands can be stored directly in two of these registers prior to the ALU operation (i.e., registers usually are preloaded with the appropriate operands). In addition, these registers are very useful for storing intermediate results to speed up operations.

4.6.3 Bottlenecks in the Basic Architecture

Two main bottlenecks can be identified in the basic architecture:

- The time taken to fetch the instructions from memory. The fetch cycle does not deliver value to the customer; the execution cycle produces the end result.
- The sequential nature of scheduling and executing instructions. One instruction should be fully completed before the microprocessor can pay attention to the next instruction.

Microprocessor designers have come up with various modifications to improve these two areas. These attempts have brought about new architectural designs. Most of them are centered around the concept of parallel processing. In parallel processing, the three architectural designs are pipelines, array processing, and multiprocessor systems.

4.6.3.1 Cache Memory

An increase in operating speed can be achieved through the use of a memory cache. A cache is a relatively small, high-speed memory buffer. A cache may be internal or external to the microprocessor. An internal memory cache is a fast set of registers available within the microprocessor. An external memory cache is a fast multiport RAM with a data and addressing bus separate from the sequencer and the ALU.

While the ALU is executing one operation, the control unit loads the next instruction into an instruction cache register, decodes the instruction, and loads all possible data operands into the data cache registers. Once in the cache registers, these instructions and data can be accessed instantly, reducing the normal memory read time. This significantly increases the processing speed.

4.6.3.2 Pipeline Processing

The time taken by a processor to complete a program is determined by three factors:

1. The number of instructions required to execute the program.
2. The average number of processor cycles required to execute an instruction.
3. The processor cycle time.

Processor performance is improved by reducing the completion time, which involves reducing one or more of these factors. Pipeline processing is a technique used by both reduced instruction set computers (RISC) and complex instruction set computers (CISC) to break the execution of individual instructions into stages and overlap the stages so several instructions can be processed in parallel.

To process instructions in a pipeline, the various steps of execution need to be performed by pipeline stages, units that independently execute the steps of different instructions. The result of each pipeline stage is communicated to the next pipeline stage via a register between the stages. Both 486 and Pentium CPUs use a five-stage pipeline:

1. Fetch an instruction from the processor cache or memory.
2. Decode the instruction.
3. Generate a memory address if the instruction includes a memory reference.
4. Execute the instruction.
5. Store, or "write back," the result.

Pipelines allow more than one microprocessor instruction to be serviced at once, which allows the microprocessor to average one clock cycle for each instruction. Under ideal conditions, each stage requires one clock cycle. When the pipeline is fully loaded, an average of one instruction per clock cycle can be produced by the pipeline. For example, the Pentium's superscalar design, incorporating two independent pipelines, nominally doubles the processor's instruction throughput.

The term *scalar processor* denotes a processor that executes one instruction at a time. A *superscalar processor* reduces the average number of cycles per instruction beyond what is possible in a scalar processor by concurrent execution of scalar instructions. Superscalar microprocessors are the next step in the evolution of microprocessors.

4.7 Using Assembly Language

The size of the instruction set and the types of instructions depend on the category of microprocessor, accumulator based or register based.

4.7.1 Accumulator-Based Microprocessors

Accumulator-based microprocessors carry an accumulator register that is involved in a majority of its internal operations. Therefore, most of the instructions in the instruction set refer to the accumulator (e.g., the 6502 is an accumulator-based microprocessor). The purpose of all the microprocessors internal registers of 6502 is predefined by the manufacturer. Therefore, these are special purpose registers. The programmer can define the use of a general purpose register. In the instruction set, many instructions perform their operations in relation to the accumulator, so the instruction set of these microprocessors is smaller.

4.7.2 Register-Based Microprocessors

The internal register bank of register-based microprocessors consists of both general purpose and special purpose registers. The programmer is free to use general purpose registers as appropriate to the specific case. For example, the

Z80's A, B, C, DE, and HL registers all are general purpose ones. B and C may be used as a pair or individually, as may the D and E and H and L registers.

In general when the number of registers in a microprocessor increases, the instruction set grows in size. Therefore, Z80, a register-based microprocessor, has a bigger instruction set than a 6502 or M6800, which are accumulator based.

4.7.3 Mnemonics and the Instruction Set

A mnemonic represents a long instruction. For example, the mnemonic LDA means "load accumulator with a specified data byte." The set of all commands to which the microprocessor responds is known as its instruction set, which consists of all the mnemonics and their operational codes. A typical instruction consists of two parts: the op code (operational code) and the operand. The op code specifies the operation to be carried out (e.g., LDA, TAX, INC). The operand specifies the location of the data to carry out this operation. Therefore, it depends on the addressing mode used in the instruction.

The instructions of any microprocessor can be broadly classified into three groups: data transfer instructions, arithmetic and logic instructions, and test and branch instructions.

4.7.3.1 Data Transfer Instructions

Data transfer instructions help the programmer to transfer data to or between internal registers, registers and memory, registers and input/output devices, or memory to memory. Data can be transferred between a microprocessor's internal registers at high speed. Short, single-byte instructions are available for this task. Some examples are TAX (transfer accumulator to register X) of 6502 and LD X, A (load X register from accumulator) of Z80.

Memory to register instructions transfer data from a memory location to an internal register and vice versa. Table 4-3 provides a few examples from a 6502 environment.

In register-based microprocessors, I/O operations are isolated from memory operations and use special instructions. This is called the *isolated I/O method*. Table 4-4 lists some examples.

TABLE 4-3 Examples of Memory to Register Transfers for the 6502 Processor

Register	Mnemonic	Typical Instruction	Remarks
A	LDA	LDA 50H	Loading A from memory location 0050H
X	LDX	LDX 80H	Loading X from memory location 0080H
Y	STY	STY F0H	Storing Y in memory location 00F0H

TABLE 4-4 Examples of Register and I/O Device Transfers with Isolated I/O

Register	Mnemonic	Typical Instruction	Remarks
A	OUT	OUT A, 02H	Sending the contents of A to port 02H
A	IN	IN A, 03H	Transferring a byte from port 03H to A

TABLE 4-5 Examples of Register and I/O Device Transfers with Memory Mapped I/O

Register	Mnemonic	Typical Instruction	Remarks
A	STA	STA 02H	Sending the contents of A to port 02H
A	LDA	LDA 03H	Transferring a byte from port 03H to A

But accumulator-based microprocessors use a *memory-mapped I/O method*, in which there is no differentiation between memory operations and I/O operations. The same instructions are used for both types of operations. Ports also are treated as memory locations. Table 4-5 lists some examples.

In this format 0002 and 0003 may be memory locations or I/O ports, so a port and a memory location cannot have the same address, unlike in the isolated I/O method. Therefore, a portion of the address map needs to be reserved for I/O ports.

Memory to memory instructions transfer data bytes directly from one memory location to another. They take more time to execute. For instance, the MOV (Move) instruction of the Motorola CPU08 can transfer data directly from memory to memory.

4.7.3.2 Arithmetic and Logic Instructions

Five subcategories can be identified in this group of instructions: arithmetic instructions, logic instructions, shift and rotate instructions, increment/decrement instructions, and comparisons.

4.7.3.2.1 Arithmetic Instructions

To explain the arithmetic instructions, let us take some examples for 6502 and Z80 instruction sets.

The 6502 has no direct addition or subtraction, only ADC (add with carry) and SBC (subtract with carry). An ADC instruction adds the contents of register A to the contents of the necessary location selected plus the value of C_{flag}. A simple method of expressing the steps of an instruction is to use a symbolic expression:

$$A + M + C_{flag} \longrightarrow A$$

This indicates that the accumulator contents, the contents of the selected memory location, and the C flag are added to obtain the result; then the result is transferred back to the accumulator. Therefore, the programmer should make sure that the C flag is 0 prior to doing an 8-bit addition. For this purpose the CLC (clear carry flag) instruction is available. The ADC should follow the CLC when doing an 8-bit addition. When adding a 16-bit number, CLC is not necessary.

In the Z80 and 8080 family, the register pair HL is used as a memory pointer. HL must point to a memory location on which the operation is to be carried out. Microprocessors like the 6800 and Z80 have a direct addition instruction as well. In the Z80, instruction LD (in format LD HL, Byte$_H$) loads HL with the necessary data byte.

```
LD HL,2400H : load HL with word 2400H
ADD (HL)    : add the byte in the memory location
              pointed at by HL to the number in the
              accumulator.
```

Similarly, when subtracting 8-bit numbers with the 6502, the programmer should remember to set the carry flag beforehand using the SEC (set carry flag) command. Operation of the SBC (subtract with carry) command can be symbolized as $A - M - C \rightarrow A$. Therefore, the SEC should precede SBC when doing 8-bit subtraction. Table 4-6 shows how several microprocessors handle some common instructions.

4.7.3.2.2 Logic Instructions

Logic instructions include AND, OR, and Exclusive OR operations. Table 4-7 summarizes different formats used by three typical microprocessors for these logic instructions.

TABLE 4-6 Comparison of Formats for Arithmetic Instructions

Instruction	6502	M6800	Z80
Add memory to accumulator: $A + M \rightarrow A$	—	ADD A label ADD B label	ADD (HL) ADD A,r
Add memory to accumulator with carry: $A + M + C \rightarrow A$	ADC label	ADC A label	ADD (HL)
Subtract memory from accumulator: $A - M \rightarrow A$	—	SUB A label	SUB (HL) SUB A,r
Subtract memory from accumulator with carry: $A - M - C \rightarrow A$	SBC label	SBC A label	SBC (HL)

Note: Label = a data byte or an address of a memory location; r = any internal register; HL = register pair used in Z80 to load memory addresses that point to the data byte.

TABLE 4-7 Comparison of Formats in Logic Instructions

Instruction	6502	M6800	Z80
AND memory with accumulator or data with accumulator: $A.M \rightarrow A$	AND label AND # data	AND A label AND B label AND A # data AND B # data	AND (HL) AND r
OR memory with accumulator or data with accumulator: $A + M \rightarrow A$	ORA label ORA # data	ORA A label ORA B label	OR (HL) OR r
Exclusive OR memory with accumulator or data with accumulator: $A \oplus M \rightarrow A$	EOR label EOR # data	EOR A label EOR B label EOR A label	XOR (HL) XOR r

A bit-by-bit logic AND operation is performed by the processor in 2 bytes. Furthermore, an AND operation can be used to mask unwanted bit positions in a byte. For example,

```
LDA   2400H
AND   #00001111B
STA   2400H
BRK
```

The byte at address 2400H is not known. But, after this operation, the first 4 bits become 0. These positions get masked:

```
Assume (2400) = 11011110
                00001111
                00001110
```

A bit-by-bit logic OR operation is performed by the processor in 2 bytes. Furthermore, an OR operation can be used to "force" logic ones into certain bit positions in a byte. For example,

```
LDA   PORTA
ORA   #00000001B
STA   PORTA
BRK
```

This will set the LSB in the byte read from PORTA without affecting other bit positions.

A bit-by-bit exclusive OR (EOR or XOR) operation is performed in 2 bytes, which is useful for comparisons. If the 2 bits are not equal, the result bit will be set. This operation can be used for checking bit positions in ports or memory locations. In the 6502, it also can be used to complement a number (negate), since the 6502 has no direct negate command.

4.7.3.2.3 Shift and Rotate Instructions

Shift and rotate instructions include shift left, shift right, rotate left, and rotate right. In shift operations, all bits are shifted to the left or right by one bit position. The bit falling out is sent to the carry flag. Table 4-8 summarizes different formats used by three typical microprocessors for these logic instructions.

In arithmetic shift operations, the sign bit is preserved during rotation. In logical shift operations, all the bits including the sign bit are shifted or rotated.

4.7.3.2.4 Increment and Decrement Instructions

Increment and decrement instructions allow us to increase or decrease the value of a selected register or a memory location by 1 (see Table 4-9).

4.7.3.2.5 Compare Instructions

The compare operation has no arithmetic result. It compares two given bytes and decides whether the first byte is less than, equal to, or greater than the second

TABLE 4-8 Comparison of Formats in Shift and Rotate Instructions

Instruction	6502	M6800	Z80
Logical shift right	LSR LSR label	LSR label LSR A LSR B	SRL (HL) SRL SRL r
Arithmetic shift right	—	ASR label ASR A B	SRA SRA (HL) SRA r
Arithmetic shift left	ASL ASL label	ASL label ASL A B	SLA SLA (HL) SLA r
Rotate right	ROR ROR label	ROR label ROE A ROR B	RR RR (HL) RR r
Rotate left	ROL ROL label	ROL label ROL A ROL B	RL RL (HL) RL r

TABLE 4-9 Comparison of Formats in Increment and Decrement Instructions

Instruction	6502	M6800	Z80
Increment memory or register: $(M + 1 \rightarrow M)$	INC label INX INY	INC label INC A INC B INX INS	INC (HL) INC r INC
Decrement memory or register: $(M - 1 \rightarrow M)$	DEC DEC label DEX DEY	DEC DEC A DEC B DEX	DEC (HL) DEC r DEC

byte. Result of the comparison is stored in the flag register. In the 6502, the CMP (compare) instruction is used. For example,

```
Load Accumulator with 50                    : LDA #50_H
Compare A with the value in address 0050 : CMP  50_H
```

After the operation read the C and Z flags to find the result. If

```
Z=1(SET) : 2 numbers are equal
C=1      ; A ≥ Other Number
C=0      ; A < Other Number
```

But, to check the $A \geqslant$ other number condition, we need to check two flags in sequence. First check the Z flag and then the C flag (i.e., $Z = 0$ and $C = 1$). In addition, the CPX (compare X register) and CPY (compare Y register) instructions compare the X and Y register contents with another number (see Table 4-10). For example,

TABLE 4-10 Comparison of Formats in Compare Instructions

Instruction	6502	M6800	Z80
Compare memory with accumulator	CMP label CPX label CPY label CMP data	CMP A label CMP B label CPX label CMP A data CMP B data	CP (HL) CP r

```
To compare X register with the value in address
2401H : CPX 2401H
```

4.7.3.3 Test and Branch Instructions

The normal sequential flow of the program can be changed using jump or branch instructions. Jumps can be unconditional or conditional (sometimes called *branches*).

Unconditional jumps can be considered "jump always" instructions. For example, assume that, at address 2400H, we write an instruction to jump unconditionally to address 3600H. The command in a 6502 will be JMP 3600H. The machine code for JMP 3600H is 4C,00,36.

A conditional jump checks a flag to decide whether or not to make the jump (see Figure 4-10). When this condition is satisfied (i.e., a flag is set or cleared), the jump occurs.

These instructions allow us to select one of two alternative courses of action, depending on the result of the test. The most important flags and the instructions available to test them and make a branching decision in a 6502 microprocessor follow:

```
Flag Z  BEQ ; Branch if equal zero (Z=1)
        BNE ; Branch if Z=0
     C  BCS ; Branch if C=set(1)
        BCC ; Branch if C=clear(0)
     N  BPL ; Branch if plus (N=0)
        BMI ; Branch if minus (N=1)(See Figure 4-11)
```

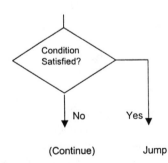

FIGURE 4-10 The typical flow of a conditional jump

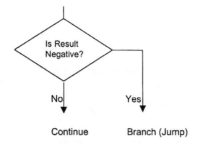

FIGURE 4-11 Flowchart for checking N flag to make the jump

```
V   BVS ; Branch if overflow set
    BVS ; Branch if V is clear
```

To jump seven locations forward if the last answer is negative, we can use

BMI 07 : Code = | 30 | 07 |

This will add 07 to the program counter (PC) and implement the jump if the N flag is found to be set. To jump seven locations forward if the last answer is negative, we can use

BMI F9 : Code = | 30 | F9 |

F9 is the two's complement of 07, which represents −7. The length of a backward jump should be calculated using two's complement numbers. Similarly, to jump two locations backward if the carry flag is not set, we can use

BCC FE : Code = | 90 | FE |

Table 4-11 shows the formats used by three microprocessors for jump and branch instructions.

4.7.4 Addressing Modes

A microprocessor could have many different modes of addressing. The format of an instruction follows the pattern

TABLE 4-11 Comparison of Formats in Jump and Branch Instructions

Instruction	*6502*	*M6800*	*Z80*
Jump and branch (unconditional)	JMP label	JMP label BRA label	JP label
Jump and branch (conditional)	BCC label, BCS BMI, BNE, BPL BVC, BVS	BCC, BCS, BEQ BGE, BGT BHI, BLE BVS, BVC BPL	JP label JP NZ, JP Z JP NC, JPC JP PO, JP PE JP P, JP M DJNZ label

```
Instruction = OPCODE  followed by OPERAND
```

The op code specifies the operation to be carried out. The operand indicates to the microprocessor where the data can be found in memory. The different ways of specifying the location of the data byte are called *addressing modes*. Some typical addressing modes are implied addressing (inherent/implicit addressing), immediate addressing, short addressing (zero page addressing/direct addressing), absolute addressing (extended addressing), indexed addressing, relative addressing, indirect addressing, or some combinations of these (e.g., indirect indexed addressing).

4.7.4.1 Implied Addressing

In implied addressing, the op code itself implies the location of the data. These generally are single-byte instructions that operate on the internal registers of the microprocessor. TAX (transfer A to X) in the 6502 implies that the data byte is in register A. In INX (increment X register), it is clear that the data byte is in the X register.

4.7.4.2 Immediate Addressing

In immediate addressing, the data byte immediately follows the op code. In the 6502, for example,

Code =	A9	24

In this example, LDA #24H. To take another example, where ADC #08H (see Figure 4-12),

Code =	69	08

This addressing mode can be used only if the data byte is known at the time of writing the program.

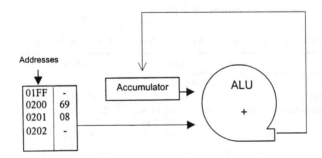

FIGURE 4-12 Implementation of the ADC #08H instruction

4.7.4.3 Zero Page Addressing

In zero page addressing (short/direct), the op code is followed by an 8-bit address (short address):

Code = | Op code | Location |

Only the location is given without a page number. Then the microprocessor seeks the data in the given location on page 00. For example, where ADC 25H,

Code = | 65 | 25 |

As another example, LDA 25H,

Code = | A5 | 25 |

The op code now will take the data byte in the 25th location of page 0 and add it to the accumulator.

4.7.4.4 Absolute Addressing

In absolute (extended) addressing the absolute address of the data byte is given in the instruction. The format is as follows:

Code = | Op code | Location | Page |

Consider an example instruction from the 6502 (see Figure 4-13):

Code = | 6D | 50 | 02 |

That is, ADC 0250. In this example, the microprocessor searches the 50th location of page 02 for data. Therefore, in absolute addressing, the full address of the data should be specified by the programmer.

Consider the following examples from the 6502. When you want to rotate the content of address 20FFH by one bit position to the right,

Code = | 6E | FF | 20 |

This is written in mnemonic form as ROR 20FFH

To compare the value of the X register with the value of the address 2050,

Code = | EC | 50 | 20 |

CPX 2050H is the mnemonic form for this instruction.

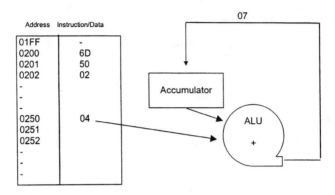

FIGURE 4-13   Implementation of absolute addressing

4.7.4.5 Relative Addressing

Relative addressing is used to make a relative jump with respect to the current position of the program counter. It is used only by test and branch instructions. In a test and branch instruction, if the test fails, the program counter (PC) remains the same and no jump is made. If the test is successful and the condition is satisfied, the displacement (jump length) value is added to the PC and the jump is made. Syntax of the instruction is,

Op code	displacement

As an example, if we take M6800, code for the instruction BNE 12H is

26	12

First, the Z flag must be checked. Then, depending on the result, one of the two courses of action will be taken. Figure 4-14 depicts relative addressing (a forward jump).

$$\text{OFFSET} = \text{EFFECTIVE ADDRESS}-\text{PROGRAM COUNTER}$$
$$= 0039_H - 0027_H$$
$$= \underline{0012}_H$$

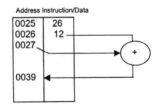

FIGURE 4-14   Relative addressing, forward jump

Here, when the jump is made, the microprocessor expects to find a new instruction in the destination. The preceding example is a forward jump of 12 locations. Next, consider a backward jump in the M6800 (see Figure 4-15): Check the Z flag and, if it is not set, jump backward 45 locations.

```
(45 Decimal = 2D Hex)

OFFSET = EFFECTIVE ADDRESS-PROGRAM CENTER
       = 0025 H - 0052 H
       = D3 H
```

4.7.4.6 Index Addressing

In index addressing the op code is followed by an offset address and the syntax is:

Op code	Offset

Index addressing uses an index register of the microprocessor. It commonly is used to sequentially access data bytes from a memory data table. In this case,

```
Effective Address = BASE ADDRESS + OFFSET.
```

Normally, the base address is stored in the index register and offset is given in the instruction. But the 6502 does it another way. It keeps the offset in the index register, since it has only a single 8-bit index register:

```
LDA 6000, X
```

Code =	BD	00	60

In the following sample program, index addressing is used to access successive memory locations in a memory table:

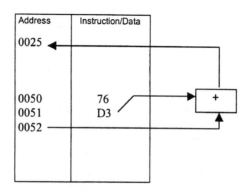

FIGURE 4-15   Relative addressing, a backward jump

```
PROG1   LDX   #16D
LOOP1   LDA   6000, X
        STA   PORTA
        DEX
        BNE   LOOP1
        BRK
```

In this program, 16 data bytes of a memory table are sent to port A in sequence. The offset in the X register is incremented in each pass through the loop.

In the 6502 microprocessor, zero page addressing and index addressing can be combined if the base address is on page 00. For example, where LDA 60, X

```
if X = 02
effective address = 0060 +
                       02
                   = 0062
```

This is to similar to normal index addressing except that the base address is on page 00. But zero page addressing saves 1 byte on the instruction, which can save a significant amount of time when program loops are running.

4.7.4.7 Indirect Addressing

Indirect addressing uses the following syntax:

OP CODE	LOCATION NUMBER	PAGE NUMBER

The address given in the instruction is the address of the data. The microprocessor reads the address of the data from the specified location and uses it to locate the data byte. Therefore, it has to read the address of the data from the address given in the instruction. Consider the following indirect jump in a 6502:

```
JUMP (6000)
```

Code =	6C	00	60

Now the microprocessor searches for the address of the data byte in memory location 6000H. Once this address is read, the data byte can be located. This method is called *memory indirection*. In register-based microprocessors, another method, register indirection, commonly is available. A memory pointer register (e.g., the HL pair in the Z80) is preloaded with the address of the data byte and the instruction refers only to this pointer register; for example, LD A, (HL) in the Z80. In this case, the microprocessor reads the data byte pointed to by the address in the memory pointer register HL.

4.7.4.8 Combinations of Basic Addressing Modes

Some combinations of the basic addressing modes provide sophisticated addressing methods. Commonly combined areas are index indirect addressing and indirect index addressing. The details of such cases are given in user manuals of microprocessors.

4.7.5 Program Execution

4.7.5.1 Stages of Execution

The microprocessor must go through two distinct steps when carrying out an instruction: a fetch cycle and an execute cycle.

The fetch cycle brings the op code of the instruction from the memory to the microprocessor. Op code is loaded to the instruction register inside the microprocessor. The execute cycle decodes and carries out the instruction. At the start of the fetch cycle, the microprocessor deposits the contents of the program counter on the address bus, then it issues a read command to the memory. The memory now searches for the location. Meanwhile, the microprocessor increments the program counter in preparation for the next cycle. After the short access time, the memory deposits the op code on the data bus. Now the microprocessor loads the op code to the instruction register.

The fetch cycle follows these steps:

1. PC → addresses bus.
2. $R/\overline{W} = 1$ (read command).
3. PC + 1 → PC.
4. The op code is deposited on the data bus by the memory subsystem.
5. The microprocessor collects the op code and loads it into the instruction register.

The execute cycle is different for each instruction. The control unit collects the instruction from the instruction register (IR) and decodes it, using its microcode decoder. By decoding the op code, it finds the following information:

What operation is requested.
What addressing mode is used.
Where to store the results.
Where the next instruction will be.

The decoded information is used by the control unit to generate a set of system control signals to carry out the operation.

4.7.5.2 Execution of Typical Instructions

Consider the following examples of an M6800 microprocessor executing some common instructions. In the first example, the load accumulator from memory is using an immediate addressing mode and LDA #30H. The fetch cycle steps are

1. PC → address bus.
2. $R/\overline{W} = 1$ (read operation).
3. PC + 1 → PC.
4. The op code comes on the data bus; the microprocessor collects and loads it to the IR.

The execute cycle steps are

1. Decode the op code and identify the operation.
2. Deposit the address in PC on address bus (PC now contains the address of the data byte)
3. Issue the read command.
4. Search the memory for the data (meanwhile, the program counter is incremented).
5. The memory subsystem deposits the data byte on the data bus.
6. The microprocessor collects it and loads it into the accumulator.

Table 4-12 shows the steps taken in executing an instruction using immediate addressing.

In the second example, the load accumulator instruction uses extended (absolute) addressing. Table 4-13 shows the steps taken in executing an instruction using absolute addressing. The instruction has 3 bytes, and LDA 0020.

TABLE 4-12 Steps in the Execution of an Instruction Using the Immediate Addressing Mode

Cycle	Address Bus	R/$\overline{W}$	Data Bus	Latched to
1. Fetch	Address of op code	1	Op code	IR of microprocessor
2. Execute	Address of data byte: (op code + 1)	1	Data byte	Accumulator of microprocessor

No. of clock cycles = 2
Time taken = $2 \times T$
$= 2 \times (1/f)$, where f is the clock frequency.
$= 2/f$ seconds

TABLE 4-13 Steps in the Execution of an Instruction Using the Absolute Addressing Mode

Cycle	Address Bus	R/$\overline{W}$	Data Bus	Latched to
1. Fetch	Address of op code	1	Op code	
2. Execute	(Op code address) + 1	1	Page number (HB) of the address of the data byte	Microprocessor
3.	(Op code address) + 2	1	Location number (LB) of the address of the data byte	Microprocessor
4.	Address of data byte (HB & LB)	1	Data byte	Accumulator of microprocessor

For the third example, the microprocessor stores the accumulator instruction using extended (absolute) addressing. Table 4-14 shows the steps taken.

```
Instruction: STA 2015H
```

During cycle no. 4, the microprocessor is internally preparing for the write operation. In the fourth example, the microprocessor executes an add instruction using the immediate addressing mode. Table 4-15 shows the steps taken in executing the instruction ADD #30.

In the fifth example, the microprocessor executes a subtract instruction using zero page (direct) addressing. Table 4-16 shows the steps taken in executing the instruction SUBA 30:

Op code	Address (zero page)

4.7.6 Program Creation

A programmer follows five steps in creating a program: defining the problem, specifying an algorithm, preparing a flowchart of the operations, writing the program in code, and testing and debugging the results.

The requirements are clearly defined at the start to ensure that the final program fulfills all the needs. The following information is unearthed at this stage:

TABLE 4-14 Steps in the Execution of a Memory Store Instruction

Cycle	Address Bus	R/$\overline{W}$	Data Bus	Latched to
1. Fetch	Address of op code	1	Op code	IR of micro-processor
2. Execute	(Op code address) + 1	1	High byte of the address of memory location to which A should be stored	Microprocessor
3.	(Op code address) + 2	1	Low byte of the address of memory location	Microprocessor
4.	Address of memory location	1	Irrelevant	—
5.	Address of memory location	1	Accumulator contents	Memory location

TABLE 4-15 Steps in the Execution of an Addition Instruction Using the Immediate Addressing Mode

Cycle	Address Bus	R/$\overline{W}$	Data Bus	Latched to
1. Fetch	Address of op code	1	Op code	IR of micro-processor
2. Execute	(Op code address) + 1	1	Data byte	Accumulator of microprocessor

TABLE 4-16 Steps in the Execution of a Subtraction Instruction Using the Direct Addressing Mode

Cycle	Address Bus	R/$\overline{W}$	Data Bus	Latched to
1. Fetch	Address of op code	1	Op code	IR of micro-processor
2. Execute	(Op code address) + 1	1	Zero page memory address	Microprocessor
3.	Address of data bytes	1	Data byte	ALU

1. The exact function of the program.
2. The number and types of program input and output.
3. The execution time restrictions.
4. The level of accuracy expected in the result.
5. What memory will be occupied.
6. What action will be taken if operator errors occur.

A program that handles mistakes very well and provides the user guidelines is considered a robust program.

The algorithm is a sequence of steps that detail the procedure to solve the problem. A problem can have more than one algorithm.

The flowchart is a graphical representation of the algorithm. Preparing a flowchart minimizes logic errors and duplications and supports documentation.

Coding is the process of translating the steps of the flowchart to instructions in the computer language.

The coded program is entered into the system, and a test run is carried out. It is difficult to write a completely error-free program in one attempt. Program errors (bugs) occur even to very experienced programmers. Finding them and fixing them are called *debugging*. Debugging tools are available to assist in this process.

4.8 Single-Chip Microcontrollers and Embedded Processor Core Applications

4.8.1 Single-Chip Microcontrollers

A microcontroller is a single-chip microcomputer. The advancement of IC manufacturing technology has made possible the integration of all three subsystems of the central controller on one silicon chip. As a result, the IC count of a system becomes smaller, reducing its cost, size, and weight and making it more compact and reliable. Like microprocessors, microcontrollers appear in the market as families. Different members of a family differ in the provision of memory, input and output ports, the speed of operation, and computing power. Therefore, it is easy to find a family member to closely fit into the requirements of a system designer. Microcontrollers are popular in dedicated microelectronic systems (e.g., industrial control systems, domestic appliances).

In a microcontroller, the following components generally are integrated on the same semiconductor chip: a processor, input/output interface circuits, clock generators, and memory circuits. Some common examples include the Intel 8035, 8048, and 8051 and the National COP420L.

Consider the internal architecture of the popular 8051 microcontroller (Figure 4-16), an accumulator-based microcontroller. The programmer has 255 instructions available to command the processor. A basic instruction cycle takes 12 clocks; however, Dallas Semiconductors has designed the instruction-execution circuitry to reduce the instruction cycle to 4 clocks.

It has four banks of 8-bit registers in its on-chip RAM, which helps context switching. These registers reside within the lower 128 bytes of internal RAM, along with a bit-operation area and scratch pad RAM. These lower bytes can be addressed directly or indirectly using an 8-bit address. The upper 128 bytes of on-chip data RAM encompass two overlapping address spaces. One space is for directly addressed special function registers (SFRs) and the other space is for indirectly addressed RAM or stack. The SFRs define peripheral operations and configurations. The 8051 also has 16-bit addressable bytes of on-chip RAM for flags or variables. The 8051 processor can directly address 64 kB of data memory. However, software tools with an external latch can extend it to any multiple of 64 kB pages. The software seamlessly handles all page transitions. Register indirection uses an 8-bit register for on-chip RAM addresses. Off-chip addresses need a 16-bit data-pointer register (DPTR). The DPTR cannot be indexed but can be incremented.

The 8051 performs extensive bit manipulation via special instructions, such as set, clear, complement, and jump on bit set or jump on bit clear, only for a 16-byte area of RAM and some SFRs. It also can execute AND or OR instructions with a carry bit. Dallas versions have variable-length move external data instructions. Math functions include add, subtract, increment, decrement, multiply, divide, complement, rotate, and swap nibbles. Some of the Siemens devices have a hardware multiplier/divider for 16-bit multiplication and 32-bit division (see Figure 4-17). Microcontrollers offer a low-cost solution especially to the microelectronic system designers who design dedicated systems.

4.8.2 Embedded Processor Core Applications

Microcontrollers may not have the right mix of internal components to perfectly match the system designer's requirements; therefore, they may not be a good fit for some applications. To solve this problem, many suppliers have gone back and stripped off all functional support blocks from the integrated CPU. This results in a derivative called the *processor core*. Then the suppliers placed the CPU and support functions in megacell libraries. Those libraries can be used to assemble the optimum single-chip microcontroller for a particular customer. The product is customized, and in many cases provides a unique solution. Furthermore, suppliers can design application-specific integrated circuits (ASIC) using the processor cores and megacell libraries. The technology has

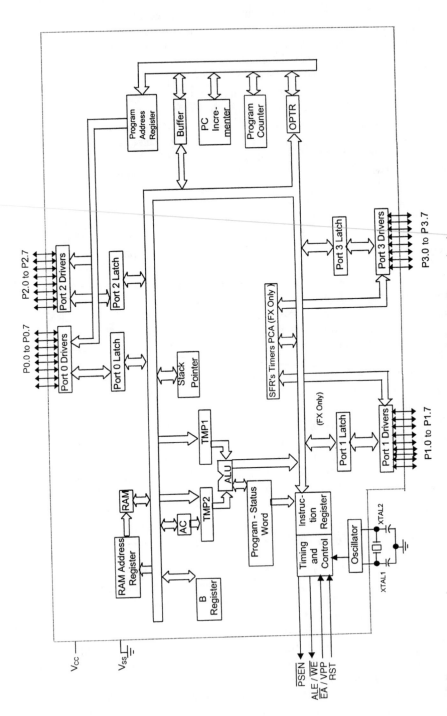

FIGURE 4-16 Intel 8051 microcontroller. (Reproduced by permission of Intel Corporation.)

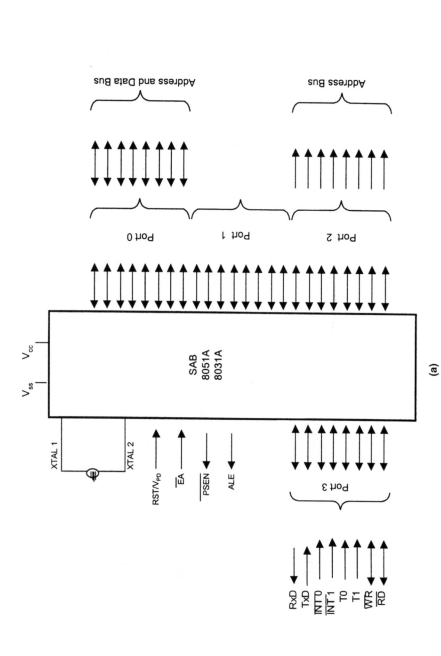

FIGURE 4-17 An Industrial version of 8051 by Siemens Corporation. (a) Pin Assignment (Source: Siemens Microcontroller Databook)
(b) Block Diagram (Reprinted from EDN Magazine, Sept 25, 1997; Copyright Cahners Business Information, 1999)

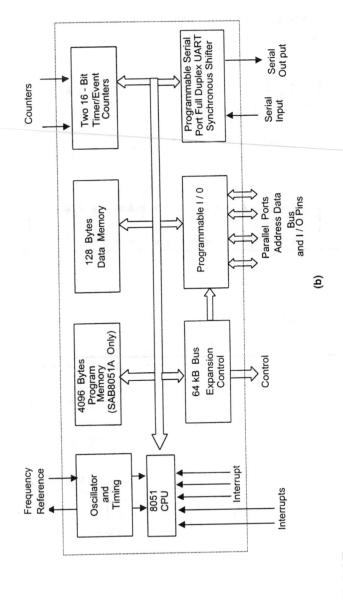

(b)

FIGURE 4-17 Continued

evolved into the construction of application-specific cores as well. A core microprocessor may be added to an ASIC. Reduced instruction set computing cores are popularly used to couple with ASICs. A core microprocessor contains only those logic elements needed to execute its instruction set, such as microcode ROM, microcode state machines, a program counter, special function registers, ALU, and interrupt logic.

An embedded application uses a processor to perform a specific task. Both microcontrollers and microprocessors can be used as embedded controllers. Microcontrollers are designed specifically to perform embedded tasks. Today, it has become easier and more convenient for a new system designer to embed a standard processor core into a new product design. An embedded processor core is a feature or component of an application. It provides standardized mechanisms for feeding and displaying data into a system. It is can provide hardware and software flexibility in the system being developed. Buying an embedded processor core is cheaper and much more convenient than designing that part. Then the compatibility with standard desktop computers is automatically available.

Before embedding a processor core into a product, the designer should decide the degree of personal computer compatibility required. The processor (and core logic chip set) chosen by the designer determines the level of personal computer compatibility in the embedded processor core. To be fully compatible, the embedded processor core must support the complete PC-AT specification. For example, the Intel 80386 is now in the embedded processor core market. A block diagram is shown in Figure 4-18.

The 386 developed a strong presence in embedded processor core applications. It has a register-based architecture with four general purpose registers and four index/pointer registers, supplemented by six 16-bit segment registers and two 32-bit status and control registers.

The main features of 80386 include

1. Segmentation, 64 kB segments to extend addressing to 1 MB.
2. Segment limits extended to the full 4 GB addressing range.
3. Protected mode addressing, the base address adds to the 32-bit effective address, producing a 32-bit linear address, which is used as a physical address or a linear-page address.
4. Six debugging registers, four code/data breakpoint registers, and two control registers. The breakpoint registers can be set with addresses for halting execution on a program or data access.
5. Power management, in which a system management mode and a power management mechanism are available. System management operating modes reduce chip power dissipation. Integrated versions of the 386 (for example, Intel's 386EX) have idle and power down modes. The idle mode discontinues CPU processing but keeps the peripherals active. The power down mode shuts down the entire chip.
6. The 386 instruction set is a superset of the 8086/186. The 386 has seven additional instructions to support the system management mode.

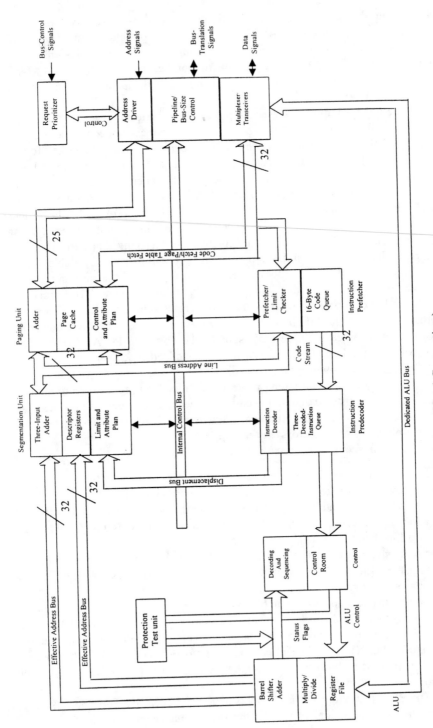

FIGURE 4-18 Intel 80386. (Reproduced by permission of Intel Corporation.)

4.9 RISC vs. CISC Microprocessor Architecture

Microprocessors fall into two main categories, according to the way they process instructions: the PowerPC is a RISC processor and the Pentium is a CISC. By reducing the number of instructions that a CPU supports and thereby the complexity of the chip, it's possible to make individual instructions execute faster and achieve a net gain in performance even though more instructions might be required to accomplish a task.

When a program runs, the microprocessor reads, or fetches, the instructions one by one and executes them. For a 233 MHz Pentium, one clock is less than 5 ns. Programs can be made to run faster by increasing the clock speed or decreasing the number of clock cycles required to execute an instruction.

Traditionally, CISC microprocessors like the x86 have required anywhere from 1 to more than 100 clocks to complete an instruction. For example, the multiply instruction on the 8088 that powered the original IBM personal computer required up to 133 clocks, to multiply on a 486 required as few as 13 clocks, and register-to-register transfer requires just 1.

The original goal of RISC was to limit the number of instructions in the instruction set. More instructions are needed to carry out a job, but if the instruction timings are low enough, the RISC chip can complete the task earlier than a CISC chip. There is a trade-off between instruction set complexity and instruction execution time. RISC seeks to strike a better balance between the two to produce a faster microprocessor. The transistors it saves on large instructions can be used for cache, pipelines, and more registers.

Today's RISC chips often have richer and more complex instruction sets than CISC chips. The PowerPC 601, for example, supports more instructions than the Pentium. Yet the 601 is considered a RISC chip, while the Pentium is definitely CISC. In reality, the instruction sets are not really reduced; the difference is not in the size of the instruction set anymore. Furthermore, the number of transistors that can be packed into a silicon wafer has steadily increased. A Pentium contains more than 3 million transistors in an area roughly a half-inch square. It executes two instructions per clock cycle, not one instruction in half a clock cycle. Actually, many instructions still require multiple clock cycles. But the use of pipelines allows instructions to overlap, so that five instructions each requiring five clock cycles can be completed in a total of five clocks — an average of one instruction per clock cycle. Pentium's two independent execution units qualify it as a superscalar microprocessor. They permit two pipelines to run in parallel. What really distinguishes RISC from CISC is more deeply rooted in the chip architecture.

4.9.1 Architectural Differences

RISC microprocessors have more general purpose registers (e.g., Pentium has just 8 general purpose registers but a PowerPC chip has 32), which help to minimize the number of times data stored in memory is accessed. Accessing registers is much faster than accessing memory addresses.

RISC microprocessors use load and store architectures. CPU instructions that operate on data in memory consume more time. RISC designs minimize the number of instructions that access memory in favor of load and store operations. For instance, when adding two numbers the instructions require one number to be loaded into a register before the other is added and then stored back to memory.

RISC microprocessors use uniform instruction lengths. On a Pentium, the length of one instruction can vary from as few as 1 byte to as many as 7. RISC favors making all instructions the same length, usually 32 bits. This simplifies the instruction fetching and decoding logic. Furthermore, an instruction can be fetched with one 32-bit memory access.

RISC microprocessors emphasize floating point performance, with high-performance floating point units built in. In general, RISC designers have been quick to adopt cutting-edge technologies such as on-chip code and data caches, superscalar designs, instruction pipelines, and branch prediction logic. But now these are available in CISC as well, so it is difficult to distinguish RISC from CISC on the basis of those features. For example, the MIPS RX000 family combines reduced instruction sets with huge internal caches and AMD, Cyrix, and NexGen incorporate RISC-like features into their Pentium clones.

The problem with CISC processors is the lack of registers. With only 16 or so registers, CISC processors allow for little flexibility. CISC processors generally dissipate much more heat and consume more power than RISC processors. This is a barrier to achieving high clock speeds with CISC-based processors. Table 4-17 lists additional differences between RISC and CISC.

TABLE 4-17 Advantages of CISC vs. RISC

CISC	RISC
Microprogramming is as easier to implement and much less expensive than hardwiring a control unit.	Since a simplified instruction set allows for a pipelined, superscalar design, RISC processors often achieve two to four times the performance of CISC processors using comparable semiconductor technology and the same clock rates.
The ease of microcoding new instructions allows CISC machines to be upwardly compatible: A new computer could run the same programs as earlier computers because the new computer would contain a superset of the instructions of the earlier computers.	Because the instruction set of a RISC processor is simpler, it uses up much less chip space than a CISC processor. Extra functions, such as memory management units or floating point arithmetic units, can be placed on the same chip.
As each instruction became more capable, fewer instructions could be used to implement a given task. This made more efficient use of the relatively slow main memory.	Since RISC processors can be designed more quickly, they can take advantage of new technological developments sooner than corresponding CISC designs.
Because microprogram instruction sets can be written to match the constructs of high-level languages, the compiler need not be very complicated.	

4.9.2 The Intel Pentium: A CISC Example

The Pentium has achieved code compatibility with earlier x86 CPUs while attaining third-generation RISC performance. It implements the complex x86 instruction set and emphasizes simple instruction execution over the more complex ones. In the Pentium, the simple, RISC-like register-to-register instructions drive the implementation, keeping the microcoded complex instructions as second priority.

The Pentium achieves a two-instruction issue peak and has two five-stage pipelines (U and V). These pipelines are not symmetric. The U pipeline has precedence over the V one. If the first instruction does not cause data interlocks, then the second instruction is scheduled for the V pipe. The U and V pipelines are fed from a common instruction fetch/align stage that fetches multiple instructions from the cache. The CPU fetches and passes a full line (256 bits) to the instruction decoder. Each pipeline has two decoder stages to decode simple and complex instructions. The wide cache-to-decoder path, coupled with a two-stage decode, enables the Pentium to decode the x86's variable-length instructions and deliver competitive performance.

It carries out superscalar dual-instruction load and store operations. Both data and instructions are cached. Pentium's floating point unit (FPU) features an eight-stage pipeline, which shares the first five stages of the U and V pipeline. Data transfers to or from the FPU use a wide 64-bit data path to the data cache to keep the FPU pipeline fed. The Pentium uses burst reads to fill its 256-bit-wide cache line. It also has burst write-back writes. The memory interface uses a pipeline, allowing a second bus cycle to set up while the first bus cycle completes. The Pentium reads or writes a 64-bit double word each cycle in burst mode.

The following key issues must be addressed when moving code from a CISC processor to a RISC processor:

1. The quality of the code. The performance of a RISC application depends critically on the quality of the code generated by the compiler. Therefore, developers have to choose the compiler carefully, based on the quality of the generated code. If a compiler does not schedule instructions properly, execution slows down significantly.
2. Debugging. Instruction scheduling makes debugging more difficult. When scheduled for execution assembly language instructions are arranged in a different order. Therefore, finding bugs and tracing the execution are difficult unless debugging is done on the unscheduled form.
3. Code expansion. Code expansion is the increase in size when a program written for a CISC processor is converted for use in a RISC machine. CISC machines perform complex actions with a single instruction. RISC machines require multiple instructions for the same action; therefore, the code expands with the conversion.
4. System design. RISC machines require instructions at a faster rate; therefore, their chips contain large first-level memory caches.

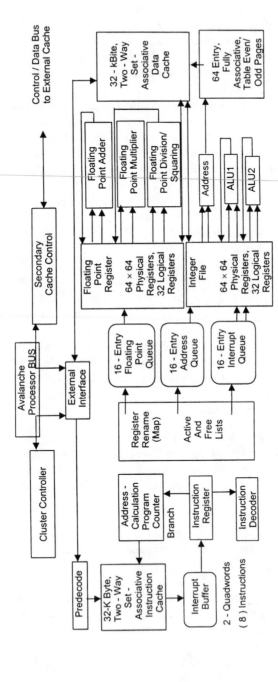

FIGURE 4-19 The internal architecture of the MIPS RI0000 RISC processor. (Reproduced by permission of the MIPS Group of Silicon Graphics.) (Reprinted from EDN Magazine, September 25, 1997; Copyright Cahners Business Information, 1999)

4.9.3 The MIPS RI0000: A RISC Example

Although the architecture of the RI0000 (Figure 4-19) is new, its designers decided that it would run R4000 code with no performance degradation. After they go through a two-stage instruction fetch-and-decode unit, the RI0000 can dispatch instructions simultaneously to five functional units: floating point add, floating point multiply, two integer operations, and a load and store. The pipeline integer unit consists of six stages: fetch, decode, issue, execute, cache access, and write back. Floating point instructions use a seventh stage attached to the integer pipe.

During sequencing, the RI0000 maintains an instruction status table to determine the instructions waiting to graduate and to put the instructions in order. When handling out-of-order execution, a completed instruction may never graduate because an exception or branch could invalidate the results. For this reason, an instruction may complete, but until graduation, its results are tentative and may be discarded.

The RI0000 also facilitates designing tightly coupled multiprocessing systems. To accomplish this goal, the CPU has a 64-bit cluster bus configuration that allows direct connection of four RI0000 processors. Attaching the RI0000 to an external agent, or cluster coordinator, creates a cluster bus that manages the flow of data within the cluster.

References

Brey, Barry B. *Microprocessors and Peripherals*, 2d ed. Columbus, OH: Merrill Publishing Company, 1989.

Bursky, Dave. "Pipelined CISC Processor Exceeds RISC Throughput." *Electronic Design* (January 1990), p. 91.

Bursky, Dave. "Embedded-Controller Architectures Suit All Needs." *Electronic Design* (January 8, 1996), p. 56.

Fujii, Masaru, and Douglas W. Fire. "Flash Memories Illustrate Progress in Technology." *JEE* (February 1994), p. 98.

Gunn, Lisa. "Closing from the Rising Tide of Controllers." *Electronic Design* (March 23, 1989), p. 51.

Hall, Douglas V. *The Microprocessor and Interfacing*. New York: McGraw-Hill, 1986.

Harold, Peter. "Core Microprocessors." *EDN* 2 (February 1989), p. 130.

Hwang, Kai, and Faye A. Briggs. *Computer Architecture and Parallel Processing*. New York: McGraw-Hill, 1984.

Levy, Markus. "Zero in on x86 Derivatives for Your Embedded PC." *EDN* (June 22, 1995).

Miller, Michael A. *The M68000 Microprocessor*. Lincolnwood, IL: Bell & Howell Company, 1988.

Mosley, J. D. "8 and 16 bit Microcontrollers." *EDN* (September 28, 1989), p. 108.

Motorola. *CPU 08 Central Processor Unit Reference Manual*. Motorola Inc., USA, 1994.

Nass, Richard. "80C51 Microcontroller Jumps to 16 Bits." *Electronic Design* (September 1994), p. 180.

Warner, William C. "Use a Single Chip /uC as the Heart of your Position Controller." *EDN* (September 1, 1988).

Zaks, Rodney, and Austin Lesea. *Microprocessor Interfacing Techniques*, 3d ed. SYBEX Inc.

Zilog Inc. Z80 CPU Technical Manual. Zilog Inc. USA.

"RISC vs RISC The Real Story." http://web.singnet.com.sg/~alphalht/risc.htm.

"CISC Pros & Cons." http://kandor.isi.edu/aliases/powerpc.programming-info/intro-to-risc/irt4-cisc4.html.

"RISC Pros & Cons." http://gatekeeper.informatik.fh-dortmund.de/person/prof/si/risc/intro-to-risc/irts-risc3.html.

CHAPTER 5

Digital Signal Processors

5.1 Introduction

The processing of information is as old as the human race. The technology associated with it dates back to before the earliest form of writing to the use of counting beads and to the time of cave drawings. However, the inventions of printing (1049), the electronic binary coding of data (1837), and more recently radio communications (1891) and the electronic digital computer (1946) have increased the speed of generation and processing of data to such an extent that information and control technology no longer is concerned just with automatic processes replacing manual ones but provides the opportunity to do entirely new things. The developments during the 50 or so years since the invention of the transistor (1948) have helped designers introduce products that can hear, talk, and even detect the motion of objects. The late 1990s have seen developments related to automatic processes that could replace human sensory and cognitive processes as well as manipulative ones.

Early generations of 4- and 8-bit CISC processors have evolved into 16-, 32-, and 64-bit components with CISC or RISC architectures. Digital signal processors can be considered special cases of RISC architecture or sometimes parallel developments of CISC systems to tackle real-time signal processing needs. Over the past several decades, the field of digital signal processing has grown from a theoretical infancy to a powerful practical tool and matured into an economical yet successful technology.

At its early stages, audio and the many other familiar signals in the same frequency band have appeared as a magnet for DSP development. The 1970s saw the implementation of special signal processing algorithms for filters and fast Fourier transforms by means of digital hardware developed for the purpose. Early sequential program DSPs are described by Jones and Watson (1990).

In the late 1990s, the market for DSPs is generated mostly by wireless, multimedia, and similar applications. As per industry estimates, by the year 2001, the market for DSPs is expected to grow to about $9.1 billion (Schneiderman, 1996). Currently, communications represent more than half of the applications for DSPs (Schneiderman, 1996). This chapter provides an essential guide for designers to understand the DSPs and briefly compares microprocessors and DSPs.

5.2 What Is a DSP?

A digital signal processor accepts one or more discrete-time inputs, $x_i(n)$, and produces one or more items of output, $y_I(n)$, for $n = \ldots, -1, 0, 1, 2, \ldots$, and $I = 1, \ldots, N$, as depicted in Figure 5-1(a). The input could represent appropriately sampled (and analog-to-digital converted) values of continuous time signals of interest, which are processed in the discrete-time domain to produce output in discrete time that could then be converted to continuous time, if necessary. The operation of the digital signal processor on the input samples could be linear or nonlinear, time invariant or time varying, depending on the application of interest. The samples of the signal are quantized to a finite number of bits, and this word length can be either fixed or variable within the processor. Signal processors operate on millions of samples per second, require large memory bandwidth, and are computationally very demanding, often requiring as many as a few hundred operations on each sample processed. These real-time capabilities are beyond

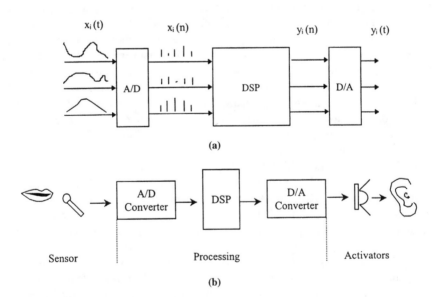

FIGURE 5-1 A digital signal processing system: (a) Mathematical representation, (b) A practical example

the capabilities of conventional microprocessors and mainframe computers. A practical example of voice processing by a DSP is shown in Figure 5-1(b).

Signal processors can be either programmable or of a dedicated nature. Programmable signal processors allow flexibility of implementation of a variety of algorithms that can use the same computational kernel, while dedicated signal processors are hardwired to a specific algorithm or class of algorithms. Dedicated processors often are faster than or dissipate less power than general purpose programmable processors, although this is not always the case.

Digital signal processors traditionally have been optimized to compute the finite impulse response convolutions (sum of products), infinite impulse response recursive filtering, and fast Fourier transform-type (butterfly) operations that typically characterize most signal processing algorithms. They also include interfaces to external data ports for real-time operation. It is interesting to note that one of the earliest digital computers, ENIAC, included characteristics of a DSP (Marven and Ewers, 1994).

5.3 Comparison Between a Microprocessor and a DSP

Following the preceding chapter's discussion of microprocessors and microcontrollers, we can compare the microprocessors and DSPs. General architectures for computers and single-chip microcomputers fall into two categories. The architectures for the first significant electromechanical computer had separate memory spaces for the program and the data, so that both could be accessed simultaneously. This is known as a *Harvard architecture*, having been developed in the late 1930s by Howard Aiken, a physicist at Harvard University. The Harvard Mark 1 computer became operational in 1944.

The first general purpose electronic computer was probably the ENIAC (electronic numerical integrator and calculator) built during 1943–1946 at the University of Pennsylvania. The architecture was similar to that of the Harvard Mark 1 with separate program and data memories. Due to the complexity of two separate memory systems, Harvard architecture has not proven popular in general purpose computer and microcomputer design.

A consultant to the ENIAC project, John von Neumann, a Hungarian-born mathematician, is widely recognized as the creator of a different, very significant architecture, published by Burks, Goldstine, and von Neumann (1946; reprinted in Bell and Newell, 1971). The so-called von Neumann architecture set the standard for developments in computer systems over the next 40 years and more. The idea was very simple, based on two main premises: that there is no intrinsic difference between instructions and data and that instructions can be partitioned into two major fields containing the operation command and the address of the operand (data to be operated on); therefore, a single memory space could contain both instructions and data.

Common general purpose microprocessors, such as the Motorola 68000 family and the Intel i86 family, share the von Neumann architecture. These and other general purpose microprocessors also have other characteristics typical of most computers over the past 40 years. The basic computational blocks are an arithmetic logic unit and a shifter. Operations such as add, move, and subtract are performed easily in a very few clock cycles. Complex instructions such as multiply and divide are built up from a series of simple shift, add, and subtract operations. Devices of this type are known as *complex instruction set computers*. CISC devices have multiply instructions, but this will simply execute a series of microcode instructions which are hard coded in on-chip ROM. The microcoded multiply operation therefore takes many clock cycles.

Figure 5-2 compares the basic differences between traditional micro-processor architecture and typical DSP architecture. Real-time digital signal processing applications require many calculations of the form

$$A = BC + D \qquad (5.1)$$

This simple equation involves a multiplication operation and an addition operation. Because of its slow multiplication, a CISC microcomputer is not very efficient at calculating it. We need a machine that can multiply and add in just one clock cycle. For this, we need a different approach to computer architecture.

Many embedded applications are well defined in scope and require only a few calculations to be performed, but they require very fast processing. Examples of such applications are digital compression of images, compact disc players, and digital telephones. In addition to these computation-intensive functions demanding the continuous processing, the processor has to perform comparatively simple functions such as menu control for satellite TV, selection of tracks for CD players, or number processing in a digital PBX, all of which require significantly less processing power.

In such applications, computation-intensive functions such as digital filtering and data compression require continuous signal processing, which requires multi-plication, addition, subtraction, and other mathematical functions. While RISC processor architectures could be optimized to handle these situations by incorpo-rating cache memory, direct access internal registers, and the like, DSP systems provide more computation-intensive functions such as fast Fourier transforms, convolutions, and digital filters. Particularly in a DSP-based system, such tasks should be performed on a real-time basis, as per Figure 5-3. This indicates that the sample period and computational latency are becoming key parameters.

5.3.1 The Importance of the Sample Period and Latency in the DSP World

The sample period (the time interval between the arrival of successive samples of the input signal) depends on the technology employed in the processor. The time interval between the arrival of input and the departure of the corresponding output sample is the computational latency of the processor. To ensure the stability of the input ports, the output samples have to depart at the same sample period as the input samples. In signal processing applications, the

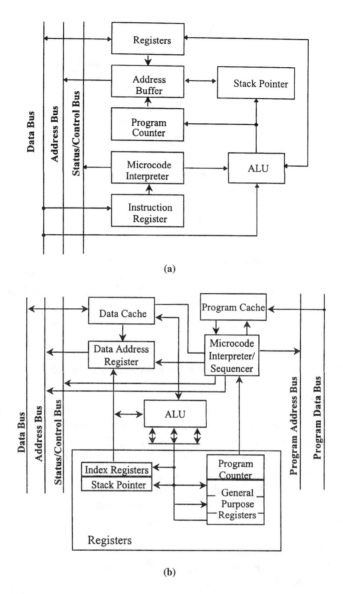

FIGURE 5-2 Comparison of microprocessor and DSP architectures: (a) Traditional microprocessor architecture, (b) Typical DSP architecture

minimum sample period that can be achieved often is more important than the latency of the circuit. Once the first output sample emerges, successive samples will emerge at the sample period rate, hiding the effects of a large latency of circuit operation. This makes sense because typical signal processing applications deal with a few million samples of data in every second of operation. For details on the relationship between these two parameters, see Madisetti (1995).

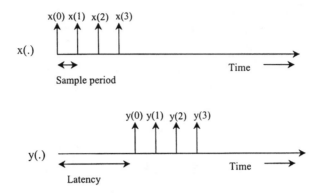

FIGURE 5-3 Sample period and latency

Other important measures are the area of the VLSI implementation and its power dissipation. These directly contribute to the cost of a DSP chip. One or more of these measures usually is optimized at the cost of others. These trade-offs again depend on the application. For instance, signal processors for portable communication require low power consumption combined with small size, usually at the cost of an increased sample period and latency.

5.3.2 The Merging of Microprocessors and DSPs

Diverse, high-volume applications such as cell phones, disk drives, antilocking brakes, modems, and fax machines require both microprocessor and DSP capability. This requirement has led many microprocessor vendors to build in DSP functionality. In some cases, such as in Siemens' Tricore architecture (Levy, 1998a), the functional merging is so complete that it is difficult to determine whether to consider the device a DSP or a microprocessor. At the other extreme, some vendors claim that their microprocessors have high-performance DSP capability, when in fact they have added only a "simple" 16×16-bit multiplication instruction.

5.4 Filtering Applications and the Evolution of DSP Architecture

Digital signal processing techniques are based on mathematical concepts familiar to most engineers. From these basic ideas spring the myriad applications of DSP, including fast Fourier transform, linear prediction, nonlinear filtering, and decimation and interpolation (see Figure 5-4). One of the most common signal processing functions is linear filtering. High-pass, low-pass, and bandpass filters, which traditionally are analog designs, can be constructed with DSP techniques. To build a linear filter using digital methods, a continuous-time input signal, $x_c(t)$, is sampled to produce a sequence of numbers, $x(n) = x_c(nT)$. This sequence is

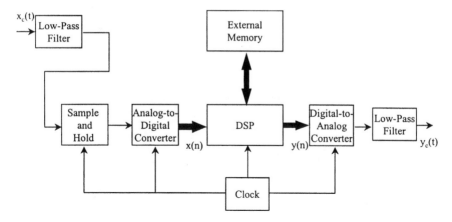

FIGURE 5-4 A DSP-based filter implementation

transformed by a discrete-time system — that is, a computational algorithm — into an output sequence of numbers, $y(n)$. Finally, a continuous-time output signal, $y_c(t)$, is reconstructed from the sequence $y(n)$. The essentials of filtering and sampling as applied to the world of DSP were discussed in Chapter 3.

5.4.1 Digital Filters

Digital filters for many years have been the most common application of digital signal processors. Digital design, of any kind, ensures repeatability. Two other significant advantages accrue with respect to filters. First, it is possible to reprogram the DSP and drastically alter the filter's gain or phase response. For example, we can reprogram a system from low pass to high pass without throwing away the existing hardware. Second, we can update the filter coefficients while the program is running; that is, build "adaptive" filters. The two basic forms of digital filter, the finite impulse response (FIR) filter and the infinite impulse response (IIR) filter, are explained next. The initial descriptions are based on a low-pass filter. It is very easy to change low-pass filters to other types: high pass, bandpass, and so forth. Parks and Burrus (1987) and Oppenheim and Schafer (1988) cover this in detail.

5.4.1.1 Finite Impulse Response Filter

The mechanics of the basic FIR filter algorithm are straightforward. The blocks labeled z^{-1} in Figure 5-5 are unit delay operators; their output is a copy of the input sample delayed by one sample period. A series of storage elements (usually memory locations) are used to simulate series of these delay elements (called a *delay line*). The FIR filter is constructed from a series of taps. Each tap includes a multiplication operation and an accumulation operation. At any given time, $n - 1$ of the most recent input samples resides in the delay line, where n is the number of taps in the filter. Input samples are designated x_k; the first input

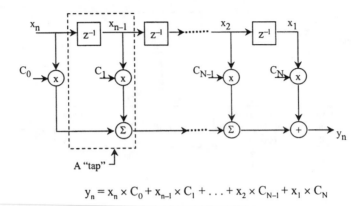

$$y_n = x_n \times C_0 + x_{n-1} \times C_1 + \ldots + x_2 \times C_{N-1} + x_1 \times C_N$$

FIGURE 5-5 Finite impulse response filter

sample is x_1, the next is x_2, and so on. Each time a new input sample arrives, the previously stored samples are shifted one place to the right along the delay line and a new output sample is computed by multiplying the newly arrived sample and each of the previously stored input samples by the corresponding coefficient. In the figure, coefficients are represented as C_n, where n is the coefficient number. The results of'each multiplication are summed together to form the new output sample, y_n. Later we discuss how DSPs are designed to help implement these.

5.4.1.2 Infinite Impulse Response Filter

The other basic form of digital filter is the infinite impulse response filter. A simple form of this is shown in Figure 5-6. Using the same notations as for the FIR, we can see that

$$y(n) = x(n) + a_1 y(n-1) + a_2 y(n-2) \tag{5.2}$$
$$= x(n) + (a_1 z^{-1} + a_2 z^{-2}) \cdot y(n)$$
$$= x(n) \frac{1}{1 - a_1 z^{-1} - a_2 z^{-2}} \tag{5.3}$$

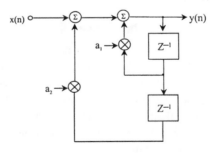

FIGURE 5-6 A simple IIR filter

Take the math for granted—it is just relatively simple substitution. Therefore, the transfer function is given by

$$H(n) = \frac{y(n)}{x(n)} = \frac{1}{1 - a_1 z^{-1} - a_2 z^{-2}} \qquad (5.4)$$

From equation (5.2) we can see that each output, $y(n)$, is dependent on the input value, $x(n)$, and two previous outputs, $y(n-1)$ and $y(n-2)$. Taking this one step at a time, let us assume that there were no previous input samples before $n = 0$, then

$$y(0) = x(0)$$

At the next sample instant,

$$y(1) = x(1) + a_1 y(0)$$
$$= x(1) + a_1 x(0)$$

Then, at $n = 2$,

$$y(2) = x(2) + a_1 y(1) + a_2 y(0)$$
$$= x(2) + a_1 [x(1) + a_1 x(0)] + a_2 x(0)$$

Then, at $n = 3$,

$$y(3) = x(3) + a_1 y(2) + a_2 y(1)$$
$$= x(3) + a_1 \{x(2) + a_1 [x(1) + a_1 x(0)] + a_2 x(0)\} + a_2 [x(1) + a_1 x(0)]$$

We already can see that any output depends on all the previous inputs and we could go on, but the equation just gets longer. An alternative way of expressing this is to say that each output depends on an infinite number of inputs. This is why this filter type is called an *infinite impulse response*.

If we look again at Figure 5-6, the filter actually is a series of feedback loops, and as with any such design, we know that, under certain conditions, it may become unstable. Although instability is possible with an IIR design, it has the advantage that, for the same roll-off rate, it requires fewer taps than FIR filters. This means that, if we are limited in the processor resources available to perform a desired function, we may have to use an IIR. We just have to be careful to design a stable filter. More advanced forms of these filters are discussed with simple explanation in Marven and Ewers (1994).

5.4.2 Filter Implementation in DSPs

To explain the filter implementation, let us take the case of a first-order recursive filter. A signal flow graph or signal flow diagram is a convenient representation of a signal processing algorithm. Consider the first-order recursive filter shown in Figure 5-7(a). The sequential computations involved are not clearly evident in the signal flow graph, since it appears as if all the operations can be

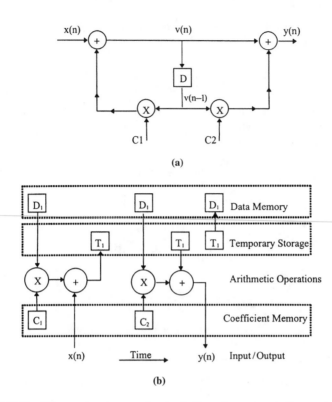

(a)

(b)

FIGURE 5-7 Filtering implementation by DSP techniques: (a) First-order IIR filter, (b) Assembler instructions at the Register Transfer level

evaluated at the same time. However, operations have to follow a certain precedence to preserve correct operation. It is also not clear where the data operands and coefficients are stored prior to their utilization in the computation. A more convenient mode of description would be the one in Figure 5-7(b), which shows the storage locations for each operand and the sequence of computations in terms of micro-operations at the register-transfer level (RTL) ordered in time from left to right. We assume that the state variable $v(n-1)$ is stored in the data memory (DM) at location D_1, while the coefficient C1 stored in a coefficient memory (CM) at location C_1. Both these operands are fetched and multiplied and the result is added to the input sample, $x(n)$, and the sum is stored in a temporary location, T_1. Then, another multiplication is performed using coefficient C2 and the product is added to the contents of T_1. The final result is the output $y(n)$. The new variable $v(n)$ is stored in memory location D_1. One may wonder why temporary location T_1 has been used. Temporary locations such as T_1 often provide a longer word length (or precision) than the word length of the memory. Repeated sums of products, as required in this example, quickly can exceed the dynamic range provided by the word length. Temporary locations provide the additional bits required to offset the deleterious effects of overflow. One also can observe that, in this example, the multiplier and adder operate in tandem and the second coefficient multiplication can utilize the same multiplier when the input sample is being added. Thus, only one multiplier and one adder are required as arithmetic units. One data memory location, two coefficient memory locations, and one temporary storage register are required for correct operation of the filter. The specification of the sequence of micro-operations required to perform the computation is called *programming in assembler.*

From the preceding discussion, any candidate signal processor architecture for the IIR filter needs a coefficient memory, a data memory, temporary registers for storage, a multiplier, an adder, and interconnection. In addition, address must be calculated for the memories as well as interpretation (or decoding) of the instruction (obtained from the program memory). The coefficient memory and the program memory can be combined into one memory (the program memory). Nothing can be written into this read-only memory (ROM). Data can be written and read from the random-access data memory (RAM). The architecture shown in Figure 5-8 is a suitable candidate architecture for this application. The program counter and the index registers are used in computing the addresses of the next instruction and the coefficients. The instruction is decoded by the instruction register (IR), where the address of the data is calculated using the adder and the base index register provided with the data memory. The program bus and the data bus are separate from each other, as are the program and data memories. This separation of data and program memories and buses characterizes the Harvard architecture for digital signal processors. The shifter is provided to allow incorporation of multiple word lengths within the data path (the multiplier and the adder) and the data and program buses. The T_1 register is configured as a higher-precision accumulator. Input samples are read in from the input buffer and written into the output buffer. The DSP can interact with a host computer via the external interface. In Figure 5-8, the integers represent the number of bits carried on each bus. For a detailed account of digital filters, see Jones and Watson (1990, Chapter 7).

The inherent advantages of digital filters are these:

1. They can be made to have no insertion loss.
2. Linear phase characteristics are possible.
3. Filter coefficients easily are changed to enable adaptive performance.
4. Frequency response characteristics can be made to approximate closely to the ideal.
5. They do not drift.
6. Performance accuracy can be controlled by the designer.
7. They can handle very low-frequency signals.

5.4.3 DSP Architecture

The simplest processor memory structure is a single bank of memory, which the processor accesses through a single set of address and data lines, as shown in Figure 5-9. This structure, which is common among non-DSP processors, is

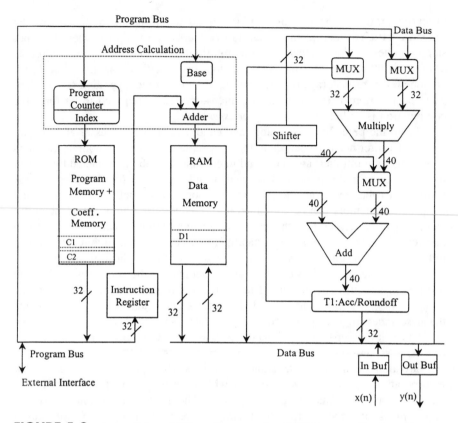

FIGURE 5-8 A candidate DSP architecture for IIR/FIR type filtering. (Source: Thompson and Tewksbury, 1982 (©1995 IEEE).)

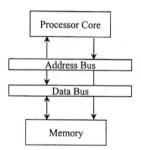

FIGURE 5-9 Von Neumann architecture for non-DSP processors

often considered a von Neumann architecture. Both program instructions and data are stored in the single memory. In the simplest (and most common) case, the processor can make one access (either a read or a write) to memory during each instruction cycle.

If we consider programming a simple von Neumann architecture machine to implement the example FIR filter algorithm, the shortcomings of the architecture become immediately apparent. Even if the processor's data path is capable of completing a multiply-accumulate operation in one instruction cycle, it will take four instruction cycles for the processor to actually perform the multiply-accumulate operation, since the four memory accesses outlined previously must proceed sequentially, with each memory access taking one instruction cycle. This is one reason why conventional processors often do not perform well on DSP-intensive applications and why designers of DSP processors have developed a wide range of alternatives to the von Neumann architecture, which we explore next.

The previous discussions indicate that parallel memories are preferred in DSP applications. In most DSPs, Harvard architecture coexists with data pipelines and instruction processors in a very efficient manner. The systems with specific addressing modes for signal processing applications could be best described as special instruction set computers (SISC). SISC architecture is characterized by a memory-oriented special purpose instruction set.

5.4.3.1 Basic Harvard Architecture

Harvard architecture refers to a memory structure in which the processor is connected to two independent memory banks via two independent sets of buses. In the original Harvard architecture, one memory bank holds program instructions and the other holds data. Commonly, this concept is extended slightly to allow one bank to hold program instructions and data, while the other bank holds data only. This "modified" Harvard architecture is shown in Figure 5-10. The key advantage of the Harvard architecture is that two memory accesses can be made during any one instruction cycle. Thus, the four memory accesses required for the example FIR filter can be completed in two instruction cycles. This type of

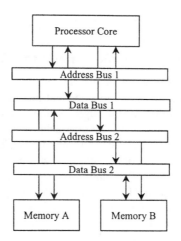

FIGURE 5-10 Harvard architecture

memory architecture is used in many DSP families including the Analog Devices ADSP21xx.

5.4.3.2 SISC Architecture

While microprocessors are based on register-oriented architecture, signal processors have memory-oriented architectures. Multiple memories for both program and data have been present even in the first-generation DSPs such as TMS320C10. Modern DSPs have as many as six parallel memories for the use of the instruction or the data processors. External memory is as easily accessible as internal memory. In addition, a rich set of addressing modes tailored for signal processing applications also are provided. We describe the architecture representative of SISC computers and expect that future generations of SISC computers will have communication primitives as part of the standard instruction set. The basic instruction cycle is a unit of time measurement in the context of signal processing architectures, in some sense, the average time required to execute an ALU instruction. The basic instruction cycle is further divided into subcycles (usually two to four). The memory cycle time is that required to access one operand from the memory. The high-memory bandwidth requirement in SISC computers can be met by either providing for memories with very low-memory cycle times or multiple memories with relatively slower cycle times. Typically, an instruction cycle is twice as long as a memory cycle for on-chip memory (and equal to the memory cycle for external memory). Clearly, this facilitates the use of operand fetch and execution pipelines of two-operand instructions with on-chip data memories. If parallel data memories are provided, then the total number of memory cycles per instruction cycle is increased. The total number of memory cycles possible within a single basic instruction cycle is defined as the demand ratio (Kogge, 1981) for a SISC machine. Higher demand ratios lead to a higher throughput of instructions:

$$\text{Demand ratio} = \frac{(\text{Basic cycle time}) \times (\text{Number of memories})}{\text{Memory cycle time}} \tag{5.5}$$

5.4.3.3 Multiple Access Memory-Based Architecture

As discussed, Harvard architecture achieves multiple memory accesses per instruction cycle by using multiple, independent memory banks connected to the processor data path via independent buses. While a number of DSP processors use this approach, there are other ways to achieve multiple memory accesses per instruction cycle. These include using fast memories that support multiple, sequential accesses per instruction cycle over a single set of buses and using "multiported" memories that allow multiple concurrent memory accesses over two or more independent sets of buses.

Achieving increased memory access capacity by use of multiported memory is becoming popular with the development of memory technology. A multiported memory has multiple independent sets of address and data connections, allowing multiple independent memory access to proceed in parallel. The most common type of multiported memory is the dual-ported variety, which provides two simultaneous accesses. However, triple- and even quadruple-ported varieties sometimes are used. Multiported memories dispense with the need to arrange data among multiple, independent memory banks to achieve maximum performance. The key disadvantage of multiported memories is that they are much more costly (in terms of chip area) to implement than standard, single-ported memories. Some DSP processors combine a modified Harvard architecture with the use of multiported memories. The memory architecture shown in Figure 5-11, for example, includes a single-ported program memory with a dual-ported data memory. This arrangement provides one program memory access and two data memory accesses per instruction word and is used in the Motorola DSP561xx processors. For a more-detailed discussion of these techniques, see Lapsley et al. (1997).

5.4.4 Modifications to Harvard Architecture

The basic Harvard Architecture can be modified into six different types. This discussion is beyond the scope of the chapter and for details, see Lee (1988, 1989).

5.5 Special Addressing Modes

In addition to general addressing modes used in microprocessor systems, several special addressing modes are used in DSPs, including circular addressing

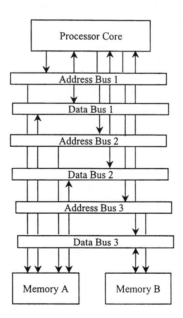

FIGURE 5-11 Modified Harvard architecture with dual-ported memory

and bit reversed addressing. For a comprehensive discussion on addressing modes, see Lapsley et al. (1997), as only circular addressing and bit reversed addressing are discussed here.

5.5.1 Circular Addressing

Many DSP applications need to manage data buffers. A data buffer is a section of memory used to store data that arrive from an off-chip source or a previous computation until the processor is ready to process the data. In real-time systems, where dynamic memory allocation is prohibitively expensive, the programmer usually must determine the maximum amount of data that a given buffer must hold and set aside a portion of memory for that buffer. The buffers generally use a first-in-first-out (FIFO) protocol, meaning that data values are read out of the buffer in the order in which they arrived.

In managing the movement of data into and out of the buffer, the programmer maintains two pointers, which are stored in registers or in memory: a read pointer and a write pointer. The read pointer points to (that is, contains the address of) the memory location containing the next data value to arrive, as illustrated in Figure 5-12. Each time a read or write operation is performed, the read or write pointer is advanced and the programmer must check to see whether the pointer has reached the last location in the buffer. When the pointer reaches the end of the buffer, it is reset to point to the first location in the buffer. Checking whether the pointer has reached the end of the buffer after each buffer operation and resetting it if it has is time consuming. For systems that use buffers extensively, this linear addressing can cause a significant performance bottleneck.

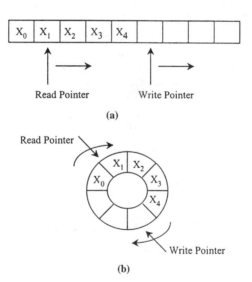

FIGURE 5-12 Comparison of linear and circular addressing: (a) A FIFO buffer with linear addressing, (b) The same data in FIFO buffer with circular addressing

To address this bottleneck, many DSPs have a special addressing capability that allows them, after each buffer address calculation, to automatically check whether the pointer has reached the end of the buffer and reset it at the buffer start location if necessary. This capability is called *modulo addressing* or *circular addressing*.

The term *modulo* refers to modulo arithmetic, where numbers are limited to a specific range. This is similar to the arithmetic used in a clock, which is based on a 12-hour cycle. When the result of a calculation exceeds the maximum value, it is adjusted by repeatedly subtracting from it the maximum representable value until the result lies within the specified range. For example, four hours after 10 o'clock is 2 o'clock (14 modulo 12).

When modulo address arithmetic is in effect, read and write pointers (address registers) are updated using pre- or postincrement register-indirect addressing (Lapsley et al., 1997). The processor's address generation unit performs modulo arithmetic when new address values are computed, creating the appearance of a circular memory layout, as illustrated in Figure 5-11(b). Modulo address arithmetic eliminates the need for the programmer to check the read and write pointers to see whether they have reached the end of the buffer and reset them once they have reached the end. This results in much faster buffer operations and makes modulo addressing a valuable capability for many applications.

In most real-time signal processing applications, such as those found in filtering, the input is an infinite stream of data samples. These samples are placed in "windows" and used in filtering applications. For instance, a sliding window of N data samples is used by an FIR filter with N taps. The data samples simulate a tapped-delay line and the oldest sample is written over by the most recent sample. The filter coefficients and the data samples are written into two circular buffers. Then, they are multiplied and accumulated to form the output sample result, which is stored. The address pointer for the data buffer is updated and the samples appear shifted by one sample period, the oldest data being written out and the most recent data is written into that location.

5.5.2 Bit-Reversed Addressing

Perhaps the most unusual of addressing modes, bit-reversed addressing is used only in very specialized circumstances. Some DSP applications make heavy use of the fast Fourier transform (FFT) algorithm. The FFT is a fast algorithm for transforming a time-domain signal into its frequency-domain representation and vice versa (Oppenheim and Schafer, 1988; Kularatna, 1996, Chapter 9). However, the FFT has the disadvantage that it either takes its input or leaves its output in a scrambled order. This dictates that the data be rearranged to or from natural order at some point.

The scrambling required depends on the particular variation of the FFT. The radix-2 implementation of an FFT, a very common form, requires reordering of a particularly simple nature, bit-reversed ordering. The term *bit reversed* refers to the observation that, if the output values from a binary counter are written in reverse order (that is, least significant bit first), the resulting sequence of counter output values will match the scrambled sequence of the FFT output data. This phenomenon is illustrated in Figure 5-13.

Because the FFT is an important algorithm in many DSP applications, many DSP processors include special hardware in their address generation units to facilitate generating bit-reversed address sequences for unscrambling FFT results. For example, the Analog Devices ADSP-210xx provides a bit-reverse mode, which is enabled by setting a bit in a control register. When the processor is in the bit-reverse mode, the output of one of its address registers is bit reversed before being applied to the memory address bus.

An alternative approach to implementing bit-reversed addressing is the use of reverse-carry arithmetic. With reverse-carry arithmetic, the address generation unit reverses the direction in which carry bits propagate when an increment is added to the value in an address register. If reverse-carry arithmetic is enabled in the AGU and the programmer supplies the base address and increment value

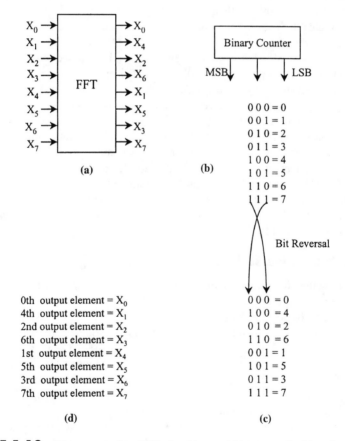

FIGURE 5-13 The output of an FFT algorithm and bit-reversed addressing: (a) FFT output and input relations, (b) Binary counter output, (c) Bit reversal, (d) Transformation of output into order

in bit-reversed order, then the resulting addresses will be in bit-reversed order. Reverse-carry arithmetic is provided in the AT&T DSP32xx, for example.

5.6 Important Architectural Elements in a DSP

Based on the preceding chapter's discussion on microprocessors, it may be relevant for us to discuss special function blocks in a DSP chip. Performing efficient digital signal processing on a microprocessor is a tricky business. Although the ability to support single-cycle multiplier/accumulators (MACs) is the most important function a DSP performs, many other functions are critical for real-time DSP applications. Executing a real-time DSP application requires an architecture that supports high-speed data flow to and from the computation units and memory through a multiport register file. This execution often involves the use of direct memory access units and address generation units that operate in parallel with other chip resources. Address generation units or AGUs, which perform address calculations, allow the DSP to bring two pieces of data per clock, which is a critical need for real-time DSP algorithms.

It is important for DSPs to have an efficient looping mechanism, because most DSP code is highly repetitive. The architecture allows for zero-overhead looping, in which no additional instructions are needed to check the completion of loop iterations. Generally, DSPs take looping a step further by including the ability to handle nested loops.

DSPs typically handle an extended precision and dynamic range to avoid overflow and minimize round-off errors. To accommodate this capability, DSPs generally include dedicated accumulators with registers wider than the nominal word size to preserve precision. DSPs also must support circular buffers to handle algorithmic functions, such as tapped delay lines and coefficient buffers. DSP hardware updates circular-buffer pointers during every cycle in parallel with other chip resources. During each clock cycle, the circular-buffer hardware performs an end-of-buffer comparison and resets the pointer with no overhead when it reaches the end of the buffer. FFTs and other DSP algorithms also require bit-reversed addressing.

5.6.1 Multiplier/Accumulator

The multiplier/accumulator provides high-speed multiplication, multiplication with cumulative addition, multiplication with cumulative subtraction, saturation, and clear-to-zero functions. A feedback function allows part of the accumulator output to be used directly as one of the multiplicands of the next cycle. To explain MAC operation, we take a real-life example from the ADSP21XX family (see Figure 5-14).

The multiplier has two 16-bit input ports, X and Y, and a 32-bit product output port, P. The 32-bit product is passed to a 40-bit adder/subtracter, which adds or subtracts the new product from the content of the multiplier result (MR) register or passes the new product directly to MR. The MR register is 40 bits

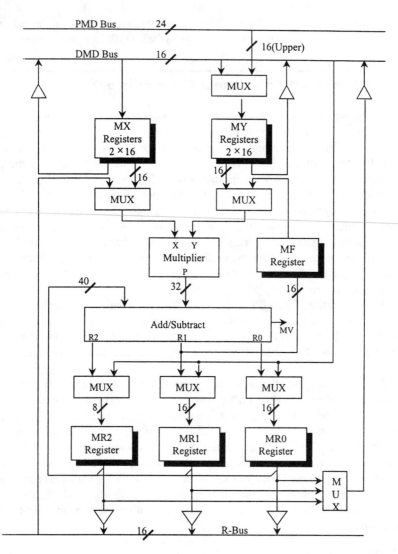

FIGURE 5-14 MAC block diagram of the ADSP-2104. (Reproduced by permission of Analog Devices Inc.)

wide. In this discussion, we refer to the entire register as MR, although it actually consists of three smaller registers: MR0 and MR1, which are 16 bits wide, and MR2, which is 8 bits wide.

The adder/subtracter is greater than 32 bits to allow for intermediate overflow in a series of multiply/accumulate operations. The multiply overflow (MV) status bit is set when the accumulator has overflowed beyond the 32-bit boundary; that is, when there are significant (nonsign) bits in the top nine bits of the MR register (based on two's-complement arithmetic). The input/output registers of the MAC section are similar to the ALU. The X input port can accept data from either the MX register file or any register on the result (R) bus. The R bus connects the output registers of all the computational units, permitting them to be used directly as input operands. Two registers in the MX register file, MX0 and MX1, can be read and written from the data memory data (DMD) bus. The MX register file output is dual ported so that one register can provide input to the multiplier while the other one drives the DMD bus.

The Y input port can accept data from either the MY register file or the MF register. The MY register file has two registers, MY0 and MY1, which can be read and written from the DMD bus and written from the program memory data (PMD) bus. The ADSP-2101 instruction set also provides for reading these registers over the PMD bus but with no direct connection; this operation uses the DMD-PMD bus exchange unit. The MY register file output also is dual ported so that one register can provide input to the multiplier while either one drives the DMD bus.

The output of the adder/ subtracter goes to either the MF register or the MR register. The MF register is a feedback register that allows bits 16–31 of the result to be used directly as the multiplier Y input on a subsequent cycle. The 40-bit adder/subtracter register (MR) is divided into three sections: MR2, MR1, and MR0. Each register can be loaded directly from the DMD bus and its output sent to either the DMD bus or the R bus.

Any register associated with the MAC can be both read and written in the same cycle. Registers are read at the beginning of the cycle and written at the end of the cycle. A register read instruction, therefore, reads the value loaded at the end of a previous cycle. A new value written to a register cannot be read out until a subsequent cycle. This allows an input register to provide an operand to the MAC at the beginning of the cycle and be updated with the next operand from memory at the end of the same cycle. It also allows a result register to be stored in memory and updated with a new result in the same cycle.

The MAC contains a duplicate bank of registers, shown in Figure 5-14 behind the primary registers. There actually are two sets of MR, MF, MX, and MY register files. Only one bank is accessible at a time. The additional bank of registers can be activated for extremely fast context switching. A new task, such as an interrupt service routine, can be executed without transferring current states to storage. The selection of the primary or alternate bank of registers is controlled by bit 0 in the processor mode states register (MSTAT). If this bit is 0, the primary bank is selected; if it is 1, the secondary bank is selected. For details, see Ingle and Proakis (1991) and New (1995).

5.6.2 Address Generation Units

Most DSP processors include one or more special address generation units dedicated to calculating addresses. Manufacturers refer to these units by various names. For example, Analog Devices calls its AGU a data address generator, and AT&T calls its a control arithmetic unit. An AGU can perform one or more complex address calculations per instruction cycle without using the processor's main data path. This allows address calculations to take place in parallel with arithmetic operations on data, improving processor performance. The differences among address generation units are manifested in the types of addressing modes provided and the capability and flexibility of each addressing mode. As an example let us take data addressing units in the ADSP-21xx family.

5.6.2.1 Data Address Units of ADSP-21xx Family: An Example

Data address generator (DAG) units contain two independent address generators so that program and data memories can be accessed simultaneously. Let us discuss the operation of the DAGs taking the ADSP-2101 as an example. The DAGs provide indirect addressing capabilities and perform automatic address modification. In the ADSP-2101, the two DAGs differ: DAG1 generates data memory addresses and provides an optional bit-reversal capability, DAG2 can generate both data memory and program memory addresses but has no bit reversal.

Figure 5-15 shows a block diagram of a single DAG. There are three register files: the modify (M) register file, the index (I) register file, and the length (L)

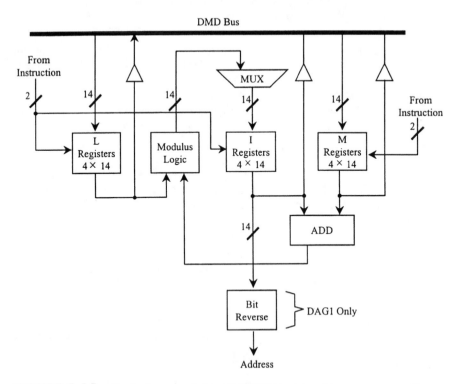

FIGURE 5-15 Block diagram of the ADSP-2101's data address generator. (Reproduced by permission of Analog Devices, Inc.)

register file. Each file contains four 14-bit registers that can be read from and written to via the DMD bus. The I registers (I0-3 in DAG1, I4-7 in DAG2) contain the actual addresses used to access memory. When data is accessed in the indirect mode, the address stored in the selected I register becomes the memory address. With DAG1, the output address can be bit reversed by setting the appropriate mode bit in the mode status register, as discussed next. Bit reversal facilitates FFT addressing.

The data address generator employs a postmodification scheme. After an indirect data access, the specified M register (M0-3 in DAG2) is added to the specified I register to generate the new I value. The choice of the I and M registers is independent within each DAG. In other words, any register in the I0-3 set may be modified by any register in the M0-3 set in any combination but not by those in DAG2 (M4-7). The modification values stored in the M register are signed numbers so that the next address can be either higher or lower. The address generators support both linear and circular addressing. The value of the L register determines which addressing scheme is used. For circular buffer addressing, the L register is initialized with the length of the buffer. For linear addressing, the modulus logic is disabled by setting the corresponding L register to 0. L registers and I registers are paired and the selection of the L register (L0-3 in DAG1, L4-7 in DAG2) is determined by the I register used. Each time an I register is selected, the corresponding L register provides the modulus logic with the length information. If the sum of the M register content and the I register content crosses the buffer boundary, the modified I register value is calculated by the modulus logic using the L register value.

All data address generator registers (I, M, and L registers) are loadable and readable from the lower 14 bits of the DMD bus. Since the I and L register content is considered unsigned, the upper 2 bits of the DMD bus are padded with zeros when reading them. The M register content is signed; when reading an M register, the upper 2 bits of the DMD bus are sign extended. The modulus logic implements automatic pointer wraparound for accessing circular buffers. To calculate the next address, the modulus logic uses the following information:

- The current location, found in the I register (unsigned).
- The modify value, found in the M register (signed).
- The buffer length, found in the L register (unsigned).
- The buffer base address.

From such input, the next address is calculated using the formula

$$\text{Next address} = (I + M - B) \text{ modulo } (L) + B \qquad (5.6)$$

where

I = current address;

M = modify value (signed);

B = base address (generated by the linker);

$$L = \text{buffer length } M+;$$

$$I = \text{modified address;}$$

and $M < L$ (which ensures that the next address cannot wrap around the buffer more than once in one operation).

5.6.3 Shifters

Shifting a binary number allows scaling. A shifter unit in a DSP provides a complete set of shifting functions, which can be divided into two categories: arithmetic and logical. A logical left shift by 1 bit inserts a 0 bit in the least significant bit, while a logical right shift by 1 bit inserts a 0 bit in the most significant bit. In contrast, an arithmetic right shift duplicates the sign bit (either a 1 or 0, depending on whether the number is negative or not) into the most significant bit. Although people use the term *arithmetic left shift*, arithmetic and logical left shifts really are identical: Both shift the word left and insert a 0 in the least significant bit.

Arithmetic shifting provides a way of scaling data without using the processor's multiplier. Scaling is especially important in fixed-point processors, where proper scaling is required to obtain accurate results from mathematical operations.

Virtually all DSPs provide shift instructions of one form or another. Some processors provide the minimum; that is, instructions to do arithmetic left or right shifting by 1 bit. Some processors may provide instructions for 2- or 4-bit shifts. These can be combined with single-bit shifts to synthesize n-bit shifts, although at a cost of several instruction cycles.

Increasingly, many DSP processors feature a barrel shifter and instructions that use the barrel shifter to perform arithmetic or logical left or right shifts by any number of bits. Examples include the AT&T DSP16xx, the Analog Devices ADSP-21xx and ADSP-210xx, the DSP Group OakDSPCore, the Motorola DSP563xx, the SGS-Thompson D950-CORE, and the Texas Instruments TMS320C5x and TMS320C54x. If you start with a 16-bit input, a complete set of shifting functions needs a 32-bit output. These include arithmetic shift, logical shift, and normalization. The shifter also derives the exponent and common exponent for an entire block of numbers. These basic functions can be combined to efficiently implement any degree of numerical format control, including full floating point representation. Figure 5-16 shows a block diagram of the ADSP-2101.

The variable shifter section in the ADSP-2100 can be divided into a shifter array, an OR/PASS logic, an exponent detector, and the exponent compare logic.

The shifter array is a 16×32 barrel shifter. It accepts a 16-bit input and can place it anywhere in the 32-bit output field, from off-scale right to off-scale left, in a single cycle. This gives 49 possible placements within the 32-bit field. The placement of the 16 input bits is determined by a control code (C) and a HI/LO reference signal.

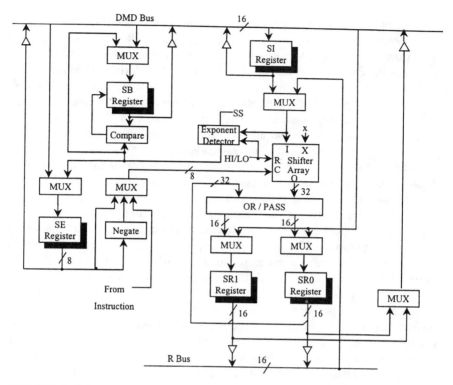

FIGURE 5-16 Block diagram of the ADSP-2101's shifter. (Reproduced by permission of Analog Devices Inc.)

The shifter array and its associated logic are surrounded by a set of registers. The shifter input (SI) register provides input to the shifter array and the exponent detector. The SI register is 16 bits wide and is readable and writable from the DMD bus. The shifter array and the exponent detector also take as inputs arithmetic, shifter, or multiplier results via the R bus. The shifter result (SR) register is 32 bits wide and divided into two 16-bit sections, SR0 and SR1. The SR0 and SR1 registers can be loaded from the DMD bus and sent to either the DMD bus or the R bus. The SR register also is fed back to the OR/PASS logic to allow double-precision shift operations. The SE (shifter exponent) register is 8 bits wide and holds the exponent during the normalize and denormalize operations. The SE register is loadable and readable from the lower 8 bits of the DMD bus. It is a two's-complement, integer value.

The SB (shifter block) register is important in block floating point operations where it holds the block exponent value; that is, the value by which the block values must be shifted to normalize the largest value. SB is 5 bits wide and holds the most recent block exponent value. The SB register is loadable and readable from the lower 5 bits of the DMD bus. It is a two's-complement, integer value.

Whenever the SE or SB registers are loaded onto the DMD bus, they are sign extended to a 16-bit value. Any of the SI, SE, or SR registers can be read and written in the same cycle. Registers are read at the beginning of the cycle and written at the end of the cycle. All register reads, therefore, read values loaded at the end of a previous cycle. A new value written to a register cannot be read out until a subsequent cycle. This allows an input register to provide an operand to the shifter at the beginning of the cycle and be updated with the next operand at the end of that cycle. It also allows a result register to be stored in memory and updated with a new result in the same cycle.

The shifter section contains a duplicate bank of registers, shown in Figure 5-16 behind the primary registers. There actually are two sets of SE, SB, SI, SR1, and SR0 registers, only one bank accessible at a time. The additional bank of registers can be activated for extremely fast context switching. A new task, such as an interrupt service routine, can be executed without transferring current states to storage. The selection of the primary or alternate bank of registers is controlled by bit 0 in the processor mode status register. If this bit is 0, the primary bank is selected; if it is 1, the secondary bank is selected.

The shifting of the input is determined by a control code (C) and a HI/LO reference signal. The control code is an 8-bit signed value that indicates the direction and number of places the input is to be shifted. Positive codes indicate a left shift (upshift) and negative codes indicate a right shift (downshift). The control code can come from three sources: the content of the shifter exponent register, the negated content of the SE register, or an immediate value from the instruction.

The HI/LO signal determines the reference point for the shifting. In the HI state, all shifts are referenced to SR1 (the upper half of the output field); and in the LO state, all shifts are referenced to SR0 (the lower half). The HI/LO reference feature is useful when shifting 32-bit values since it allows both halves of the number to be shifted with the same control code. HI/LO reference signal is selectable each time the shifter is used.

The shifter fills any bits to the right of the input value in the output field with zeros, and bits to the left are filled with the extension bit (X). The extension bit can be fed by three possible sources depending on the instruction being performed: the MSB of the input, the AC bit from the arithmetic status register, or a zero.

The OR/PASS logic allows the shifted sections of a multiprecision number to be combined into a single quantity. When PASS is selected, the shifter array output is passed through and loaded into the shifter result register unmodified. When OR is selected, the shifter array is bitwise ORed with the current contents of the SR register before being loaded there.

The exponent detector derives an exponent for the shifter input value. The exponent detector operates in one of three ways, which determine how the input value is interpreted. In the HI state, the input is interpreted as a single precision number or the upper half of a double precision number. The exponent detector determines the number of leading sign bits and produces a code that indicates how many places the input must be upshifted to eliminate all but one of the sign bits. The code is negative so that it can become the effective exponent for the mantissa formed by removing the redundant sign bits.

In the HI-extend state (HIX), the input is interpreted as the result of an add or subtract performed in the ALU section, which may have overflowed. Therefore, the exponent detector takes the arithmetic overflow (AV) status into consideration. If AV is set, then $a + 1$ exponent becomes output to indicate an extra bit is needed in the normalized mantissa (the ALU carry bit); if AV is not set, then HI-extend functions exactly like the HI state. When performing a derive exponent function in HI or HI-extend modes, the exponent detector also sends out a shifter sign (SS) bit, which is loaded into the arithmetic status register. The sign bit is the same as the MSB of the shifter input except when AV status is set; when AV status is set in the HI-extend state, the MSB is inverted to restore the sign bit of the overflow value. In the LO state, the input is interpreted as the lower half of a double precision number. In the LO state, the exponent detector interprets the SS bit in the arithmetic status register as the sign bit of the number. The SE register is loaded with the output of the exponent detector only if SE contains P15. This occurs only when the upper half — which must be processed first — contains all sign bits. The exponent detector output also is offset by P16 to indicate that the input actually is the lower half of a 32-bit value.

The exponent compare logic is used to find the largest exponent value in an array of shifter input values. The exponent compare logic in conjunction with the exponent detector derives a block exponent. The comparator compares the exponent value derived by the exponent detector with the value stored in the shifter block exponent register and updates the SB register only when the derived exponent value is larger than the value in the SB register.

Shifters in different DSPs have different capabilities and architecture. For example, the TMS320C25 scaling shifter shifts to the left from none to 16 bits. Two other shifters can shift data coming from the multiplier left 1 bit or 4 bits or can shift data coming from the accumulator left from none to 7 bits. These two shifters add the advantage of being able to scale data during the data move instead of requiring an additional shifter operation.

5.6.4 Loop Mechanisms

DSP algorithms frequently involve the repetitive execution of a small number of instructions, so-called inner loops or kernels. FIR and IIR filters, FFTs, matrix multiplication, and a host of other application kernels are performed by repeatedly executing the same instruction or sequence of instructions. DSPs have evolved to include features to efficiently handle this sort of repeated execution. To understand the evolution, we look at the problems associated with traditional approaches to related instruction execution. First, a natural approach to looping uses a branch instruction to jump back to the start of the loop.

Second, because most loops execute a fixed number of times, the processor must use a register to maintain the loop index; that is, the count of the number of times the processor has been through the loop. The processor's data path must be used to increment or decrement the index and test to see if the loop condition has been met. If not, a conditional branch brings the processor back to the top of the loop. All of these steps add overhead to the loop and use precious registers.

DSPs have evolved to avoid these problems via hardware looping, also known as *zero-overhead looping*. Hardware loops are special hardware control constructs that repeat between hardware loops and software loops so that hardware loops lose no time incrementing or decrementing counters, checking to see if the loop is finished, or branching back to the top of the loop. This can result in considerable savings. To explain how a loop mechanism improves the efficiency, we once again use the ADSP-2101 as an example (see Figure 5-17).

The ADSP-2100A program sequencer supports zero overhead DO UNTIL loops. Using the count stack, loop stack, and loop comparator, the processor can determine whether a loop should terminate and the address of the next instruction (either the top of the loop or the instruction after the loop) with no overhead cycle.

A DO UNTIL loop may be as large as program memory size permits. A loop may terminate when a 16-bit counter expires or when any other arithmetic condition occurs. The following example shows a three-instruction loop that is to be repeated 100 times:

```
CNTR = 100
Do Label UNTIL CE
First instruction of loop
Second instruction of loop
Label: Last instruction of loop
        First instruction outside loop
```

The first instruction loads the counter with 100. The DO UNTIL instruction contains the address of the last instruction in the loop (in this case the address represented by the identifier, Label) and the termination condition (in this case the count expiring, CE). The execution of the DO UNTIL instruction causes the address of the first instruction of the loop to be pushed on the program counter stack and the address of the last instruction of the loop to be pushed on the loop stack (see Figure 5-17).

As instruction addresses are sent to the program memory address bus and the instruction is fetched, the loop comparator checks to see if the instruction is the last instruction of the loop. If it is, the program sequencer checks the status and condition logic to see if the termination condition is satisfied. The program sequencer then either takes the address from the program counter stack (to go back to the top of the loop) or simply increments the program counter (to go to the first instruction outside the loop).

The looping mechanism of the ADSP-2100A is automatic and transparent to the user. As long as the DO UNTIL instruction is specified, all stack and counter maintenance and program flow is handled by the sequencer logic with no overhead. This means that, in one cycle, the last instruction of the loop is being executed and, in the very next cycle, the first instruction of the loop is executed or

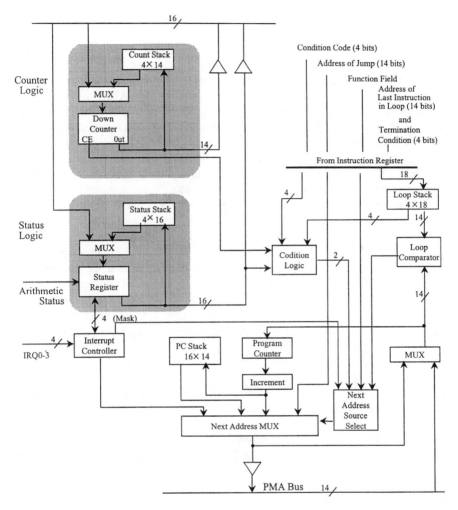

FIGURE 5-17 The ADSP-2100A program sequencer architecture. (Reproduced by permission of Analog Devices Inc.)

the first instruction outside the loop is executed, depending on whether the loop terminated or not. For further details of program sequencer and loop mechanisms of the ADSP-2100A, see Ingle and Proakis (1991) and Fine (●●●).

5.7 Instruction Set

Generally, a DSP instruction set is tailored to the computation-intensive algorithms common to DSP applications. This is possible because the instruction set allows data movement between various computational units with minimum overhead. For example, sustained single-cycle multiplication/accumulation operations are possible.

Again, we use the ADSP-2101 as an example. The instruction set provides full control of the ADSP-2101's three computation units: the ALU, MAC, and shifter. Arithmetic instructions can process single-precision 16-bit operands directly with provisions for multiprecision operations. The ADSP-2101 assembly language uses an algebraic syntax for arithmetic operations and data moves. The sources and destinations of computations and data moves are written explicitly, eliminating cryptic assembler mnemonics. There is no performance penalty for this; each program statement assembles into one 24-bit instruction, which executes in one cycle. There are no multicycle instructions in the ADSP-2101 instruction set. Some 50 registers surrounding the computational units are dual purpose, available for general purpose on-chip storage when not used in computation. This saves many memory access cycles and provides excellent freedom in coding. The control instructions provide conditional execution of most calculations and, in addition to the usual JUMP and CALL, support a DO UNTIL looping instruction. Return from Interrupt (RTI) and the Return from Subroutine (RTS) also are provided. These services are made compact and speedy by the single-cycle content save. The contents of the primary register set are held constant while the alternate set is enabled for subroutine and interrupt services. This eliminates the cluster of PUSHes and POPs of stacks common in general purpose microprocessors.

The ADSP-2101 also provides an IDLE instruction for idling the processor until an interrupt occurs. IDLE puts the processor into a low-power state while waiting for interrupts. Two addressing modes are supported for memory fetches. Direct addressing uses immediate values; indirect addressing uses the two data address generators.

The 24-bit instruction word allows a high degree of parallelism in performing operations. The instruction set allows for a single-cycle execution of any of the following combinations:

- Any ALU, MAC, or shifter operation (may be conditional).
- Any register-to-register move.
- Any data memory read or write.
- A computation with any data register/data register move.
- A computation with any memory read or write.
- A computation with a read from two memories.

The instruction set provides moves from any register to any other register or from most registers to and from either memory. For combining operations, almost any ALU, MAC, or shifter operation may be combined with any register-to-register moves or with a register move to or from either internal or external memory.

There are five basic categories of instruction: computational instructions, data move instructions, multifunction instructions, program flow control instructions, and miscellaneous instructions, all of which are described in the next several

TABLE 5-1 Notation Used in the Instruction Set of the ADSP-21xx Family. (Reproduced by permission of Analog Devices, Inc.)

Symbol	Meaning
+, −	Add, subtract
*	Multiply
a = b	Transfer into a the contents of b
,	Separates multifunction instruction
DM(addr)	The contents of data-memory at location "addr"
PM(addr)	The contents of program memory at location "addr"
[option]	Anything within square brackets is an optional part of the instruction statement
|option a|	List of parameters enclosed by parallel vertical lines require the choice of one parameter from among the available list
CAPITAL LETTERS	Capital letters denote reserved words. These are instruction words, register names, and operand selections
Lower-case letters	Parameters are shown in small letters and denote an operand in the instruction for which there are numerous choices
< data >	These angle brackets denote an immediate data value
< addr >	These angle brackets denote an immediate value of an address to be coded in the instruction
;	End of instruction

sections, with tables summarizing the syntax of each instruction category. The notation used in an instruction is shown in Table 5-1.

As it is beyond the scope of a chapter of this kind to explain the whole group of instructions, the computation instructions of the ADSP-2101 are described in a summary form. A more-detailed version instruction set can be found in Ingle and Proakis (1991) and the ADSP literature.

5.7.1 Computation Instructions: A Summary of the ADSP-21xx Family

The computation group executes all ALU, MAC, and shifter instructions. There are two functional classes: standard instructions, which include the bulk of the computation operations, can be executed conditionally (IF condition . . .), test the ALU status register, and may be combined with a data transfer in single-cycle multifunction instructions; and special instructions, which form a small subset and must be executed individually. Table 5-2 indicates permissible conditions for computation instructions, and Table 5-3 describes the computational input/output registers.

5.7.1.1 MAC Functions

Standard MAC instructions include multiply, multiply/accumulate, multiply/subtract, transfer AR conditionally, and clear. As an example, consider a MAC instruction for multiply/accumulate in the form:

TABLE 5-2 Permissible Conditions for Computation Instructions of ADSP-2101. [Reproduced by permission of Analog Devices, Inc.]

Condition	Keyword
ALU result is	
equal to zero	EQ
not equal to zero	NE
greater than zero	GT
greater than or equal to zero	GE
less than zero	LT
less than or equal to zero	LE
ALU carry status:	
carry	AC
not carry	NOT AC
x-input sign:	
positive	POS
negative	NEG
ALU overflow status:	
overflow	AV
not overflow	NOT AV
MAC overflow status:	
overflow	MV
not overflow	NOT MV
Counter status:	
not expired	NOT CE

TABLE 5-3 Computational Input/Output Registers. [Reproduced by permission of Analog Devices, Inc.]

Source for X input (xop)	Source for Y input (yop)	Destination*
ALU		
AX0, AX1, AR	AY0, AY1	AR
MR0, MR1, MR2	AF	AF
SR0, SR1		
MAC		
MX0, MX1, AR	MY0, MY1	MR (MR2, MR1, MR0)
MR0, MR1, MR2	MF	MF
SR0, SR1		
Shifter		
SI, SR0, SR1		SR (SR1, SR0)
AR		
MR0, MR1, MR2		

*Destination for output port R for ALU and MAC or destination for shifter output.

```
[IF Condition] MR  =  MR + xop * yop   (SS)  ;
               MF                        SU
                                         US
                                         UU
                                         RND
```

If the options MR and UU are chosen; if xop and yop are the contents of MXO and MYO, respectively; and if MAC overflow condition is chosen, then a conditional instruction would read

```
IF NOT  MV  MR  =  MR + MXO * MYO (UU) ;
```

The conditional expression, IF NOT MV, tests the MAC overflow bit. If the condition is not true, an NOP is executed. The expression MR = MR + MXO * MYO is the multiply/accumulate operation: The multiplier result register gets the value of itself plus the product of the X and Y input registers selected. The modifier selected in parentheses (UU) treats the operands as unsigned. Only one such modifier can be selected from the available set: (SS) means both are signed, (US) and (SU) mean that either the first or second operand is signed; (RND) means to round the (implicitly signed) result.

Accumulator saturation is the only MAC special function:

```
IF  MV  SAT  MR  ;
```

The instruction tests the MAC overflow bit (MV) and saturates the MR register (for only one cycle) if that bit is set.

5.7.1.2 ALU Group Functions

Standard ALU instructions include add, subtract, logic (AND, OR, NOT, exclusive-OR), pass, negate increment, decrement, clear, and absolute value. The − function does two's-complement subtraction while NOT obtains a one's-complement. The PASS function passes the listed operand but tests and stores status information for later sign/zero testing. As an example, consider an ALU addition instruction for add/add-with-carry in the form

```
[IF Condition]  AR  =  xop   + ypo    ;
                AF           + c
                             + yop + c
```

Instructions are in similar form for subtraction and logical operations. If the options AR and + yop + C are chosen, and if xop and yop are the contents of AXO and AYO, respectively, the unconditional instruction would read

```
AR = AXO + AYO + C;
```

This algebraic expression means that the ALU result register gets the value of the ALU x-input and y-input registers plus the value of the carry-in bit. This shortens the code and speeds execution by eliminating many separate register-move instructions.

When an optional IF condition is included, and if ALU carry bit status is chosen, then the conditional instruction would read

```
IF  AC  AR = AXO + AYO + C ;
```

The conditional expression, IF AC, tests the ALU carry bit. If there is a carry from the previous instruction, this instruction executes; otherwise, an NOP occurs and execution continues with the next instruction.

Division is the only ALU special function. It is executed in two steps: DIVS computes the sign, then DIVQ computes the quotient. A full divide of a signed 16-bit divisor into a signed 32-bit quotient requires a DIVS followed by 15 DIVQs.

5.7.1.3 Shifter Group Functions

Shifter standard functions include arithmetic and logical shift as well as floating point and block floating point scaling operations, derive exponent, normalize, denormalize, and block exponent adjust. As an example, consider a shifter instruction for normalize:

```
IF   NOT CE   SR   =   SR   OR   NORM   SI   (HI)   ;
```

The conditional expression, IF NOT CE, tests the "not counter expired" condition. If the condition is false, an NOP is executed. The destination of all shifting operations is the shifter result register. (The destination of the exponent detection instructions is SE or SB.) In this example, SI, the shifter input register, is the operand. The amount and direction of the shift are controlled by the signed value in the SE register in all shift operations except an immediate shift. Positive values cause left shifts; negative values cause right shifts.

The SR OR modifier (which is optional) logically ORs the result with the current contents of the SR register; this allows the user to construct a 32-bit value in SR from two 16-bit pieces. NORM is the operator and (HI) is the modifier that determines whether the shift is relative to the HI or LO (16-bit) half of SR. If SR OR is omitted, the result is passed directly into the SR.

Shift-immediate is the only shifter special function. The number of places (exponents) to shift is specified in the instruction word.

5.7.2 Other Instructions

Other instructions in a DSP could be grouped as in Table 5-4. The details could depend on the DSP family and hence Table 5-4 should be considered only a guideline.

TABLE 5-4   Instruction Set Groups (Using the ADSP 21xx Family as an Example)

Instruction Type	Purpose
Data move instructions	Move data to and from data registers and external memory
Multifunction instructions	Exploit the inherent parallelism of a DSP by combinations of data moves and memory writes/reads in a single cycle
Program flow control instructions	Directs the program sequence. In normal order, the sequence automatically fetches the next contiguous instruction for exertion. This flow can be altered by these instructions
Miscellaneous instruction	Such as NOP (no operation), PUSH/POP, and the like

5.8 Development Systems

Although a development system is needed only initially (when the application is being designed) and not in the final product, a designer most likely will be working with development tools. Therefore, understanding the capabilities of these tools is as essential as understanding the architecture of the DSP itself.

The development process begins with the task of defining the target system hardware environment. The *system builder* is used to define the hardware environment. The *system specification file* includes the target hardware information. The system builder reads this file and creates an *architecture description file* that passes information about the target hardware to the *linker, simulator*, and *emulator*.

Code generation begins by creating assembly source code modules. An assembly module is a unit of source code, such as a calling program, subroutine, data buffer declaration section, or any combination. Each assembly code module is assembled separately by the assembler. Several modules then are linked to form an executable program.

The linker needs the target hardware information located in the architecture description file to determine placement of the code and data fragments. In the assembly modules, we have the option of specifying each code or data fragment as completely relocatable, relocatable within a defined memory segment, or placed at an absolute address. Absolute code or data modules are placed at the specified base address, provided the specified memory area has the correct attributes. Relocatable objects are placed in memory by the linker.

Using the architecture description file and the assembler output files, the linker determines the placement of relocatable code and data segments (including circular buffers) and places all segments in memory locations with the correct attributes (CODE or DATA, RAM or ROM). The linker generates an executable image file, which may be loaded into the simulator and emulator for debugging.

The simulator provides windows that display different aspects of the hardware environment. To replicate the target hardware environment, the simulator configures its memory according to the system builder output and simulates I/O ports according to user-entered simulator commands. This simulation provides the capability to debug the system and analyze performance before committing to a hardware prototype.

After debugging with the simulator, the emulator is used in the prototype target system to debug hardware, timing, and real-time software problems. It provides overlay memory to replace target system off-chip memory, including boot memory, if desired.

The *PROM splitter* translates the executable memory image file (linker output) into a file compatible with a PROM burner. Once the ADSP-2101 code is burned into PROM and an ADSP-2101 is plugged into the target board, the prototype is ready to run.

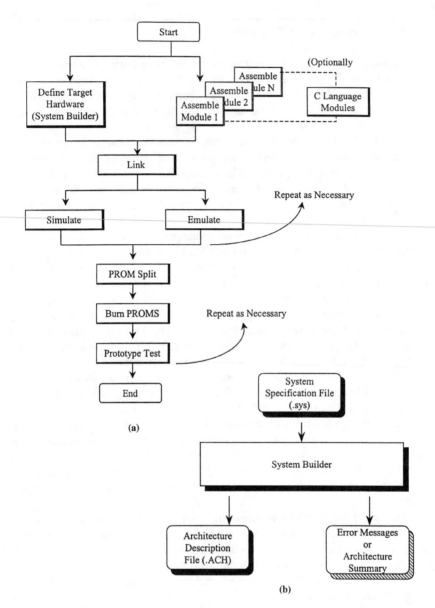

FIGURE 5-18 ADSP-2101 system development: (a) Development flow, (b) System builder I/O. (Reproduced by permission of Analog Devices, Inc.)

Figure 5-18(a) shows a flowchart of the ADSP-2101 development cycle. Figure 5-18(b) shows the system builder I/O. All the steps in the preceding development process except emulation are carried out by the software development system, while the hardware development consists of the emulator and the prototype target system.

5.9 Interface Between DSPs and Data Converters

Advances in semiconductor technology have given DSPs fast processing capabilities and data converter ICs have the conversion speeds to match the faster processing speeds. This section considers the hardware aspects of practical design.

5.9.1 Interface Between ADCs and DSPs

Precision sampling analog/digital converters generally have either parallel data output or a single serial output data link. We consider these separately.

5.9.1.1 Parallel Interfaces with ADCs

Many parallel output sampling ADCs offer three-state output that can be enabled or disabled using an output enable pin on the IC. While it may be tempting to connect the three-state output directly to a back plane data bus, severe performance-degrading noise problems will result. All ADCs have a small amount of internal stray capacitance between the digital output and the analog input (typically $0.1-0.5$ μF). Every attempt is made during the design and layout of the ADC to keep this capacitance to a minimum. However, if there is excessive overshoot and ringing and possibly other high-frequency noise on the digital output lines (as would probably be the case if the digital output were connected directly to a back plane bus), this digital noise will couple back into the analog input through the stray capacitance. The effect of this noise would be to decrease the overall ADC SNR and ENOB. Any code-dependent noise also will tend to increase the ADC harmonic distortion.

The best approach to eliminating this potential problem is to provide an intermediate three-state output buffer latch located close to the ADC data output. This latch isolates the noisy signals on the data bus from the ADC data outputs, minimizing any coupling back into the ADC analog input.

The ADC data sheet should be consulted regarding exactly how the ADC data should be clocked into the buffer latch. Usually, a signal called *conversion complete* or *busy* from the ADC is provided for this purpose.

It also is a good idea not to access the data in the intermediate latch during the actual conversion time of the ADC. This practice will further reduce the possibility of corrupting the ADC analog input with noise. The manufacturer's data sheet timing information should indicate the most desirable time to access the output data.

Figure 5-19 shows a simplified parallel interface between the AD676–16 bit, 100 kSPS ADC (or the AD7884) and the ADSP-2101 microcomputer. (Note that the actual device pins shown have been relabeled to simplify the following general discussion. In a real-time DSP application (such as in digital filtering), the processor must complete its series of instructions within the ADC sampling interval. Note that the entire cycle is initiated by the sampling clock edge from the sampling clock generator. Even though some DSP chips offer the capability

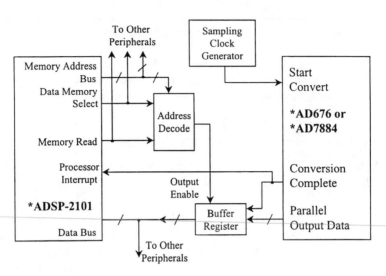

FIGURE 5-19 Generalized DSP-to-ADC parallel interface. (Reproduced by permission of Analog Devices Inc.)

to generate lower-frequency clocks from the DSP master clock, the use of these signals as precision sampling clock sources is not recommended due to the probability of timing jitter. It is preferable to generate the ADC sampling clock from a well-designed low noise crystal oscillator circuit as has been previously described.

The sampling clock edge initiates the ADC conversion cycle. After the conversion is completed, the ADC conversion complete line is asserted, which in turn interrupts the DSP. The DSP places the address of the ADC that generated the interrupt on the data memory address bus and asserts the data memory select line. The read line of the DSP then is asserted. This enables the external three-state ADC buffer register outputs and places the ADC data on the data bus. The trailing edge of the read pulse latches the ADC data on the data bus into the DSP internal registers. At this time, the DSP is free to address other peripherals that may share the common data bus.

Because of the high-speed internal DSP clock (50 MHz for the ADSP-2101), the width of the read pulse may be too narrow to access properly the data in the buffer latch. If this is the case, adding the appropriate number of programmable software wait states in the DSP will both increase the width of the read pulse and cause the data memory select and the data memory address lines to remain asserted for a correspondingly longer period of time. In the case of the ADSP-2101, one wait state is one instruction cycle, or 80 ns.

5.9.1.2 Interface Between Serial Output ADCs

ADCs that have a serial output (such as the AD677, AD776, and AD1879) have interfaces to the serial port of many DSP chips, as shown in Figure 5-20.

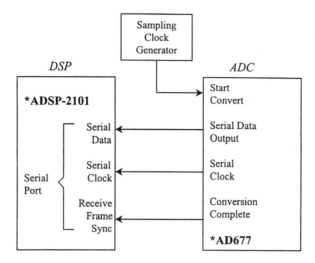

FIGURE 5-20 Generalized serial DSP-to-ADC interface. (Reproduced by permission of Analog Devices Inc.)

The sampling clock is generated from the low-noise oscillator. The ADC output data is presented on the serial data line one bit at a time. The serial clock signal from the ADC is used to latch the individual bits into the serial input shift register of the DSP serial port. After all the serial data are transferred into the serial input register, the serial port logic generates the required processor interrupt signal. The advantages of using serial output ADCs are a reduction in the number of interface connections as well as reduced noise because fewer noisy digital program counter tracks are close to the converter. In addition, SAR and Σ-Δ ADCs are inherently serial-output devices. The number of peripheral serial devices permitted is limited by the number of serial ports available on the DSP chip.

5.9.2 Interfaces with DACs

5.9.2.1 Parallel Input DACs

Most of the principles previously discussed regarding interfaces with ADCs also apply to interfaces with DACs. A generalized block diagram of a parallel input DAC is shown in Figure 5-21(a). Most high-performance DACs have an internal parallel DAC latch that drives the actual switches. This latch deskews the data to minimize the output glitch. Some DACs designed for real-time sampling data DSP applications have an additional input latch so that the input data can be loaded asynchronously to the DAC latch strobe. Some DACs have an internal reference voltage that can be either used or bypassed with a better external reference. Other DACs require an external reference.

The output of a DAC may be a current or a voltage. Fast-video DACs generally are designed to supply sufficient output current to develop the required signal levels across resistive loads (generally 150 Ω, corresponding to a 75 Ω source and

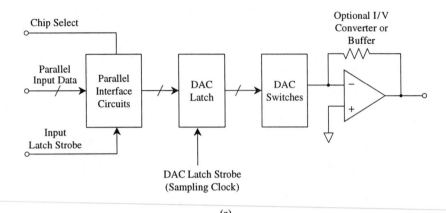

(a)

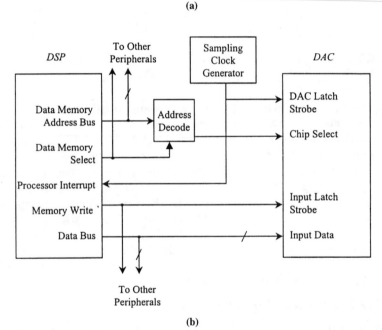

(b)

FIGURE 5-21 Interface between DSPs and parallel DACs: (a) Parallel input DAC, (b) DSP and parallel DAC input

load-terminated cable). Other DACs are designed to drive a current into a virtual ground and require a current-to-voltage converter (which may be internal or external). Some high-impedance voltage-output DACs require an external buffer to drive reasonable values of load impedance.

A generalized parallel DSP-to-DAC interface is shown in Figure 5-21(b). The operation is similar to that of the parallel DSP-to-ADC interface described earlier. In most DSP applications, the DAC is operated continuously from a stable sampling clock generator external to the DSP. The DAC requires double-buffering because of the asynchronous interface to the DSP. The sequence of events as follows. Asserting the *sampling clock generator* line clocks the word contained in the DAC *input latch* into the DAC *latch* (the latch that drives the DAC switches). This causes the DAC output to change to the new value. The sampling clock edge also interrupts the DSP, which then addresses the DAC, enables the DAC *chip select*, and writes the next data into the DAC *input latch* using the *memory write* and data bus lines. The DAC now is ready to accept the next sampling clock edge.

5.9.2.2 Serial Input DACs

A block diagram of a typical serial input DAC is shown in Figure 5-22(a). The digital input circuitry consists of a serial-to-parallel converter driven by a

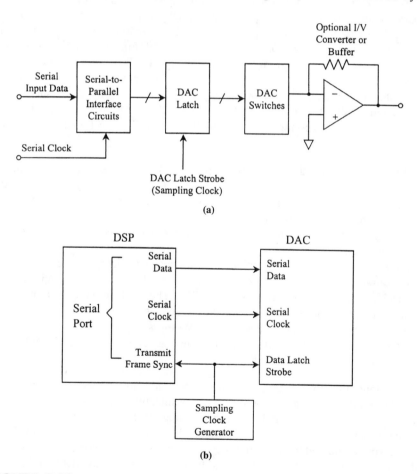

FIGURE 5-22 Interface between DSPs and serial DACs: (a) Serial input DAC, (b) DSP and parallel DAC input

serial data line and a serial clock. After the serial data is loaded, the DAC latch strobe clocks the parallel DAC latch and updates the DAC switches with a new word. Interface between DSPs and serial DACs is quite easy using the DSP serial port (Figure 5-22(b)). The serial data transfer process is initiated by the assertion of the sampling clock generator line. This updates the DAC latch and causes the serial port of the DSP to transmit the next word to the DAC using the serial clock and the serial data line.

5.10 Practical Components and Recent Developments

During 1997 and 1998, incredible developments took place in the DSP components world. Vendors were focusing on several key aspects of the DSP architecture. The most obvious architectural improvements were in the increased "parallelism": the number of operations the DSP can perform in an instruction cycle. An extreme example of parallelism is Texas Instruments' C6x very-long-instruction-word (VLIW) DSP with eight parallel functional units. Although Analog Devices' super Harvard architecture (SHARC) could perform as many as seven operations per cycle, the company and other vendors were working feverishly to develop their own VLIW-ized DSPs. In contrast to superscalar architectures, VLIW simplifies a DSP's control logic by providing independent control for each processing unit. During 1997 the following important developments were achieved (Levy, 1997):

- While announcing the first general purpose VLIW DSP, Texas Instruments also announced the end of the road for the C8x DSP family. The company emphasized the importance of the compilers for DSPs with the purchase of DSP-compiler company Tartan.
- Analog Devices broke the $100 price barrier with its SHARC floating-point architecture.
- Lucent Technologies discontinued new designs incorporating its 32-bit, floating-point DSP. The company also focused its energy on application-specific rather than general purpose DSPs. The application-specific products target modems and other communication devices.
- Motorola's DSP Division became the Wireless Signal Processing Division, although the company still supports many general purpose DSP and audio applications.

Among the hottest architectural innovations during 1998 was the move to dual multiply/accumulate units. The architecture of these MACs allows performing twice the digital/signal processing as before. TI kicked off this evolution with its VLIW-based C6x. Meanwhile, engineers designing with DSPs need a simple method to compare processor performance. Unfortunately, as types of processor architecture diversify, traditional metrics such as MIPS and MOPS have become less relevant. Alternatively, Berkeley Design Technology (BDTI, www.bdit.com) has become well known in the DSP industry for providing DSP benchmarks. Instead of using full-application benchmarks, BDTI has adopted a benchmark methodology based on DSP-algorithm kernels, such as FFTs and FIR filters. BDTI implements its suite of 11 kernel-based benchmarks on a variety of processors. You can find the results of these benchmarks in the company's *Buyer's Guide to DSP Processors* at Berkeley's web site.

To see the developments over the past ten years, compare Cushman (1987) with Levy (1997, 1998b).

References

Bell, C. G., and A. Newell. "Computer Structures," McGraw Hill, New York, USA, 1971.

Cushman, Robert H. "μP-Like DSP Chips." *EDN* (September 3, 1987), pp. 155–186.

Fine, Bob. "Considerations for Selecting a DSP Processor-ADSP 2100A vs. TMS 320C25." Application note. Analog Devices, Inc., USA.

Ingle, Vinay K., and John G. Proakis. *Digital Signal Processing Laboratory Using the ADSP-2101 Microcomputer.* Englewood Cliffs, NJ: Prentice-Hall/Analog Devices, 1991.

Jones, N. B., and J. D. M. Watson: *Digital Signal Processing-Principles, Devices and Applications.* London: Peter Peregrinus/IEE, 1990.

Kogge, P. M. *The Architecture of Pipe-Lined Computers.* New York: Hemisphere Publishing Co./McGraw-Hill, 1981.

Kularatna, N. *Modern Electronic Test and Measuring Instructions.* London: IEEE, 1996.

Lapsley, Phil, Jeff Bier, Amit Shoham, and Edward A. Lee. *DSP Processor Fundamentals: Architecture and Features.* Piscataway, NJ: IEEE Press, 1997

Lee, E. A. "Programmable DSP Architectures: Parts I." *IEEE Transactions on Acoustics, Speech, and Signal Processing.* ASSP Magazine (October 1988), pp. 4–19.

Lee, E. A. "Programmable DSP Architectures: Parts II." *IEEE Transactions on Acoustics, Speech, and Signal Processing.* ASSP Magazine (January 1989), pp. 4–14.

Levy, Markus. "EDN's 1997 DSP-Architecture Directory." *EDN* (May 8, 1997), pp. 43–107.

Levy, Markus. "Microprocessors and DSP Technologies Unite for Embedded Applications." *EDN* (March 2, 1998), pp. 73–81.

Levy, Markus. "EDN's 1998 DSP-Architecture Directory." *EDN* (April 23, 1998), pp. 40–111.

Madisetti, Vijay K. *VLSI Digital Signal Processors.* Boston: Butterworth-Heinemann, 1995.

Marven, Craig, and Gillan Ewers. *A Simple Approach to Digital Signal Processing.* Texas Instruments, 1994.

New, Bernie. "A Distributed Arithmetic Approach to Designing Scalable DSP Chips." *EDN* (August 17, 1995), pp. 107–114.

Oppenheim, A. V., and R. W. Schafer. *Digital Signal Processing.* Englewood Cliffs, NJ: Prentice-Hall, 1988.

Parks, T. W., and C. S. Burrus. *Digital Filter Design.* New York: John Wiley & Sons, 1987.

Schneiderman, Ron. "Faster, More Highly Integrated DSPs—Designed for Designers." *Wireless Systems Designs* (November 1996), pp. 12–13.

Thompson, J., and S. Tewksbury. "LSI Signal Processor Architecture for Telecommunications Applications." *IEEE Transactions on Acoustics, Speech, and Signal Processing* ASSP-30, no. 4 (August 1982), pp. 613–632.

CHAPTER 6

Optoisolators

6.1 Introduction

Since its introduction over 20 years ago, the classic optically coupled pair consisting of a photo emitter and a photo detector in one package has found its way into a multitude of applications. The ubiquitous "opto" provides galvanic isolation against kilovolts of input/output voltage differential and does so without the complexity and other problems of electromechanical or magnetic components.

Optocouplers are available in several possible output configurations, including LDRs (light-dependent resistors), various transistor types, logic elements, thyristors and their variations, and even photovoltaic output elements. This chapter presents the basic characteristics and applications of optos in practical circuits.

6.2 Light-Emitting Diodes and Photosensors

6.2.1 Light-Emitting Diodes

Key to the operation of an optocoupler is the emitter (normally, a light-emitting diode), which generates the light energy, and the photosensitive silicon detector, acting as the output device. Present light-emitting diodes (LEDs) are fabricated from gallium arsenide (GaAs), gallium arsenide phosphide (GaAsP), gallium phosphide (GaP), or gallium aluminum arsenide (GaAlAs). LEDs are mass produced in red, super-red, yellow, and green. Blue light-emitting diodes based on silicon carbide (SiC) are entering the market gradually. GaAs emits infrared radiation around 900 nm; GaAlAs emits red light between 650 and 670 nm; GaP emits green light between 520 and 570 nm or red light between 630 and 790 nm; GaAsP emits light over a broad range from green to infrared depending on the percentage of phosphorus in the material. SiC is the only material that allows reproducible P and N doping and possesses a suitable bandgap for the emission of blue light. Very recently white LEDs suitable for illumination applications have been released by Infineon Technologies (former Siemens) based on GaN technology (OSRAM, 1999).

6.2.2 Photosensors

The basic types of photosensors used in optoisolators are photoconductive bulk effect, photoconductive junction, and photovoltaic.

6.2.2.1 Photoconductors

Photoconductive bulk effect cells normally are made of cadmium sulfide (CdS) or cadmium selenide (CdSe). They have no junctions. The entire layer of material changes in resistance when it is illuminated. In this respect it is analogous to a thermistor, except that the heat is replaced by light. The photoconductive cell decreases in resistance as the light level increases and increases in resistance as the light level decreases. The absolute value of resistance of a particular cell at a specific light level depends on the photosensitive material being used, cell size, electrode geometry, and the spectral composition of the incident light. Cadmium sulfide and cadmium selenide are the two materials most widely used in photoconductive cells.

6.2.2.2 Photoconductive Junction Sensors

Photodiodes and phototransistors represent the junction-type photoconductors. The resistance across the semiconductor junction changes as a function of light falling on it. They are very fast in response but limited in sensitivity due to the small area of the junction.

Photodiodes are similar to solar cells in that, when light strikes the PN junction, the junction develops a voltage and therefore a current when it is connected to a circuit. The current and voltage output vary with the light intensity, and the output is very linear over several decades of light intensity. The silicon diode is sensitive through the visible spectrum and into the near infrared; however, its greatest sensitivity is in the infrared range.

A phototransistor is basically a photodiode with another junction added. This results in a light-sensitive solid-state unit that permits the transistor to be biased to detect a light signal at a specific level or permits the use of speed-up circuitry to increase the speed of response. The speed of response of a phototransistor is a lot slower than that of a regular transistor because the phototransistor has a larger area for the collection of light than the regular transistor. Thus, the additional capacitance of the larger area reduces the response time.

A photo-Darlington is basically a phototransistor internally coupled to a second transistor. The emitter of the phototransistor feeds the base of the second transistor to gain increased sensitivity. However, the gain in sensitivity results in a slower response, which can make the Darlington phototransistor too slow for switching logic circuits.

6.2.2.3 Photovoltaic Sensors

The photovoltaic type generates a voltage across a PN junction as a function of the photons impinging on it. This class is usually made of selenium or silicon and is the only self-generating type, requiring no external power supply. International Rectifier's PVA and PVD series are examples (International Rectifier, 1996). This series is designed to replace electromechanical relays.

6.3 Optoisolators

An optoisolator combines a photoconductor or a phototransistor with a high-quality, long-life light source in an encapsulated package that is light tight. The combination of various photosensors and light sources is available in a wide variety of packages. The main advantage to the use of an optical coupling device is that switching or variations in a circuit can be made without generating electrical noise. For example, if the light output of the source is varied by a potentiometer, the noise generated by the wiper of the potentiometer is not transmitted through the light beam. Figure 6-1 shows different optocoupler configurations and the relative efficiency of silicon detectors versus different emitters.

Optocouplers such as IL1 from Siemens consist of a GaAs infrared-emitting diode and a silicon phototransistor mounted in a single package.

When forward current (I_F) is passed through the GaAs diode, it emits infrared radiation peaking at about 900 nm wavelength. This radiant energy is transmitted through an optical coupling medium and falls on the surface of the NPN phototransistor.

Phototransistors are designed to have a large base-collector junction area and a small emitter area. Some fraction of the photons that strike the base area cause the formation of electron-hole pairs in the base region. This fraction is called the *quantum efficiency* of the photodetector. If we ground the base and emitter (Figure 6-2(a)) and apply a positive voltage to the collector of the phototransistor, the device operates as a photodiode. The high field across the collector base junction quickly draws the electrons across into the collector region. The holes drift toward the base terminal, attracting electrons from the terminal. Therefore, a current flows from collector to base, causing a voltage drop across the load resistance (R_L). The high junction capacitance, C_{cb}, results in an output circuit time constant $R_L C_{cb}$, with a corresponding output voltage rise time. The output in this configuration is quite small and hence this connection normally is not used.

The most common circuit configuration is an open base connection (Figure 6-2(b)). With this connection, the holes generated in the base region cause the base potential to rise forward biasing the base-emitter junction. Electrons then are injected into the base from the emitter, trying to neutralize the excess holes. Because of the close proximity of the collector junction, the probability of an electron recombining with a hole is small and most of the injected electrons are immediately swept into the collector region. As a result, the total collector current

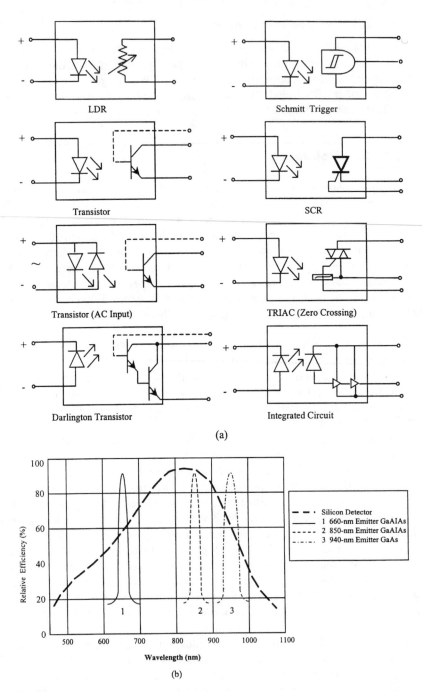

FIGURE 6-1 Different kinds of optoisolators and the relationship between efficiency and wavelength in emitters and silicon detectors: (a) a variety of optocoupler configurations, (b) relative efficiency and wavelength

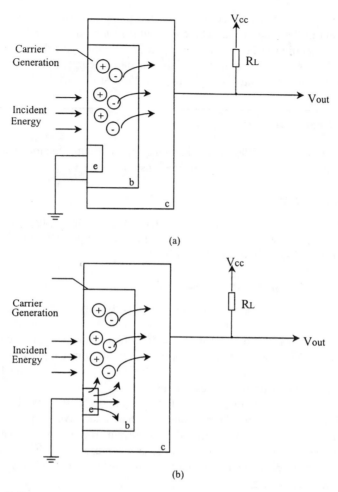

FIGURE 6-2 Effect of incident energy on the phototransistor: (a) emitter and base grounded, (b) base open

is much higher than the photogenerated current and in fact is amplified β times. The total collector current is several hundred times greater than for the previous connection.

This gain comes with a penalty of much slower operation. Any drop in collector voltage is coupled to the base due to the collector/base capacitance which tends to turn off the injected current. The only current available to charge this junction capacitance is the original photocurrent. Thus, the rate of change of the output voltage is the same for both the diode and transistor connections. In the latter case, the voltage swing is β times as great, so the total rise time is β times as great as for the diode connection. Therefore, the effective output time constant is $\beta R_L C_{cb}$. For the IL1, a typical two-wavelength rise time for 100 Ω results.

The ratio of the output current from the phototransistor (I_C or I_E) to the input current in the diode is called the *current transfer ratio* (CTR). For the IL1, the CTR is specified at 20% minimum with 35% being typical at $I_F = 10$ mA. So, for a 10 mA input current the minimum output current is 2 mA. Another important parameter is that V_F typically is 1.25 V at 60 mA I_F.

6.4 Practical Circuits

Optos have many different applications. The following sections examine the most common digital and analog applications.

6.4.1 Digital Interfaces

In digital circuits, most common applications are output sensing circuits and input driving elements. We discuss some of the most important considerations in output sensing circuits and the input drivers.

6.4.1.1 Output Sensing Circuits

The output of the phototransistor can directly drive the input of standard logic circuits such as the 7400 TTL families. The worst case input current for the 74 series gate is -1.6 mA for $V_{IN} = 0.4$ volts. This easily can be supplied by the output of an optocoupler such as the IL1 with 10 mA input to the infrared diode. Figure 6-3(a) shows a case with active level low circuit, and for higher speed, a smaller pull-up resistor can be used.

It is more difficult to operate into TTL gates in the active level high configuration. Figure 6-3(b) shows the best method when a negative supply is available. The circuit in Figure 6-3(c) requires 10 mA current from the optotransistor, with some sacrifice of the noise margin. The case in Figure 6-3(d) has a high sensitivity, needs extra parts, and still sacrifices the noise margin. Figure 6-3(e) also is a high-sensitivity circuit with extra parts.

Several optocoupler output transistors can be connected to perform logical functions. Figure 6-4 shows logical OR and logical AND connections, respectively.

6.4.1.2 Input Driving Circuits

The input side of an optocoupler such as the IL1 has a diode characteristic like that shown in Figure 6-5. The forward current must be controlled to provide the desired operating condition. The input can be conveniently driven by integrated circuit logic elements in a number of different ways. A few examples are given in Figure 6-6. The series resistor in Figure 6-6(a) can be omitted for about 15 mA into the diode.

Obviously, many other ways may be used to drive the device with logic signals but the most common needs can be met with these circuits. All provide 10 mA into the LED, yielding a 2 mA minimum out of the phototransistor. The

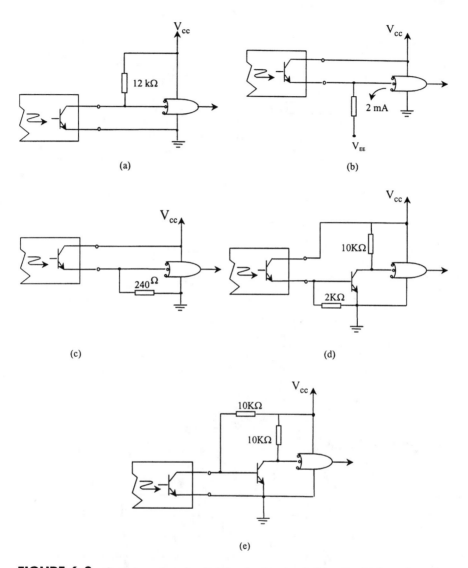

(a)

(b)

(c)

(d)

(e)

FIGURE 6-3 Output sensing circuit: (a) active low level, (b) active high configuration when a negative rail is available, (c), (d), and (e) single power rail cases

1 V diode knee and its high capacitance (typically 100 pF) provide good noise immunity. The rise time and propagation delay can be reduced by biasing the diode to about 1 mA forward current, but the noise performance will be worse. These circuits have various advantages over other ways of doing the task, such as the use of relays, pulse transformers, and integrated circuit line drivers or receivers. Further details can be found in Siemens (1995–1996).

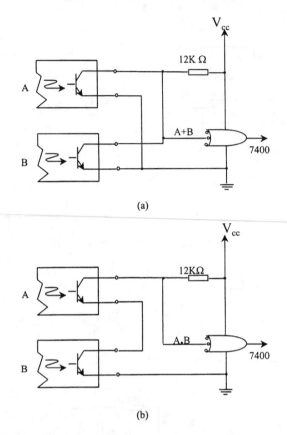

(a)

(b)

FIGURE 6-4 Logical functions: (a) OR connection, (b) AND connection

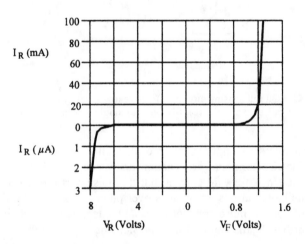

FIGURE 6-5 Input diode characteristics of an optoisolator

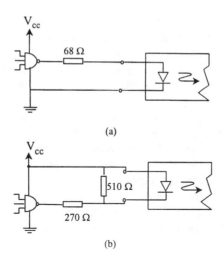

(a)

(b)

FIGURE 6-6 Input drive circuits: (a) active high type, (b) active low type

6.4.2 Linear Applications

The curve of input current vs. output current for optoisolators is somewhat nonlinear, because of the variation of β with current for the phototransistor and the variation of infrared radiation out vs. forward current in the GaAs diode. The useful range of input current is about $1-100$ mA (for devices such as the IL1 from Siemens), but higher currents may be used for short-duty cycles. For linear applications, the LED must be forward biased to some suitable current (usually $5-20$ mA). Modulating signals then can be impressed on this DC bias. As shown in Figure 6-7, a differential amplifier is a good way to accomplish this.

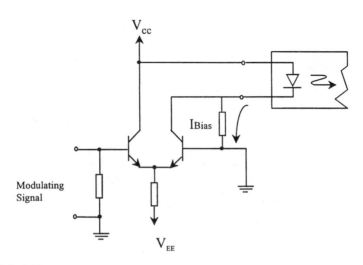

FIGURE 6-7 A differential amplifier to drive the input diode

Sensing in-linear applications can be done in several ways, depending on the requirements. For high-frequency performance, the phototransistor should be operated into a low-impedance input current amplifier. The simplest such scheme is a grounded base amplifier, as in Figure 6-8. The circuit will work equally well either way with a phase inversion between the two. Obviously, a PNP transistor would work as well.

A feedback amplifier also could be used to get a low-impedance input, as in Figure 6-9. For example, if $R1 = 900\ \Omega$, $R2 = 100\ \Omega$, and $V_{CC} = 5$ V, we would have a current gain of 10 and an input impedance of about 6.3 Ω. This would give a considerable speed improvement over a 100 Ω load.

A high-speed operational amplifier could be used to give excellent performance, as in Figure 6-10. Note that, in all cases, the output can be taken from either the collector or the emitter of the phototransistor, depending on the polarity desired. The operating speed is the same in either case.

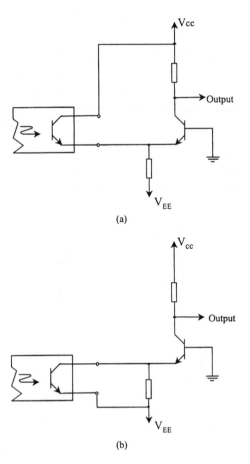

(a)

(b)

FIGURE 6-8 A phototransistor used with grounded base amplifier

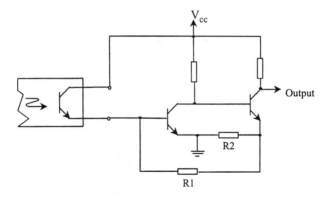

FIGURE 6-9   Use of a feedback amplifier

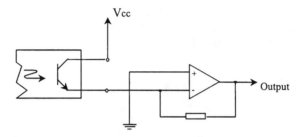

FIGURE 6-10   Use of a high-speed operational amplifier

The design engineer will see many ways to expand on these circuits to achieve his or her goals. The devices are extremely versatile and can provide better ways to share many system problems than competing components. Special designs are possible to optimize certain parameters, such as coupling capacitance or transfer ratio.

6.5 Driving High-Level Loads with Optocouplers

Frequently, a load to be driven by an optocoupler requires more current, voltage, or both than an optocoupler can provide at its output. Available optocoupler output current is found by multiplying the input LED current by the CTR. For worst-case design, the minimum specified value should be used. Temperature derating usually is unnecessary over the 0–60°C range because the LED light output and transistor β have approximately compensating coefficients (for components similar to IL1). Multiplying the minimum CTR by 0.9 would ensure a safe design over this temperature range. More margin would be required over a wider range.

The LED source current is limited by its rated power dissipation. Table 6-1 shows the maximum allowable I_F vs. maximum ambient temperature. The values

TABLE 6-1 Maximum forward diode current vs. temperature for the IL1, based on a derating factor of 1.33 mW/°C. (Reproduced by permission of Infineon Technologies.)

Maximum Temperature (°C)	I_F Maximum (mA)
40	50
60	35
80	17

for Table 6-1 are based on a 1.33 mW/°C derate from the 100 mW at 25°C power rating.

Based on the information in Table 6-1 and allowing a 10% margin for temperature effects, the minimum available output current for the IL1 will be 6.3 mA.

If the IL1 is being operated from logic with a 5 V driving transistor and 0.2 V V_{CE} saturation is assumed for the driving transistor, a R_{IF} resistor will provide the 48 mA. The forward voltage of the IR-emitting LED is about of $75R$ 1.2 V. Figure 6-11 shows two such drive circuits.

A "buffer gate" such as the SN7440 provides a very good alternative to discrete transistor drivers. Figure 6-12 shows how this is done. Note that the gate is used in the "current-sinking" rather than the "current-sourcing" mode. In other words, conventional current flows into the buffer gate to turn on the LED, because the TTL gate will sink more current than it will source. The SN7440 is specified to drive thirty 1.6 mA loads or 48 mA. Changing R_{IF} from 75 Ω to 68 Ω adjusts for the higher saturation voltage of the monolithic device.

6.5.1 Higher-Load Current

For load currents greater than 6.3 mA, a current amplifier is required. Figure 6-13 shows two single-transistor current amplifier circuits.

Since the transistor in the optocoupler is treated as a two-terminal device, no operational difference exists between the NPN and the PNP circuits. R_b provides a return path for I_{CBO} of the output transistor. Its value is $R_b = 400$ mV/$I_{CBO}(T)$, where $I_{CBO}(T)$ is found for the highest junction temperature expected.

It is necessary to use the maximum dissipated power, the specified maximum junction-to-ambient thermal resistance, and the maximum design ambient temperature in conjunction with the specified maximum I_{CBO} at 25°C to calculate $I_{CBO}(T)$, assuming that leakage currents double every ten degrees.

As an example, suppose a 2N3568 is used to provide a 100 mA load current. Also assume a maximum steady-state transistor power dissipation of 100 mW and a 60°C maximum ambient temperature. The transistor junction-to-ambient thermal resistance is 333°C/W, so a maximum junction temperature of 93°C (60°C + 33°C) is expected. This is about seven decades above 25°C. Therefore, $I_{CBO}(T) = I_{CBO}(\text{max}) \times 27 = 50$ mA $\times 128 = 6.5$ μA. A safe value for R_b is 400 mV/6.5 μA = 62 kΩ.

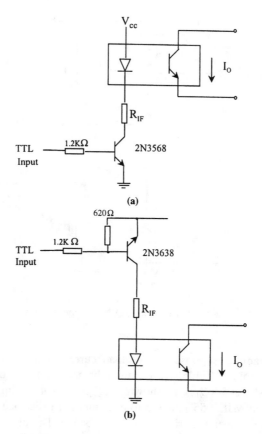

FIGURE 6-11 Driving the emitter from TTL circuits: (a) NPN driver, (b) PNP driver

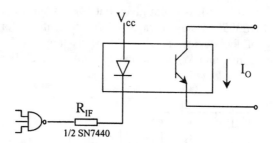

FIGURE 6-12 A buffer gate drive

Working backward, the maximum base current under load will be $I_O/h_{FE}(\text{min}) = 100 \text{ mA}/100 = 1 \text{ mA}$. The current in R_b is $V_{BE}/R_b = 600 \text{ mV}/60 \text{ k}\Omega = 10 \text{ μA}$, which is negligible. An IL1 with a 9 mA drive would operate effectively. If the load requires more current than can be obtained with the highest β transistor available, then more than one transistor must be used in

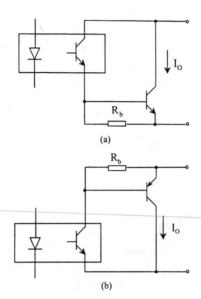

(a)

(b)

FIGURE 6-13   Current booster circuits: (a) NPN type, (b) PNP type

cascade. For example, suppose 3 A of load current and 10 W dissipation are needed. A Motorola MJE3055 (Q_2) might be used for the output transistor, driven by an MJE205 (Q_1), as shown in Figure 6-14. Using a 5°C/W heat sink and the rated MJE3055 junction-to-case thermal resistance of 1.4°C/W, we find that junction temperature rise is 64°C (6.4 × 10). Therefore maximum junction temperature is 124°C. This is ten decades above 25°C, making $I_{CBO}(T) = 2^{10}I_{CBO}(\text{max}) \simeq 10^3 I_{CBO}(\text{max})$.

$I_{CBO}(\text{max})$ at 30 V or less is not given but I_{CEO} is specified at a maximum of 0.7 mA for MJE3055. Using a value of 20 (for safety) for the minimum low-current h_{FE} of the device, I_{CBO} could be as large as $I_{CEO}/20 = 35\mu\text{A}$. Then $I_{CBO}(T)$ is 35 mA and $R_{b2} = 400 \text{ mV}/35 \text{ mA} = 11 \ \Omega$. For I_b use I_O/h_{FE}

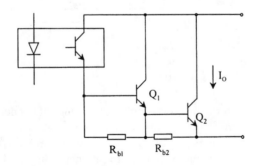

FIGURE 6-14   Two NPN current boosters

(min at 4 A) = 3 A/20 = 150 mA. I_{Rb2} = 600 mV/10 Ω = 60 mA, so $I_{E(Q1)}$ = 210 mA.

Maximum power in Q_1 will be about 1/14th the power in Q_2, since its current is lower by that ratio and the two collector-to-emitter voltages are nearly the same. This means Q_1 must dissipate 700 mW. Assuming a small "flag-type" heat sink having 50°/W thermal resistance, we find the junction temperature at about 95°C. The 150°C case temperature I_{CBO} rating for this device is 2 mA, so one can work backward and assume about 1/30 of this value, or 70 μA. On the other hand, the 25°C-rated I_{CBO} is 100 μA. Choosing the larger of these contradictory specifications, R_{b1} = 400 mV/0.1 mA = 4k ≈ 3.9 k. The Q_1 base current is $I_{E(Q1)}/h_{FE(Q1-min)}$ = 210 mA/50 = 4.2 mA. Total current is $I_{b(Q1)}$ + I_{Rb1} = 4.2 + 0.24 = 4.5 mA. An IL1 could be used here.

6.5.2 Higher-Load Voltages

All the current gain circuits shown so far have one common feature: The load voltage is limited by the voltage rating of the optocoupler and not by the voltage or power rating of the transistor(s). Figure 6-15(a) shows a method of overcoming this limitation. This circuit will stand off BV_{CEO} of Q_1. The voltage rating of the phototransistor is irrelevant since its maximum collector-emitter voltage is the base-emitter voltage of Q_1 (about 0.7 V). Unlike the Darlington configurations shown previously, this circuit operates "normally-on." When no current flows in the LED and the phototransistor is off, R_1 current is allowed to flow into the base of Q_1, turning on Q_1. When the optocoupler is energized, its phototransistor "shorts out" the R_1 current, turning off Q_1.

The value of R_1 depends only on the load-supply voltage ($V_{cc+} - V_{cc-}$) and the maximum required base current for Q_1. This is derived from the minimum β of Q_1 at minimum temperature and the load current. The required current-drive capability is the same as I_{R1}, since I_{R1} changes negligibly when the circuit goes between its on and off states.

In some applications either more current gain will be required than one transistor can provide or the power dissipated in R_1 will be objectionable. In these cases, use of Darlington high-voltage boosters (as shown in Figure 6-16(a)) is a solution. If more than one load is being driven and the negative terminals must be in common, it is necessary to use the PNP circuit, as per Figure 6-16(b). Otherwise, the NPN is better because the transistors cost less. Performance characteristics of the NPN and PNP versions are identical if the device parameters are the same.

6.5.3 Higher Speed

Figure 6-17 shows a typical circuit employing an optocoupler to transmit logic signals between electrically isolated parts of a system. In the circuit shown, the optocoupler must "sink" the current from one TTL load plus a pull-up resistor to V_{CC}. The resistor in series with the LED of the optocoupler must supply the worst-case load current divided by the CTR of the optocoupler. If an optocoupler

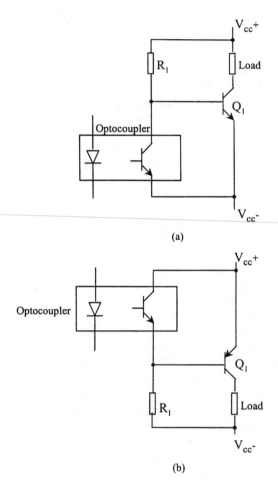

FIGURE 6-15 High-voltage boosters: (a) NPN, (b) PNP

is used with a minimum CTR of 0.2 and 80% variation in the load is allowed, 8.1 mA is required. This is supplied by the 430 Ω resistor.

The maximum repetition rate at which this circuit will operate is only about 8 kHz. This severe limitation is due entirely to the characteristics of the photo-transistor half of the optocoupler. The device has a large base-collector junction area and a very thick base region to make it sensitive to light. C_{ob} typically is 25 pF. This capacitance in the circuit of Figure 6-17 is effectively multiplied by a large factor due to the "Miller effect." Also, because the base region volume is large, base storage time is large.

A very simple method of reducing both effects is to add a resistor between the base and the emitter, as shown in Figure 6-18. This resistor helps by reducing the time constant due to C_{ob} and removing stored charge from the base region

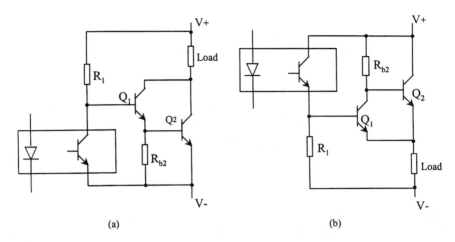

(a)   (b)

FIGURE 6-16   Darlington high-voltage boosters: (a) NPN, (b) PNP

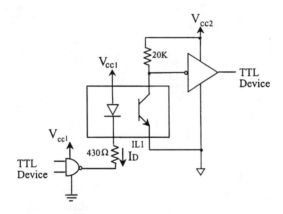

FIGURE 6-17   Low-speed isolation stage for TTL applications. (Reproduced by permission of Infineou Technologies.)

faster than recombination. When a base-emitter resistor is used, the required LED drive is increased, since much of the photo current generated in the base-collector junction is deliberately "dumped."

Using this method usually does not result in a large power supply current drain, since the average repetition rate is low in most applications. As the drive is increased and R_{BE} is reduced, turn-on and turn-off time both decrease. The total amount of charge stored can be reduced by decreasing the LED drive pulse duration. Also, as higher drive levels are used, the load resistance, R_L, can be reduced to further enhance the speed of the circuit. These parameters are related to each other such that all should be changed together for best results. One important generalization can be made concerning their interdependence. The LED drive pulse duration, $-T_{in}$, output fall time (t_f), output rise time (t_r), and propagation

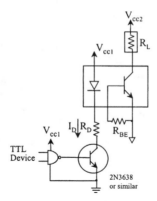

FIGURE 6-18 A higher-speed isolation stage for TTL applications. (Reproduced by permission of Infineou Technologies.)

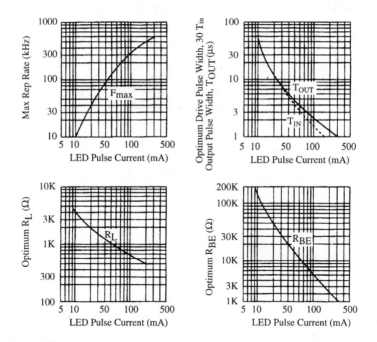

FIGURE 6-19 Parameters vs. LED pulse current. (Reproduced by permission of Infineou Technologies.)

delay (t_p) should occur in a 1.5:1:1:1 ratio, approximately. If this relationship does not occur, the circuit will not operate at as high a repetition rate as it could at the same drive level. Output pulse duration, T_{out}, equals T_{in} at low currents but stretches out at high currents.

Figure 6-19 shows graphs relating the important parameters for a typical optoisolator such as the IL1. The optimum values of T_{in}, R_{BE}, and R_L are shown vs. LED pulse current, as are the resulting output pulse width and maximum full-swing frequency. Rise, fall, and propagation times can be read as two-thirds of T_{in}. Figure 6-19 shows that increasing drive to 200 mA and using optimum R_{BE} and R_L will increase the maximum repetition rate from 3 kHz to 500 kHz, a 167:1 improvement.

Lower-grade optocouplers will behave similarly if the LED drive level is scaled appropriately to allow for a lower CTR. Another method of increasing speed is to operate the phototransistor as a photodiode. In this method, the bias voltage is supplied between the collector and base terminal, the emitter being unused. Operation to at least 1.0 MHz is possible this way, but external amplification is necessary. Figure 6-20 is a graph showing peak output current vs. drive pulse duration for 200 mA peak drive current.

Since output current is small, some type of wide bandwidth amplifier must be employed to drive TTL loads. One simple solution for intermediate-speed

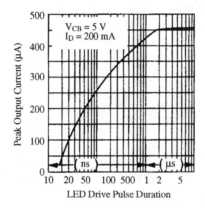

FIGURE 6-20 Diode mode output current vs. drive pulse duration. (Reproduced by permission of Infineou Technologies.)

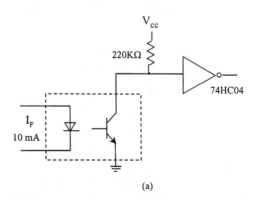

FIGURE 6-21 Circuits for intermediate-speed operation: (a) use of a MOS inverter, (b) comparator based

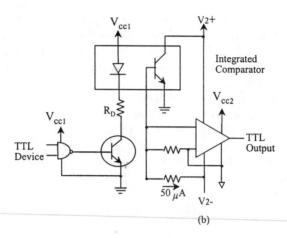

(b)

FIGURE 6-21 Continued

operation is the use of a MOS inverter (1/6 74HC04), as per Figure 6-21(a). Another device that will provide a good interface is an integrated comparator amplifier, as in Figure 6-21(b).

6.6 Photovoltaic Devices

6.6.1 Photovoltaic Isolators

A photovoltaic isolator (PVI) generates an electrically isolated voltage on a receipt of an input signal. Conventional photocouplers merely modulate the resistance of an output device such as a transistor, diode, or resistor. Such photocouplers require a separate voltage source to detect the presence of an input signal. In contrast, a PVI actually transmits (and transforms) energy across the isolation barrier and directly generates an output voltage. This DC voltage, available at a 2500 VAC isolation level, gives circuit designers a new and uniquely useful electronic component.

The input of the PVI is an LED optically coupled to, but electrically isolated from, the output. A GaAlAs LED is used for high output and maximum stability. The infrared emission from the LED energizes, by photovoltaic action, a series connection of silicon PN junctions. A unique alloyed junction stack, which is edge illuminated, is used to form the output photovoltaic generators. This novel structure produces extremely high operating efficiency.

A PVI can serve as an isolator, coupler, or isolated voltage source. As an isolator, the PVI can be the key component in a solid-state relay circuit. The PVI is ideally suited for driving power MOSFETs or sensitive gate SCRs to form solid-state relays. As a coupler, the PVI can sense a low-level DC signal and transmit a voltage signal to an electrically remote circuit. As a voltage source, the PVI can function as a "DC transformer" by providing an isolated, low-current

DC source for basing or supplying power to low quiescent current electronic devices.

An example of these are the PVI series microelectronic isolators from International Rectifier. The PVI 5100, PVI 5050, and PVI 1050 are typical devices. These units are available with single (PVI 5100, PVI 5050) or dual (PVI 1050) 5 V output, which can be series connected to produce 10 V. Figure 6-22 depicts characteristics of these devices.

Recent developments in semiconductor technology have led to the design of a new type of solid-state relay, which combines photovoltaic isolation with MOSFET power-integrated circuit techniques.

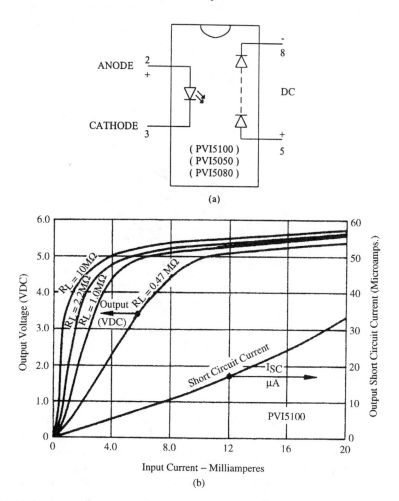

(a)

(b)

FIGURE 6-22 PVI device characteristics: (a) schematic diagrams, (b) typical output characteristics (PVI 5100), (c) typical variation of output with temperature, (d) typical response time (PVI 5100), (e) photovoltaic relays. (Reproduced by permission of International Rectifier.)

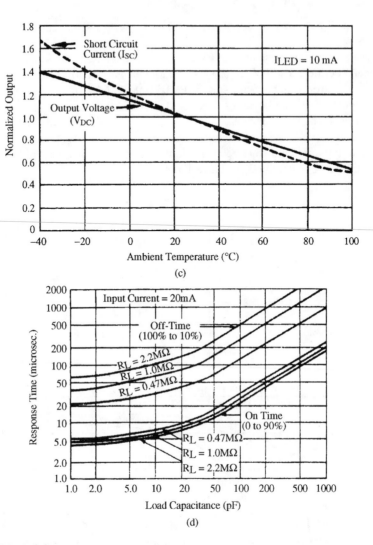

(c)

(d)

FIGURE 6-22 Continued

This new topology, photovoltaic relay (PVR), which evolved recently, is illustrated in Figure 6-23. The PVR topology achieves electro-optical isolation by means of a light-emitting diode energizing a photovoltaic generator (PVG) consisting of a series connection of silicon P-N junctions. The signal from the photovoltaic generator in turn activates a bidirectional MOSFET configuration. A PVR configuration achieves a unique combination of operating advantages not present in any other relay. The PVR has the solid-state advantages of long switching life, high operating speed, low pickup power, bounce-free operation, noninductive input, insensitivity to position and magnetic fields, extreme shock

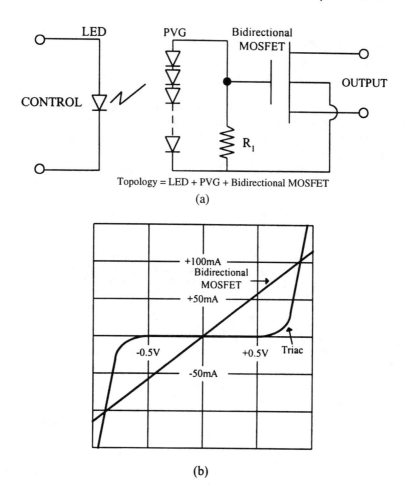

Topology = LED + PVG + Bidirectional MOSFET

(a)

(b)

FIGURE 6-23 Photovoltaic relay: (a) basic schematic, (b) solid-state output characteristics. (Reproduced by permission of International Rectifier.)

and vibration resistance, and miniaturization. In addition, modern MOSFET technology provides a much better equivalent (of an analog electromechanical) switch than thyristor or bipolar transistor technology used dominantly as the output contact in previous solid-state relays. Relative to thyristors, the MOSFET displays a linear on-resistance rather than a 0.6 V threshold in forward conduction, as shown in Figure 6-23(b). An inverse series connection of two MOSFETs can switch DC or AC at frequencies well into the RF range. Static and commutating dv/dt effects are not inherent and turn-off can be instantaneous. Relative to bipolar transistors, MOSFETs display lower on-state offset voltages, much lower off-state leakages, and, most important, have essentially infinite static forward current gain (i.e., MOSFETs are voltage controlled).

6.7 Conclusion

This chapter is an introduction to the use of optocouplers as a low-cost solution to achieve galvanic isolation between different circuits and provide sufficient isolation voltage between circuits. Information is based mostly on literature from the Seimens' (1995–1996) Optoelectronics Division (Infineon Technologies is the present name of the company). Using the fundamentals presented here, a designer should be able to use many different versions of optoisolator circuits for different applications and speeds. The following references provide more detail.

References

Clairex Electronics. *Optoelectronic Designers Handbook*. Clairex Electronics, New York, USA, 1986.
Collins, Bill. "The Photovoltaic Relay: A New Solid State Control Device." Application Note AN-104. International Rectifier-Microelectronic Relays, 7th ed., 1996, pp. E–9 to E–14.
International Rectifier. *Microelectronic Relays Data Book*, 7th ed. International Rectifier, USA, 1996.
Osram,: "LW T676. Hyper TOPLED/White LED Datasheet"; (1999-05-03).
Pryce, Dave. "Advances in Speed and Voltage Ratings Enhance Applications for Optocouplers." *EDN* (April 28, 1988), pp. 75–82.
Siemens. *Opto Electronics Data Book*. Siemens, CA, USA, 1995–1996.
Woodward, W. S. "The Many Analog Uses for Optical Isolators." *Electronic Design* (April 17, 1997), pp. 101–108.

CHAPTER 7

Sensors

7.1 Introduction

Sensors convert information about the environment such as temperature, pressure, force, or acceleration into an electrical signal. With the development of microelectronic technology with silicon as the base material in the 1970s, sensors using the properties of silicon entered the component market. Silicon's physical properties make it an ideal building material for mechanical devices. Silicon has the hardness of steel, the thermal conductivity of diamond, piezoresistive properties, a light weight, and low thermal expansion; also, it is relatively inert. It is free of hysteresis and its crystalline structure is well suited to the fabrication of miniature precision products. Silicon micromechanical products have several advantages over their conventionally manufactured counterparts: They generally are much smaller, their performance is higher because of the precise dimensional control in the fabrication, and costs are lower due to the possibility of mass-scale production.

Silicon micromachining is a powerful outgrowth of semiconductor process technology, whereby integrated circuit manufacturing techniques are supplemented by the silicon etching process to create very precise micromechanical structures. These silicon microstructures can have electronic features that allow conversion of physical input into electrical signals. Similarly, electrical signals can be applied to these devices to provide control functions. Initially developed in the 1950s and 1960s at leading semiconductor pioneers including Fairchild and National Semiconductor, the technology was further advanced in the 1970s at universities throughout the world. Commercial activities picked up in the early 1980s, with a number of startups located in the Silicon Valley area.

By the beginning of the 1980s, designers were able to incorporate integrated circuits on a single die with the sensor elements. Although this complicates

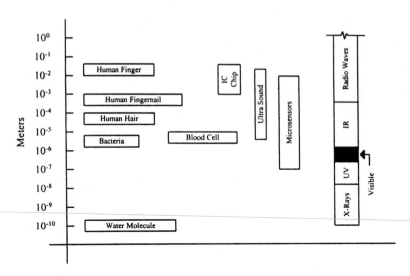

FIGURE 7-1 Comparative scale of microsensors

the fabrication process and can limit the operating-temperature range for the sensor, it often leads to superior performance at an acceptable cost. These integrated microsensors can provide a more linear output than that of the sensor itself or an output having a digital format that can readily be handled by associated data-logging or display systems. By the late 1980s, microsensors for measuring pressure, temperature, and the like were readily available, while silicon accelerometers and so forth were entering the market. Figure 7-1 shows the comparative scale of microsensors.

Nowadays, miniaturization is the aim of many research laboratories and companies. As a part of microsystem technology, sensors also will play a major role in the future and sensor interfaces and related standards are getting ready for this developing component sector. Many producers in Japan, Europe, and the United States forecast growth rates for sensors above 10% beyond the year 2000.

This chapter is a summary of modern semiconductor sensors, their characteristics, and applications with some representative devices.

7.2 The Properties of Silicon and Their Effects on Sensors

Silicon is a suitable material for sensor technology as it manifests sufficient physical and chemical effects of an acceptable strength to use in uncomplicated structures across a wide range of temperatures. Table 7-1 presents the most important effects and their applications for sensor technology.

The use of silicon has a number of implications for sensors. First, the physical properties of silicon can be used directly to measure the desired dimension, as indicated in Table 7-1. However, the range of possibilities is limited. Beyond

TABLE 7-1 The effect of silicon used in sensors

Physical Dimension	Effect	Application
Radiation	Photoresistive	Photoresistor
	Photointerface	Photodiode, phototransistor
	Ionization	Nuclear radiation sensor
	Photocapacitive	Photocapacitance
Mechanical	Piezoresistive, piezojunction, and piezotunnel	Piezoresistive power and pressure sensors, piezoelectric diode and transistor
Thermal	Thermal resistance	Resistance temperature sensors
	Thermojunction	Temperature sensors (diode, transistor)
	Thermoelectric	Thermopile
	Pyroelectric	Pyroelectric sensor
Magnetic signals	Magnetoresistive	Magnetoresistive sensors
	Hall	Hall generator
	Magnetic interface	Magnetic diode and transistor
Chemical signals	Charge-sensitive field	ISFET

this, for example, silicon can be extremely useful when used as the substrate for thin-film sensors, even when information processing electronics are integrated. For details, see Hauptmann (1991).

7.3 Micromechanics

The term *micromechanics*, with its obvious similarity to the term *microelectronics*, is used to describe a completely new discipline. Its objective is the construction of complex microsystems consisting largely of integrated sensors, a logical signal processing stage, and actuators. In this connection, *micromechanics* refers to the fabrication of mechanical structures whose geometrical size, at least in one dimension, is so small that it no longer is sensible to use the methods of fine mechanics. Depending on the boundary conditions imposed by the desired function or the properties of the material, this limit may be located anywhere between the millimeter and the submicrometer range (see Figure 7-2). In contrast to microelectronics, micromechanics is concerned with the production of three-dimensional structures.

Modern micromechanics make it possible to produce micropumps, microvalves, micro-loudspeakers, and microphones; therefore, it is of interest to disciplines other than sensor technology (Hauptmann, 1991; Guckel, 1992).

7.4 Temperature Sensors

The most common electronic temperature measurement devices currently available include the thermocouple, the resistance temperature detector, the thermistor, and the integrated circuit temperature transducer. All have

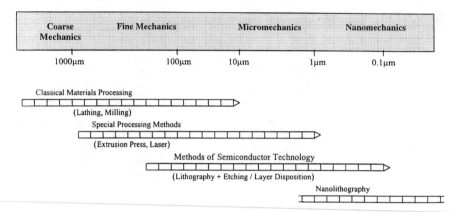

FIGURE 7-2 The size of micromechanics

associated application benefits and limitations, which are delineated in Table 7-2 (Microswitch, 1997a, pp. A1–A10).

7.4.1 Resistance Temperature Detectors

Resistance temperature detectors (RTD) are wire windings or thin-film serpentines that exhibit changes in resistance with changes in temperature. While metals such as copper, nickel, and nickel-iron often are used, the most linear, repeatable, and stable RTDs are constructed from platinum.

7.4.2 Negative Temperature Coefficient Thermistors

Negative temperature coefficient (NTC) thermistors are composed of metal oxide ceramics, low in cost, and the most sensitive temperature sensors. They also, however, are the most nonlinear and have a negative temperature coefficient. Thermistors are offered in a huge variety of sizes, base resistance values, and $R - T$ curves to facilitate both packaging and output linearization schemes.

7.4.3 Thermocouples

Thermocouples consist of two dissimilar metal wires welded together at both ends to form two junctions. Temperature differences between the junctions cause a thermoelectric potential (i.e., a voltage) between the two wires. By holding the reference junction at a known temperature and measuring this voltage, the temperature of the sensing junction can be deduced.

Thermocouples have very large operating temperature ranges despite a very small size. However, they have low output voltages, susceptibility to noise pickup by the wire loop, and relatively high drift. Silicon integrated circuits are available for interface with the thermocouples. Some examples are AD-594 and AD-595 from Analog Devices (Le Fort and Ries, AN-274; Marcin, AN-369).

TABLE 7-2 A comparison of thermal sensors. (Reproduced by permission of Microswitch Honeywell Inc.)

Characteristic	Platinum RTD		Thermistor	Thermocouple	Silicon
	Thin-Film Type	*Wire Wound Type*			
Active material	Platinum thin film	Platinum, wire wound	Metal oxide ceramic	Two dissimilar metals	Silicon transistor cascade
Relative sensor cost	Moderate to low	Moderate	Low to moderate	Low	Low
Relative system cost	Moderate	Moderate	Low to moderate	High	Low
Temperature range	$-200°C$ to $750°C$ ($560°C$ max. typ.)	$-200°C$ to $850°C$ ($600°C$ max. typ.)	$-100°C$ to $500°C$ ($125°C$ max. typ.)	$-270°C$ to $1800°C$	$-40°C$ to $125°C$
Changing parameter	Resistance	Resistance	Resistance	Voltage	Voltage
Base value	100Ω to 200Ω	100Ω	$1\ k\Omega$ to $1\ M\Omega$	$< 10\ \mu V$ at $25°C$	$750\ mV$ at $25°C$
Interchangeability	$\pm1\%, \pm3°C$	$\pm0.06\%, \pm0.2°C$	$\pm10\%, \pm2°C$ typ.	$\pm0.5\%, \pm2°C$	$\pm1\%, \pm3°C$
Stability	Excellent	Excellent	Moderate	Poor	Moderate
Sensitivity	$0.39\%/°C$	$0.39\%/°C$	$-4\%/°C$	$40\ \mu V/°C$	$10\ mV/°C$
Relative sensitivity	Moderate	Moderate	Highest	Low	Moderate
Linearity	Excellent	Excellent	Logarithmic, poor	Moderate	Moderate
Slope	Positive	Positive	Negative	Positive	Positive
Noise susceptibility	Low	Low	Low	High	Low
Lead resistance errors	Low	Low	Low	High	Low
Special requirements	—	Lead compensation	Linearization	Reference junction	—

7.4.4 Silicon Temperature Sensors

Temperature sensors that utilize the temperature-dependent properties of silicon are appearing in the market in a wide variety of types, and their prices are reasonably low. Practical integrated circuits available in the market basically are either voltage output or current output temperature sensors.

Figure 7-3(a) shows the use of the temperature dependence of the PN junction voltage to provide a temperature-dependent voltage output, V_{be}. The voltage is related to the temperature and other parameters by the equation

$$V_{be} = \frac{2kT}{q} \ln \left(\frac{I_F}{I_S} \right) \tag{7.1}$$

where

I_F = forward current of transistor;

I_S = saturation current of transistor;

q = elementary charge;

k = Boltzmann's constant.

If the ratio I_F/I_S is kept constant, then the result would be a sensor exhibiting ideal linear temperature-dependence of the forward voltage.

A similar relationship is found in transistors. If the collector and base are held at the same potential (Figure 7-3(a)), then the relationship of the base-emitter voltage, V_{be}, to the collector current, I_c, is given by

$$V_{be} = \frac{kT}{q} \ln \left(\frac{I_c}{I_s} \right) \tag{7.2}$$

Here again the saturation current, I_s, is influenced by the temperature dependence of a number of parameters, similar to the case of diodes. Despite this, if the collector current, I_c, is held constant and the components are carefully selected, it is possible to obtain approximately linear behavior for temperatures between -50 and $150°C$.

Motorola's MTS 10X series silicon temperature sensors are .a classic example of this technique. The device family allows temperature measurement precisely in the range -40 to $150°C$.

Modern temperature management ICs range from purely analog voltage vs. temperature devices to mixed-signal VLSI chips containing logic and ADCs. Most ICs rely on a bandgap reference with a known temperature coefficient to provide temperature information.

7.4.4.1 Simple Current Output Temperature Transducers

Most simple current output temperature transducers use more practical forms of the circuit in Figure 7-3. The AD-590 from Analog Devices is an example

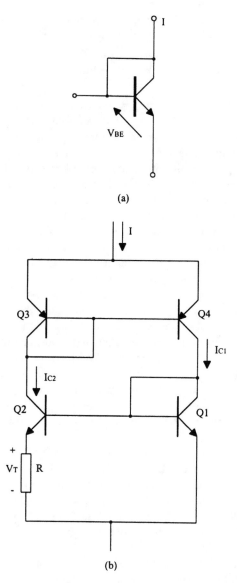

FIGURE 7-3 Temperature sensing using the PN junction properties: (a) transistor used as temperature sensor, (b) integrated temperature sensor

of this. Referring to Figure 7-3(b), the difference in base-emitter voltages of transistors Q1 and Q2 is given by

$$\Delta V_{be} = \frac{kT}{q} \ln \left(\frac{I_{c2}}{A_2} \bigg/ \frac{I_{c1}}{A_1} \right)$$

(7.3)

The temperature dependence of V_{be} therefore is solely dependent on the ratio, r, of the two collector current densities:

$$\Delta V_{be} = \frac{kT}{q} \ln(r) \qquad (7.4)$$

Provided that this ratio can be kept constant, ΔV_{be} is directly proportional to the absolute temperature. There are two ways of keeping the ratio constant. First, it is possible to operate two transistors with the same geometric dimensions on a single chip using two collector currents $(I_{c1} \neq I_{c2})$. The alternative is for a constant collector current to flow through two transistors with different emitter areas $(A_1 \neq A_2)$. The second variant has been of greater practical relevance because of the simpler circuitry involved. An example of this type of integrated sensor is presented in the basic circuit diagram in Figure 7-3(b). Transistors Q1 and Q2 perform the detection function. The identical transistors, Q3 and Q4, act as current mirrors. This causes a splitting of current I into two equal collector currents, I_{c1} and I_{c2}. The emitter area of Q2 should be r times that of Q1. Its collector current density therefore is only $1/r$ that of T1. The difference of ΔV_{be} causes a current, I_{c2}, that is proportional to the temperature to flow across a resistor, R. Because of the current mirroring, the value of I also must be proportional to the absolute temperature. Laser alignment of the resistance R makes it possible to adjust the constant of proportionality in equation (7.4) to 1 μAK^{-1}. If the circuit is changed to allow for a voltage output signal, then temperature coefficients of a few millivolts per Kelvin can be achieved.

In the AD-590, this ΔV_{be}, directly proportional to absolute temperature (PTAT), is converted to a PTAT current by low-temperature-coefficient thin-film resistors. The total current of the device is then forced to be a multiple of this PTAT current.

Figure 7-4(a) is a schematic diagram of the AD-590. Q8 and Q11 are the transistors that produce the PTAT voltage. R5 and R6 convert the voltage to current. Q10, whose collector current tracks the collector currents in Q9 and Q11, supplies all the bias and substrate leakage current for the rest of the circuit, forcing the total current to be PTAT. R5 and R6 are laser trimmed on the wafer to calibrate the device at +25°C. Figure 7-4(b) shows the typical $V - I$ characteristic of the circuit at +25°C and the temperature extremes.

The device features a 1 $\mu A/K$ linear current output over the −50 to 150°C temperature range. Some applications and accuracy are discussed in Klonowski (• • •) and Analog Devices (• • •).

Current output temperature transducers have a number of advantages:

- They are based on a linear relationship and are highly repeatable.
- The current is independent of voltage drops, voltage noise, common-mode voltage, and practically independent of excitation voltage.
- The current can be translated to a voltage at a remote destination via an appropriate value of resistance $(V = IR)$; simple offsetting circuitry may be used when necessary.

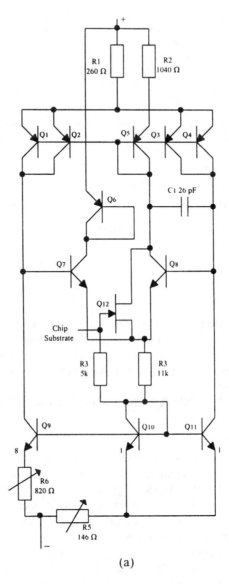

(a)

FIGURE 7-4 The AD-590: (a) schematic diagram, (b) $V - I$ characteristics. (Reproduced by permission of Analog Devices, Inc.)

- They are easy to use; they require no linearization circuitry, high-precision voltage amplifiers, resistance-measuring circuitry, or cold-junction compensation.

  Current output temperature sensors are widely used for cold-junction compensation of thermocouple circuitry.

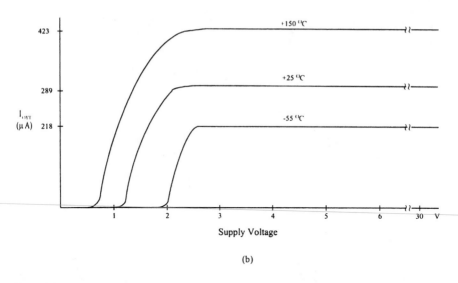

(b)

FIGURE 7-4 Continued

When voltage drops and noise are not an important consideration, it may be more convenient to work with a voltage output temperature transducer. These provide a direct output to an analog-to-digital converter or a comparator set point. Many practical components provide a voltage output as well as other functions.

7.4.4.2 AD-22100: A Ratiometric Voltage Output Temperature Sensor

The AD-22100 is a ratiometric temperature sensor IC whose output voltage is proportional to the power supply voltage. The heart of the sensor is a proprietary temperature-dependent resistor, similar to an RTD, built into the IC. Figure 7-5(a) is a simplified block diagram of the AD-22100.

The temperature-dependent resistor, R_T, exhibits a change in resistance that is nearly linearly proportional to temperature. This resistor is excited with a current source proportional to the power supply voltage (V_+). The resulting voltage across R_T therefore is both supply voltage proportional and linearly varying with the temperature (T_A). The remainder of the AD-22100 consists of an op amp signal conditioning block that takes the voltage across R_T and supplies the proper gain and offset to achieve the following output voltage function:

$$V_{out} = \left(\frac{V_+}{5}\right) \bullet [1.375 + (22.5 \times T_A)] \tag{7.5}$$

Due to its ratiometric nature, the device offers a cost-effective solution when used as an interface to an analog-to-digital converter. This is accomplished by using the ADC's +5 V power supply as a reference to both the ADC and the AD-22100 (see Figure 7-5(b)), eliminating the need for a precision reference.

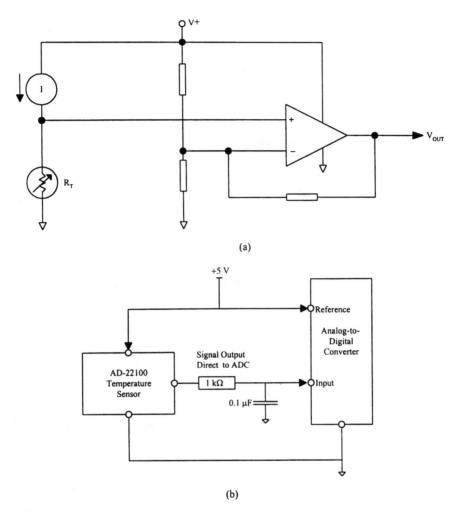

(a)

(b)

FIGURE 7-5 AD-22100 voltage output temperature sensor: (a) simplified block diagram, (b) an application. (Reproduced by permission of Analog Devices, Inc.)

The devices such as AD-22100 provide low-cost temperature measurement for microprocessor- and microcontroller-based systems. Many inexpensive 8-bit microprocessors now offer an onboard 8-bit ADC capability at a modest cost. Total "cost of ownership" becomes a function of the voltage reference and analog signal conditioning necessary to mate the analog sensor with the microprocessor ADC.

Such devices can provide a low-cost system by eliminating the need for a precision voltage reference and any additional active components. The ratiometric nature of the device allows the microprocessor to use the same power supply as its ADC reference. Variations in the supply voltage have little effect, as the sensor and the ADC use the supply as their reference. For details, see Analog Devices (1994).

7.4.5 Temperature Management ICs

Silicon temperature sensors easily can be combined with other circuit blocks for temperature control (Travis, 1996; Freeman, 1993). Overtemperature alarms, faulty circuitry shutdown, or initiation of corrective actions in a thermal feedback loop are ways in which temperature management ICs can prevent catastrophic failures. Devices commercially available include temperature controllers, airflow and temperature sensors, serial digital output, thermostat ICs, and programmable thermostat ICs. These devices are produced on a mass scale, using the standard IC production processes and prices vary from $0.50 to $4. Most of these ICs rely on a bandgap reference with a known temperature coefficient, coupled with other analog and digital circuitry, which may include the logic and ADCs as well. Two modern trends in temperature management ICs are increasing incorporation of digital circuitry and incorporation of more management functions (Travis, 1996).

Temperature control ICs include a temperature sensor that generates a voltage output proportional to the absolute temperature and a control signal from one or two outputs when the device is below or above a specified temperature range. An example of these devices is TMP01 from Analog Devices.

The TMP01 consists of a bandgap voltage reference combined with a pair of matched comparators. The reference provides both a constant 2.5 V output and a voltage proportional to absolute temperature (VPTAT), which has a precise temperature coefficient of 5 mV/K and is 1.49 V (nominal) at $+25°C$. The comparators compare the VPTAT with the externally set temperature trip points and generate an open-collector output signal when one of these thresholds has been exceeded. Figure 7-6(a) is a functional block diagram of the TMP01.

Hysteresis also is programmed by the external resistor chain and determined by the total current drawn out of the 2.5 V reference. This current is mirrored (Figure 7-6(b)) and used to generate a hysteresis offset voltage of the appropriate polarity after a comparator has been tripped. The comparators are connected in parallel, which guarantees no hysteresis overlap and eliminates erratic transitions between adjacent trip zones.

The device utilizes proprietary thin-film resistors in conjunction with production laser trimming to maintain a typical temperature accuracy of $\pm2°C$ over the rated temperature range, with excellent linearity. The open-collector outputs are capable of sinking 20 mA, enabling the TMP01 to drive control relays directly.

The TMP01 is a very linear voltage-output temperature sensor, with a window comparator that can be programmed by the user to activate one of two open-collector outputs when a predetermined temperature set point voltage has been exceeded. A low drift voltage reference is available for set point programming (see Figure 7-7).

In many temperature sensing and control applications, some type of switching is required. The open collector outputs (over and under) of TMP01 can be used to turn on a heater or switch off a motor. In such applications, the

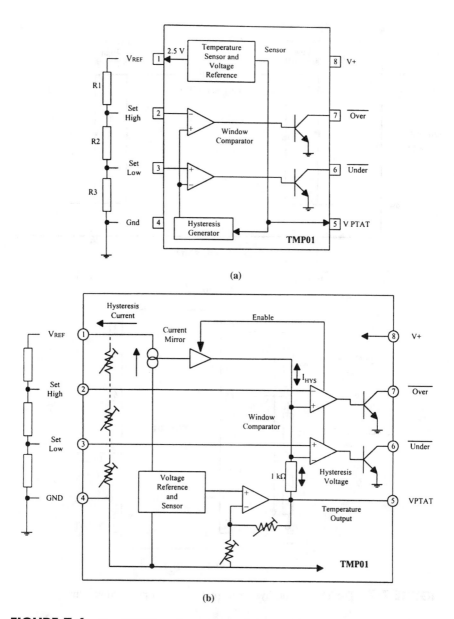

FIGURE 7-6 The TMP01, a low-power, programmable temperature controller: (a) functional block diagram, (b) detailed block diagram. (Reproduced by permission of Analog Devices, Inc.)

switches need to handle large currents, usually much more than 20 mA, which is the rated current of the output. In such cases, external switching devices such as relays, power MOSFETs, thyristors, IGBTs, or Darlington transistors can be used, as shown in Figure 7-8. For further details, see Analog Devices (1994).

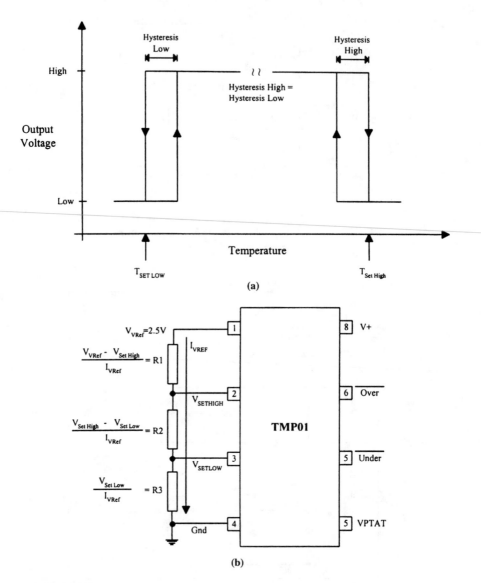

FIGURE 7-7 The TMP01: (a) hysteresis profile, (b) set point programming

7.4.6 Serial Digital Output Thermometers

Several manufacturers offer basic sensor devices coupled with analog-to-digital converters. These ICs allow a series digital output from the ADC proportional to the temperature. Examples are TMP03/04 from Analog Devices, LM 75 from National, and DS 1621 from Dallas Semiconductor.

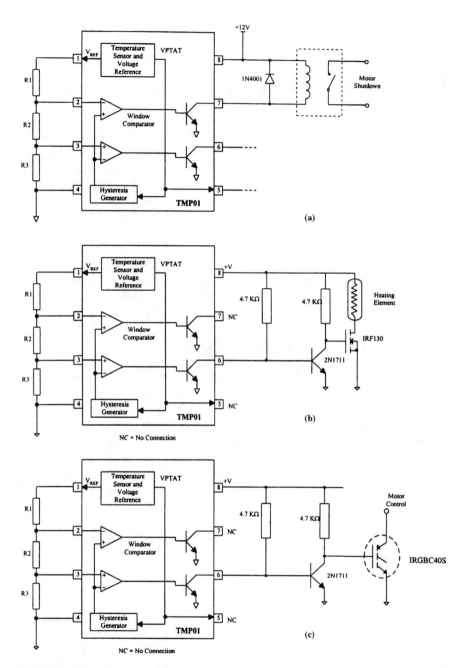

FIGURE 7-8 Switching loads with the open-collector output of the TMP01: (a) Reed relay drive, (b) driving an N-channel MOSFET, (c) driving an IGBT. (Reproduced by permission of Analog Devices, Inc.)

7.4.6.1 TMP03/04

The TMP03/TMP04 is a monolithic temperature detector generating a modulated serial digital output that varies in direct proportion to the temperature of the device. An onboard sensor generates a voltage precisely proportional to absolute temperature, which is compared to an internal voltage reference and entered into a precision digital modulator.

The sensor output is digitized by a first-order Σ-Δ modulator (see Figure 7-9(a)). This type of converter utilizes time-domain oversampling and a high-accuracy comparator to deliver 12 bits of effective accuracy in an extremely compact circuit. Figure 7-9(a) is a basic functional block diagram, and Figure 7-9(b) describes the first-order modulator interacting with the VPTAT and the voltage reference source.

The modulated output of the comparator is encoded using a circuit technique that results in a serial digital signal with a mark-space ratio format easily decoded by any microprocessor into either Centigrade or Fahrenheit degrees and readily transmitted or modulated over a single wire. Most important, the encoding method neatly avoids major error sources common to other modulation techniques, as it is clock independent.

7.4.6.1.1 Output Encoding

The TMP03/04 is designed as a low-cost three-terminal device with the output format shown in Figure 7-9(c). This patented design avoids an accurate external clock or high accuracy, low-drift types of internal clock systems within the IC. The modulation and encoding techniques within the TMP03/04 achieve this by using a simple, compact onboard clock and an oversampling digitizer that are insensitive to sampling rate variations. The digitized signal is encoded into a ratiometric format in which the exact frequency of the clock is irrelevant, and the effects of clock variations are effectively canceled on decoding by the digital filter.

The output of the TMP03/TMP04 is a square wave with a nominal frequency of 35 Hz ($\pm$20%) at +25°C. The output format is readily decoded by the user as per Figure 7-9(c):

$$\text{Temperature (}^\circ\text{C)} = 235 - \frac{400 \times T_1}{T_2} \qquad (7.6)$$

The time periods T_1 (high period) and T_2 (low period) are values easily read by a microprocessor timer/counter port, with the preceding calculations performed in software. Since both periods are obtained consecutively using the same clock, performing the division indicated in these formulas results in a ratiometric value independent of the exact frequency of, or drift in, either the originating clock of the TMP03/TMP04 or the user's counting clock. Figure 7-10 shows the output frequency and T_1/T_2 values versus temperature.

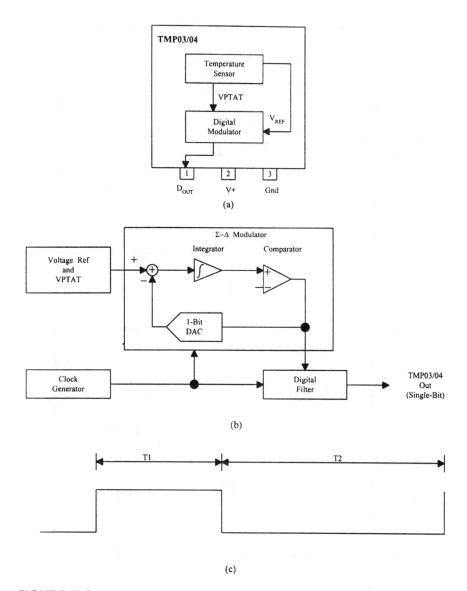

FIGURE 7-9 TMP03/04 serial digital output thermometer: (a) functional block diagram, (b) block diagram showing Σ-Δ modulator, (c) output format. (Reproduced by permission by Analog Devices, Inc.)

7.4.6.1.2 Application Considerations

These types of components are quite useful in applications such as isolated sensors, environmental control systems, computer thermal monitoring, thermal protection, and industrial process control and power system monitors. The low-voltage power supply (4.5–7 V) of these devices, low-cost three-pin package,

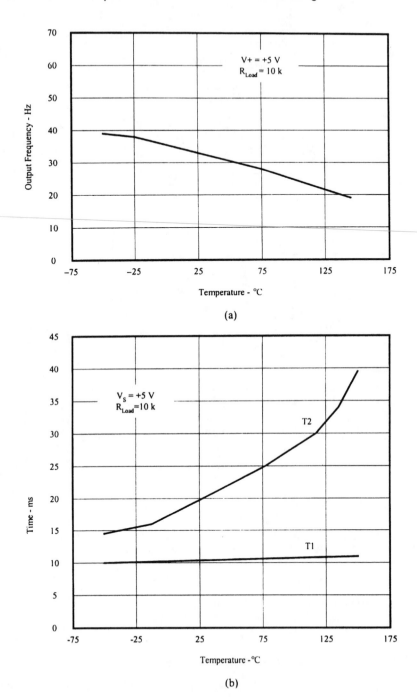

FIGURE 7-10 TMP03/04 output vs. temperature: (a) output frequency vs. temperature, (b) T_1 and T_2 vs. temperature. (Reproduced by permission of Analog Devices, Inc.)

low-power consumption, and the flexible open-collector output (TMP03) or CMOS/TTL-compatible output (TMP04) are the useful features of these devices for the wide variety of applications proposed.

Precision analog products such as the TMP series require a well-filtered power source. Since the TMP03/04 devices operate from a single +5 V supply, it is convenient to use the logic supply. Unfortunately, the logic supply often is a switchmode design, which generates noise in the 20 kHz–1 MHz range. In addition, fast logic gates can generate glitches of hundreds to a few millivolts in amplitude due to wiring resistance and inductance.

To minimize the noise affecting the operation, the circuit arrangement in Figure 7-11(a) is proposed. Even if a separate power supply trace is not available, however, generous supply bypassing will reduce supply-line-induced errors. Local supply bypassing consisting of a 10 μF tantalum electrolytic in parallel with a 0.1 μF ceramic capacitor (Figure 7-11(b)) is recommended. As the quiescent power supply current of the device typically is 900 μA, a simple RC filter network as in Figure 7-11(c) could be used when the device drives a light load, such as a CMOS gate.

The TMP03 (Figure 7-12(a)) has an open-collector NPN output suitable for driving a high-current load, such as an optoisolator. Since the output source current is set by the pull-up resistor, output capacitance should be minimized in TMP03 applications. Otherwise, unequal rise and fall times will skew the pulse width and introduce measurement errors.

The TMP04 has a "totem pole" CMOS output (Figure 7-12(b)) and provides a rail-to-rail output drive for logic interfaces. The rise and fall times of the TMP04 output are closely matched, to minimize errors caused by capacitive loading. If load capacitance is large (for example, when driving a long cable), an external buffer may improve accuracy. For more details on output configurations and interfaces to low-voltage logic and the like, see Analog Devices (1995a).

7.4.6.1.3 Microcontroller and DSP Interfaces

Here is an example of an 80C51 interface. The TMP03/TMP04 output easily is decoded with a microcomputer. The microcomputer simply measures the T_1 and T_2 periods in software or hardware and calculates the temperature using equation (7.6).

Since the TMP03/TMP04's output is ratiometric, precise control of the counting frequency is not required. The only timing requirements are that the clock frequency be high enough to provide the required measurement resolution and that the clock source be stable. The ratiometric output of the TMP03/TMP04 is an advantage because the microcomputer's crystal clock frequency often is dictated by the serial baud rate or other timing considerations.

Pulse width timing usually is done with the microcomputer's on-chip timer. A typical example, using the 80C51, is shown in Figure 7-13. This circuit requires only one input pin on the microcomputer, which highlights the efficiency of the TMP04's pulse width output format. Traditional serial input protocols, with data line, clock, and chip select, usually require three or more I/O pins.

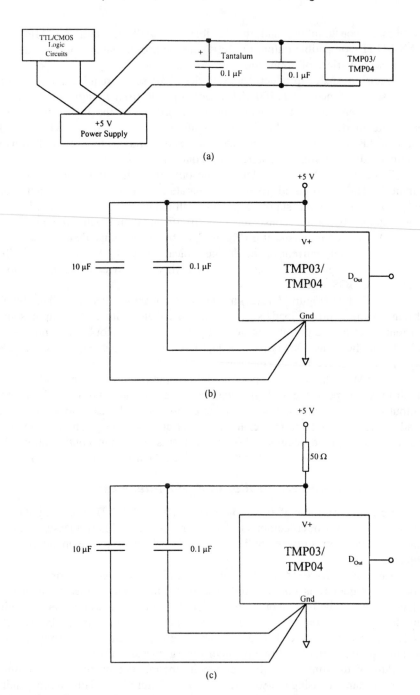

FIGURE 7-11 Supply bypassing techniques for the TMP03/04: (a) use of separate supply traces, (b) simple capacitor bypassing, (c) RC filter using a 50 Ω resistor and capacitors. (Reproduced by permission of Analog Devices, Inc.)

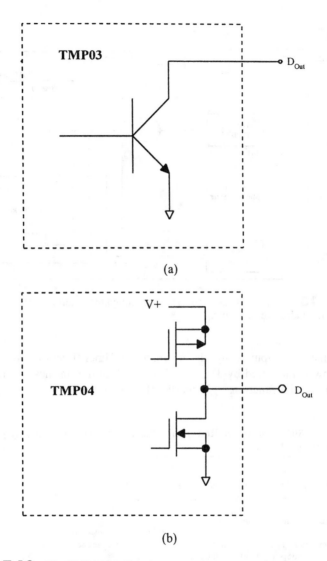

(a)

(b)

FIGURE 7-12 The TMP03/04 digital output structures: (a) the TMP03 open-collector output, (b) the TMP04 totem pole CMOS output

The 80C51 has two 16-bit timers. The clock source for the timers is the crystal oscillator frequency divided by 12. Therefore, a crystal frequency of 12 MHz or greater will provide resolution of 1 μs or less. The 80C51 timers are controlled by two dedicated registers. The TMOD register controls the timer mode of operation, while TCON controls the start and stop times. Both the TMOD and TCON registers must be set to start the timer.

Software for the interface is shown in Listing 7-1. The program monitors the TMP04 output and turns the counters on and off to measure the duty cycle.

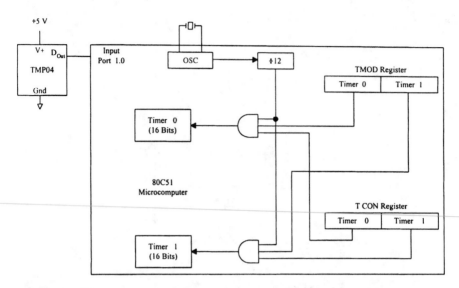

FIGURE 7-13 A TMP04 and 80C51 microcomputer interface. (Reproduced by permission of Analog Devices, Inc.)

The time that the output is high is measured by Timer 0, and the time that the output is low is measured by Timer 1. When the routine finishes, the results are available in special function registers 08AH through 08DH.

Listing 7.1. An 80C51 Software Routine for the TMP04. [Reproduced by permission of Analog Devices, Inc.]

```
;
; Test of a TMP04 interface to the 80C51,
; using timer 0 and timer 1 to measure the duty cycle
;
; This program has three steps:
;   1. Clear the timer registers, then wait for a low-to-
;      high transition on input P1.0 (which is connected
;      to the output of the TMP04).
;   2. When P1.0 goes high, timer 0 starts. The program
;      then loops, testing P1.0.
;   3. When P1.0 goes low, timer 0 stops & timer 1 starts. The
;      program loops until P1.0 goes low, when timer 1 stops
;      and the TMP04's T1 and T2 values are stored in Special
;      Function registers 8AH through 8DH (TL0 through TH1).
;
; Primary controls
$MOD51
$TITLE(TMP04 interface, Using T0 and T1)
$PAGEWIDTH (80)
$DEBUG
$OBJECT
```

Listing 7.1. Continued

```
;
;    Variable declarations
;
PORT1           DATA       90H              ;SFR register for port 1
;TCON           DATA       88H              ;timer control
;TMOD           DATA       89H              ;timer mode
;THO            DATA       8CH              ;timer 0 hi byte
;TH1            DATA       8DH              ;timer 1 hi byte
;TL0            DATA       8AH              ;timer 0 low byte
;TL1            DATA       8BH              ;timer 1 low byte
;
;
                ORG        100H             ;arbitrary start
;
READ_TMP04:     MOV        A,#00            ;clear the
                MOV        THO,A            ; counters
                MOV        TH1,A            ;  first
                MOV        TL0,A            ;
                MOV        TL1,A            ;
WAIT_LO:        JB         PORT1.0,WAIT_LO  ;wait for TMP04 output to go low
                MOV        A,#11H           ;get ready to start timer 0
                MOV        TMOD,A
WAIT_HI:        JNB        PORT1.0,WAIT_HI  ;wait for output to go high
;
;Timer 0 runs while TMP04 output is high
;
                SETB       TCON.4           ;start timer 0
;WAITTIMERO:    JB         PORT1.0,WAITTIMERO
                CLR        TCON.4           ;shut off timer 0
;
;Timer 1 runs while TMP04 output is low
;
                SETB       TCON.6           ;start timer 1
WAITTIMER1:     JNB        PORT1.0,WAITTIMER1
                CLR        TCON.6           ;stop timer 1
                MOV        A,#OH            ;get ready to disable timers
                MOV        TMOD,A
                RET
                END
```

When the READ_TMP04 routine is called, the counter registers are cleared. The program sets the counters to their 16-bit mode, and then waits for the TMP04 output to go high. When the input port returns a logic high level, Timer 0 starts. The timer continues to run while the program monitors the input port. When the TMP04 output goes low, Timer 0 stops and Timer 1 starts. Timer 1 runs until the TMP04 output goes high, at which time the TMP04 interface is complete. When the subroutine ends, the timer values are stored in their respective SFRs and the TMP04's temperature can be calculated in software.

Since the 80C51 operates asynchronously to the TMP04, there is a delay between the TMP04 output transition and the start of the timer. This delay can vary between 0 μs and the execution time of the instruction that recognized the transition. The 80C51's "jump on port-bit" instructions (JB and JNB) require 24 clock cycles for execution. With a 12 MHz clock, this produces an uncertainty of 2 μs (24 clock cycles/12 MHz) at each transition of the TMP04 output. The worst-case condition occurs when T1 is 4 μs shorter than the actual value and T2 is 4 μs longer. For a 25°C reading ("room temperature"), the nominal error caused by the 2 μs delay is only about ±0.5°C.

The TMP04 also easily interacts with digital signal processors, such as the ADSP-210x series. Again, only a single I/O pin is required for the interface (Figure 7-14).

The ADSP-2101 only has one counter, so the interface software differs somewhat from the 80C51 example. The lack of two counters is no limitation, however, because the DSP architecture provides very high execution speed. The ADSP-2101 executes one instruction for each clock cycle, versus one instruction for 12 clock cycles in the 80C51, so the ADSP-2101 actually produces a more-accurate conversion while using a lower oscillator frequency.

The timer of the ADSP-2101 is implemented as a down counter. When enabled by a software instruction, the counter is decremented at the clock rate divided by a programmable prescaler. Loading the value $n-1$ into the prescaler register will divide the crystal oscillator frequency by n.

For the circuit of Figure 7-14, therefore, loading 4 into the prescaler will divide the 10 MHz crystal oscillator by 5 and thereby decrement the counter at a 2 MHz rate. The TMP04 output is ratiometric, of course, so the exact clock frequency is not important.

A typical software routine for an interface between the TMP04 and the ADSP-2101 is shown in Listing 7-2. The program begins by initializing the prescaler and loading the counter with 0FFF. The ADSP-2101 monitors the FI flag input to establish the falling edge of the TMP04 output and starts the counter. When the TMP04 output goes high, the counter is stopped. The counter value is subtracted from 0FFFh to obtain the actual number of counts, and the count is saved. Then, the counter is reloaded and runs until the TMP04 output goes low. Finally, the TMP04 pulse widths are converted to a temperature using the scale factor of equation (7.6).

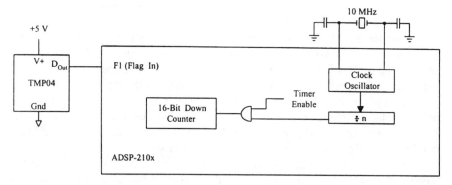

FIGURE 7-14 Interface between the TMP04 and the ADSP-210x. (Reproduced by permission of Analog Devices, Inc.)

Listing 7.2. Software Routine for the TMP04 to ADSP-210x Interface. [Reproduced by permission of Analog Devices, Inc.]

```
;
{ ADSP-21XX Temperature Measurement Routine           TEMPERAT.DSP

        Altered Registers:          ax0,ay0,af,ar,
                                    si,sr0,
                                    my0,mr0,mr1,mr2.
        Return value:               ar -> temperature result in 14.2 format
        Computation time:           2 * TMP04 output period
}
.MODULE/RAM/BOOT=0    TEMPERAT;       { Beginning TEMPERAT Program }
.ENTRY TEMPMEAS;                      { Entry point of this subroutine }
.CONST PRESCALERS=4;
.COSNT TIMFULSCALE=0xffff;
TEMPMEAS:    si=PRESCALER;            { For timer prescaler }
             sr0=TIMFULSCALE;         { Timer counter full scale }
             dm(0x3FFB)=si;           { Timer Prescaler set up to 5 }
             si=TIMFULSCALE;          { CLKin=10MHz,Timer Period=32.768ms }
             dm (0x3FFC)=si;          { Timer Counter Register to 65535 }
             dm (0x3FFD)=si;          { Timer Period Register to 65535 }
             imask=0x01;              { Unmask Interrupt timer }
TEST1:       if not fi jump TEST1;    { Check for FI=1 }
TEST0:       if fi jump TEST0;        { Check for FI=0 to locate transition }
             ena timer;               { Enable timer,count at a 500ns rate }
COUNT2:      if not fi jump COUNT2;   { Check for FI=1 to stop count }
             dis timer;
             ay0=dm(0x3FFC);          { Save counter=T2 in ALU register }
             ar=sr0-ay0;
             ax0=ar;
             dm(0x3FFC)=si;           { Reload counter at full scale }
             ena timer;
COUNT1:      if fi jump COUNT1;       { Check for FI=0 to stop count }
             dis timer;
             ay0=dm(0x3FFC);          { Save counter=T1 in ALU register }
             ar=sr0-ay0;
             my0=400;
             mr=ar*my0(uu);           { mr=400*T1 }
             ay0=mr0;                 { af=MSW of dividend,ay04=LSW }
             ar=mr1; af=pass ar;      { ax0=16-bit divisor }
COMPUTE:     astat=0;                 { To clear AQ flag }
             divq ax0; divq ax0;      { Division 400*T1/T2 }
             divq ax0; divq ax0;      { with 0.3 < T1/T2 < 0.7 }
             divq ax0; divq ax0;
             divq ax0; divq ax0;
             divq ax0; divq ax0;
             divq ax0; divq ax0;
             divq ax0; divq ax0;
             divq ax0; divq ax0;
             divq ax0; divq ax0;      { Result in ay0 }
             ax0=0x03AC;              { ax0=235*4 }
             ar=ax0-ay0;              { ar=235-400*T1/T2,result in φC }
             rts;                     { format 14.2 }
.ENDMOD;                              { End of the subprogram }
```

7.4.6.1.4 Miscellaneous Other Applications

Sensors similar to TMP03/04 can be used for many other useful applications. One such use is to monitor the temperature of a high-power microprocessor. The TMP04 interface depicted in Figure 7-15 could be used to measure the output pulse widths with a resolution of ±1 µs. The TMP04 sensors T1 and T2 periods are measured with two cascaded 74HC520 counters. The counters, accumulating clock pulses from a 1 MHz external oscillator, have a maximum period of 65 ms.

The circuit shown in Figure 7-15 can be an ASIC application (as part of the system ASIC) so that the microprocessor would not be burdened with the overhead of timing the output pulse width. For details, see Analog Devices (1995a).

Another example of using such an IC to monitor a high power dissipation ULSI is shown in Figure 7-16. The device, in a surface mounted package, is mounted directly beneath the device's pin grid array (PGA) package. In a typical application, the device's output could be connected to an ASIC, where the pulse width could be measured (Figure 7-15 is a suitable interface.) The TMP04 pulse output provides a significant advantage in this application because it produces a linear temperature output while needing only one I/O pin and no A/D converter.

7.4.7 Precision Temperature Sensors and Airflow Temperature Sensors

7.4.7.1 Low-Voltage Precision Temperature Sensors

Low-voltage temperature sensors which provide a voltage output directly proportional to the temperature for temperature monitoring and thermal control systems; for example, the TMP35, TMP36, and TMP37 from Analog Devices (Figure 7-17) or the LM35 and LM36 from National Semiconductor. These devices require no external calibration to provide a typical accuracy of ±1°C at 25°C and ±20°C over the −40 to 125°C temperature range. For application details, see Analog Devices (1996a).

7.4.7.2 Airflow Temperature Sensors

Modern electronic systems and products need be incorporated with suitable airflow temperature control systems. For such systems, such as low-cost fan controllers and overtemperature protection, commercial silicon sensors measure the airflow temperature. These devices consist of a bandgap element (with a voltage reference source and a VPTAT) and a heating element plus the associated circuitry. An example is the TMP12, an airflow and temperature sensor from Analog Devices (Figure 7-18).

The TMP12 incorporates a heating element, temperature sensor, and two user-selectable set point comparators on a single substrate. By generating a known amount of heat and using the set point comparators to monitor the resulting temperature rise, the TMP12 can indirectly monitor the performance of a system's

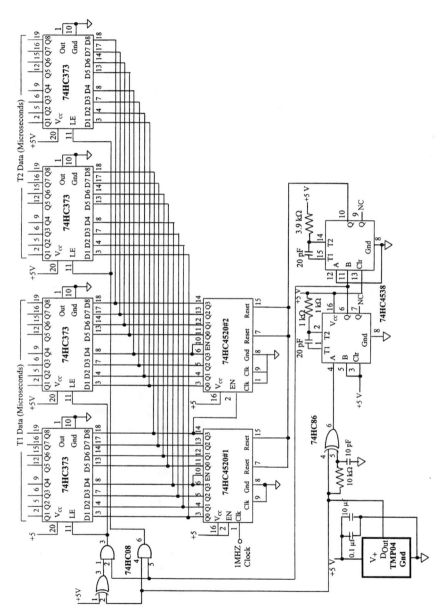

FIGURE 7-15 A hardware interface for the TMP04. (Reproduced by permission of Analog Devices, Inc.)

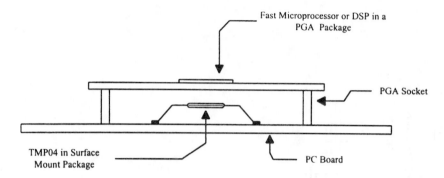

FIGURE 7-16 Monitoring the temperature of a ULSI using a surface mount sensor device

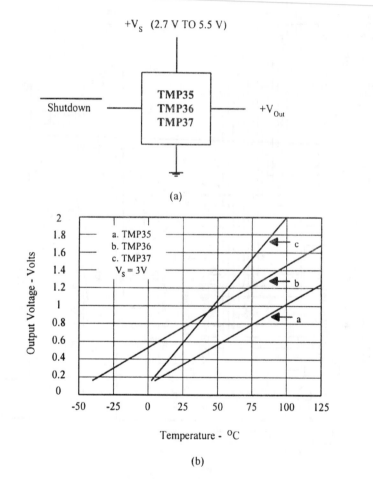

FIGURE 7-17 TMP3x series: (a) functional diagram, (b) output voltage vs. temperature. (Reproduced by permission of Analog Devices, Inc.)

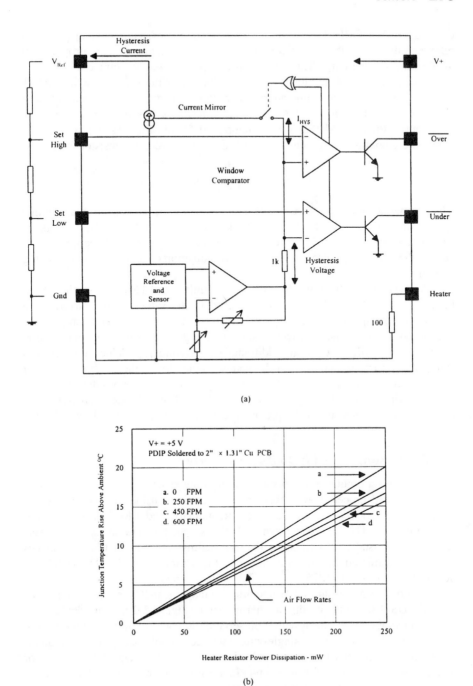

(a)

(b)

FIGURE 7-18 The TMP12 airflow and temperature sensor: (a) functional block dia-gram, (b) temperature rise vs. heater dissipation for a plastic dual inline (DIL) package. (Reproduced by permission of Analog Devices, Inc.)

cooling fan. The TMP12 temperature sensor section consists of a bandgap voltage reference that provides both a constant 2.5 V output and a voltage proportional to absolute temperature. The VPTAT has a precise temperature coefficient of 5 mV/K and is 1.49 V (nominal) at +25°C. The comparators compare the VPTAT with the externally set temperature trip points and generate an open-collector output signal when one of the respective thresholds has been exceeded. The heat source for the TMP12 is an on-chip 100 Ω thin-film resistor with a low temperature coefficient. When connected to a 5 V source, this resistor dissipates:

$$P_D = \frac{V^2}{R} = \frac{5^2}{100} = 0.25 \text{ W} \tag{7.7}$$

which generates a temperature rise of about 32°C in still air for the small outline SO packaged device. With an airflow of 450 feet per minute (FPM), the temperature rise is about 22°C. By selecting a temperature set point between these two values, the TMP12 can provide a logic-level indication of problems in the cooling system.

A typical application for devices similar to TMP12 is shown in Figure 7-19(a). The airflow sensor is placed in the same cooling airflow as a high-power dissipation IC. The sensor's internal resistor produces a temperature rise proportional to the airflow, as shown in Figure 7-19(b). Any interruption in the airflow will produce an additional temperature rise. When the sensor's chip temperature exceeds a user-defined set point limit, the system controller can take corrective action, such as reducing clock frequency, shutting down unused peripherals, or turning on an additional fan. These devices have hysteresis profiles similar to discussions in previous sections. For further details, see Analog Devices (1995b).

7.4.8 Sensors with Built-in Memories

To store set points for temperature monitoring systems, some manufacturers have developed processes to embed memory devices inside the sensor ICs; for example, the DS 1621 from Dallas Semiconductor and LM75 from National Semiconductor. The DS 1621 provides 9 bit (serial) temperature data and user-settable thermostatic set points. As the user settings are nonvolatile, the devices can be programmed before insertion into the system. For details, see Dallas Semiconductor (1994–1995).

One of the more unusual temperature management ICs is Dallas Semiconductor's DS 1820 with its digital thermometer (Figure 7-20). This device is a multidrop temperature sensor with 9 bit serial digital output. Information is sent to and from the DS 1820 over a single-wire interface, so only one wire (and ground) needs to be connected from a central microprocessor to a DS 1820. Power for reading, writing, and performing temperature conversions can be derived from the data line itself with no need for an external power source.

Because each DS 1820 contains a unique silicon serial number, multiple DS 1820s can exist on the same single-wire bus. This allows placing temperature sensors in many different places. Applications where this feature is useful

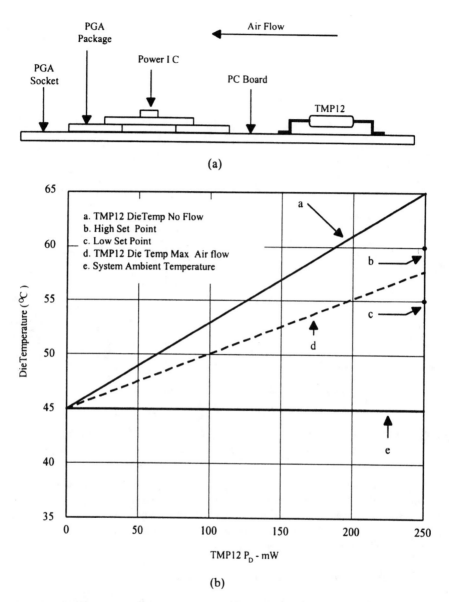

FIGURE 7-19 The TMP12: (a) typical application, (b) choosing temperature set points. (Reproduced by permission of Analog Devices, Inc.)

include HVAC environmental controls; sensing temperatures inside buildings, equipment, or machinery; and process monitoring and control. The block diagram of Figure 7-20(b) shows the major components of the DS 1820. The DS 1820 has three main data components: a 64 bit ROM, a temperature sensor, and nonvolatile temperature alarm triggers, TH and TL. The device derives its power from the

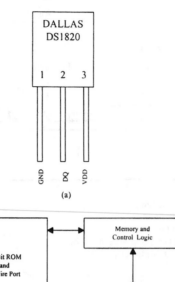

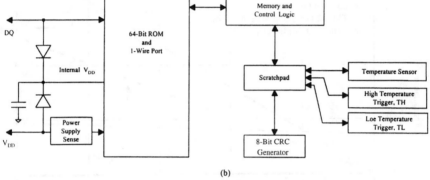

FIGURE 7-20 The DS 1820: (a) device in its 3 terminal package, (b) block diagram package. (Reproduced by permission of Dallas Semiconductor.)

one-wire communication line by storing energy on an internal capacitor during periods of time when the signal line is high and continues to operate off this power source during the low times of the one-wire line until it returns high to replenish the parasite (capacitor) supply. As an alternative, the DS 1820 can be powered from an external 5 V supply.

7.4.9 Thermal Response Time of Sensors

The time required for a temperature sensor to settle to a specified accuracy is a function of the thermal mass of and thermal conductivity between the sensor and the object being sensed. Thermal mass often is considered equivalent to capacitance. Thermal conductivity, commonly represented by the symbol θ, can be thought of as thermal resistance. It commonly is specified in units of degrees per watt of power transferred across the thermal joint.

The time required for the sensor IC to settle to the desired accuracy depends on the package selected, the thermal contact established in the particular application, and the equivalent power of the heat source. In most applications, the settling time probably is best determined empirically. Thermal time constants for thermal sensors can vary from a few seconds to over 100 seconds, depending on the package and socket used, air velocity, and other factors. Practical techniques for maximum accuracy from sensors are discussed in Steele (1996).

7.5 Silicon Pressure Sensors

7.5.1 Background on Piezoresistive Effect

The roots of silicon micromachining technology date back to Bell Laboratories. The research team developing the basics of semiconductor technology discovered a piezoresistive effect in silicon and germanium. The piezoresistive effect creates a resistance change in the semiconductor material in response to stress. This change was approximately two orders of magnitude larger than the equivalent resistance change of metals (used previously for strain gauge applications), promising an attractive option for sensors. The high sensitivity, or gauge factor, is perhaps 100 times that of wire strain gauges. Piezoresistors are implanted into a homogeneous single crystalline silicon medium. The implanted resistors thus are integrated into the silicon force sensing member. Typically, other types of strain gauges are bonded to force sensing members of dissimilar material, resulting in thermoelastic strain and complex fabrication processes. Most strain gauges are inherently unstable due to degradation of the bond, as well as temperature sensitivity and hysteresis caused by the thermoelastic strain. Silicon is an ideal material for receiving the applied force because it is a perfect crystal and does not become permanently stretched. After being strained, it returns to the original shape. Silicon wafers are better than metal for pressure sensing diaphragms, as silicon has extremely good elasticity within its operating range. Silicon diaphragms normally fail only by rupturing.

7.5.2 Piezoresistive Effect-Based Pressure Sensor Basics

The most popular silicon pressure sensors are piezoresistive bridges that produce a differential output voltage in response to pressure applied to a thin silicon diaphragm. The sensing element of a typical solid-state pressure sensor consists of four nearly equal piezoresistors buried in the surface of a thin circular silicon diaphragm (see Figure 7-21).

A pressure or force causes the thin diaphragm to flex, inducing a stress or strain in the diaphragm and the buried resistors. The resistor values will change, depending on the amount of strain they undergo, which depends on the amount of pressure or force applied to the diaphragm. Therefore, a change in pressure (mechanical input) is converted to a change in resistance (electrical output). The resistors can be connected in either a half-bridge or a full-Whetstone-bridge arrangement. For a pressure or force applied to the diaphragm using a full-bridge arrangement, the resistors can be approximated theoretically as shown in Figure 7-21 (nonamplified units). Here, $R + \Delta R$ and $R - \Delta R$ represent the actual

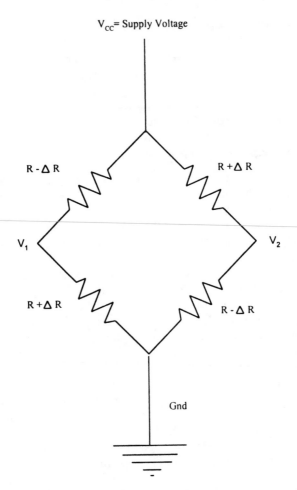

FIGURE 7-21 Four-element bridge used in a piezoresistive pressure sensor

resistor values at the applied pressure or force. R represents the resistor value for the undeflected diaphragm (pressure is zero) where all four resistors are nearly equal in value. And ΔR represents the change in resistance due to an applied pressure or force. All four resistors will change by approximately the same value. Note that two resistors increase and two decrease depending on their orientation with respect to the crystalline direction of the silicon material. The signal voltage generated by the full-bridge arrangement is proportional to the amount of supply voltage (V_{cc}) and the amount of pressure or force applied that generates the resistance change, ΔR. In a practical pressure sensor such as the Motorola MPX2100, the Whetstone bridge as shown in Figure 7-22 is used.

Bridge resistors RP1, RP2, RV1, and RV2 are arranged on a thin silicon diaphragm such that when pressure is applied RP1 and RP2 increase in value while RV1 and RV2 decrease a similar amount. Pressure on the diaphragm,

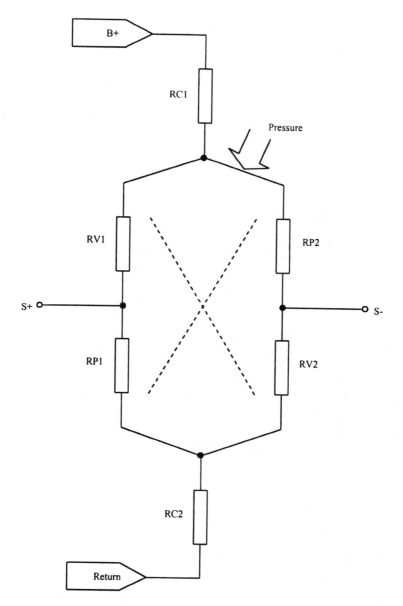

FIGURE 7-22 Sensor equivalent circuit. (Reproduced by permission of Motorola Inc.)

therefore, unbalances the bridge and produces a differential output signal. A fundamental property of this structure is that the differential output voltage is directly proportional to bias voltage, $B+$. This characteristic implies that the accuracy of the pressure measurement depends directly on the tolerance of the bias supply. It also provides a convenient means for temperature compensation. The bridge resistors are silicon resistors that have positive temperature coefficients. Therefore, when they are placed in series with zero T_C temperature compensation resistors, RC1 and RC2, the amount of voltage applied to the bridge increases with temperature. This increase in voltage produces an increase in electrical sensitivity, which offsets and compensates for the negative temperature coefficient associated with piezoresistance.

Since RC1 and RC2 are approximately equal, the output voltage common mode is very nearly fixed at 1/2B+. In a typical MPX2100 sensor, the bridge resistors are nominally 425 Ω; RC1 and RC2 are nominally 680 Ω. With these values and 10 V applied to B+, a ΔR of 1.8 Ω at full-scale pressure produces 40 mV of differential output voltage.

7.5.3 Pressure Sensor Types

Most pressure sensor manufacturers support three types of pressure measurements: absolute pressure, differential pressure, and gauge pressure. These are illustrated in Figure 7-23.

Absolute pressure is measured with respect to a vacuum reference, an example of which is the measurement of barometer pressure. In absolute devices, the P2 port is sealed with a vacuum representing a fixed reference. The difference in pressure between the vacuum reference and the measured amount applied at the P1 port causes the deflection of the diaphragm, producing the output voltage change (Figure 7-23(a)).

Differential pressure is the difference between two pressures. For instance, the measurement of pressure dropped across an orifice or venturi used to compute the flow rate. In differential devices, measurements are applied to both ports (Figure 7-23(b)).

Gauge pressure is a form of differential pressure measurement in which atmospheric pressure is used as the reference. Measurement of auto tire pressure, where a pressure above atmosphere is needed to maintain tire performance characteristics, is an example. In gauge devices, the P1 port is vented to atmospheric pressure and the measured amount is applied to the P2 port (Figure 7-23(c)).

7.5.4 Errors and Sensor Performance

In practical applications, when calculating the total error of a pressure sensor, several defined errors should be used. To determine the degree of specific errors for the pressure sensor selected, it is necessary to refer to the sensor's specification sheets. In specific customer applications, some of the published specifications can be reduced or eliminated. For example, if a sensor is used over half the specified temperature range, then the specific temperature error can be reduced by half. If an auto-zeroing technique is used, the null offset and null shift errors can be eliminated. The major factor affecting high-performance applications is the temperature dependence of the pressure characteristics. Some of the error parameters are these.

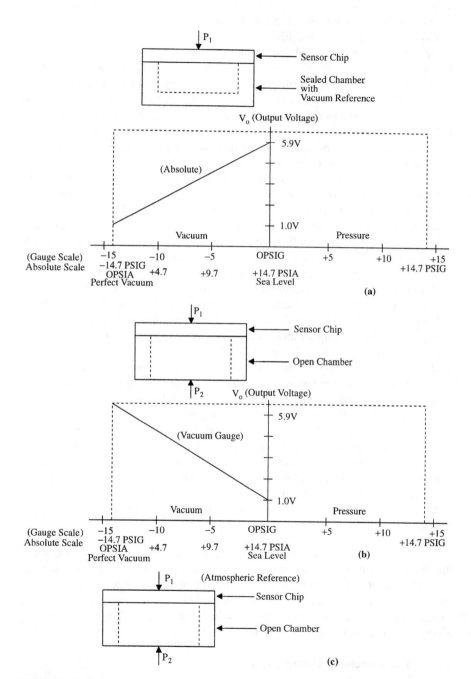

FIGURE 7-23 Different pressure measurements: (a) absolute, (b) differential, (c) gauge. (Reproduced by permission of Microswitch, Honeywell, Inc., USA.)

- *Null offset.* Null offset is the electrical output present when the pressure or force on both sides of the diaphragm is equal.
- *Span.* Span is the algebraic difference between the output end points. Normally, the end points are null and full scale.
- *Null temperature shift.* Null temperature shift is the change in null resulting from a change in temperature. Null shift is not a predictable error because it can shift up or down from unit to unit. A change in temperature will cause the entire output curve to shift up or down along the voltage axis (Figure 7-24(a)).
- *Sensitivity temperature shift.* Sensitivity temperature shift is the change in sensitivity due to a change in temperature. A change in temperature will cause a change in the slope of the sensor output curve (Figure 7-24(b)).
- *Linearity error.* Linearity error is the deviation of the sensor output curve from a specified straight line over a desired pressure range. One method of computing linearity error is least squares, which mathematically provide a best fit straight line to the data points (Figure 7-24(c)). Another method is terminal-base linearity or end point linearity, which is determined by drawing a straight line (L1) between the end data points on the output curve. Next a perpendicular line is drawn from line L1 to a data point on the output curve. The data point is chosen to achieve the maximum length of the perpendicular line. The length of the perpendicular line represents terminal-base linearity error (Figure 7-24(d)).
- *Repeatability error.* Repeatability error is the deviation in output readings for successive applications of any given input pressure or force with other conditions remaining constant (Figure 7-24(e)).
- *Hysteresis error.* Hysteresis error usually is expressed as a combination of mechanical hysteresis and temperature hysteresis. Some manufacturers, such as Microswitch, express hysteresis as a combination of the two effects (Figure 7-24(f)). Mechanical hysteresis is the output deviation at a certain input pressure or force when that input is approached first with increasing pressure or force and then with decreasing pressure or force. Temperature hysteresis is the output deviation at a certain input, before and after a temperature cycle.
- *Ratiometricity error.* Ratiometricity implies the sensor output is proportional to the supply voltage with other conditions remaining constant. Ratiometricity error is the change in this proportion and usually is expressed as a percent of span.

When choosing a pressure or force sensor, the total error contribution is important. Two methods take into account the individual errors and the unit-to-unit interchangeability errors: the root sum squared using maximum values and the worst-case error. The root sum squared method gives the most realistic value for accuracy. With the worst-case error method, the chances of one sensor having all errors at the maximum are very remote.

7.5.5 Practical Components

Pressure sensing is one of the most established and well-developed areas of sensor technology. One reason for its popularity is that it can be used to measure

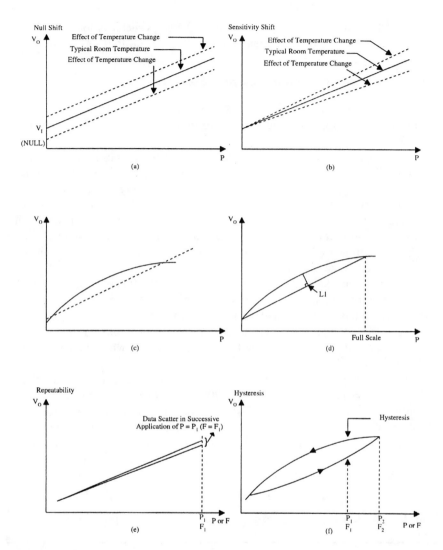

FIGURE 7-24 Typical error curves for pressure sensors: (a) null shift error, (b) sensitivity error, (c) best-fit straight-line linearity, (d) terminal-base linearity, (e) repeatability, (f) hysteresis

various real-world phenomena, like flow, fluid level, and acoustic intensities, in addition to pressure. In the automotive industry alone, for example, pressure sensors have been identified for use in ten different applications. In guidance control and industrial control systems, pressure sensors long have been used for a number of precision pressure measurement.

Practical components available from manufacturers could be basically divided into several categories: basic uncompensated types, calibrated and temperature compensated types, and signal conditioned types.

The standard pressure ranges, from manufacturers such as Motorola, Honeywell, and IC Sensors, vary between none to a few psi up to 0–5000 psi.

7.5.5.1 Basic Uncompensated Sensors

Most of the basic uncompensated pressure sensor devices are silicon piezoresistive strain gauge designs. Some examples of these devices are listed in Table 7-3.

These uncompensated basic sensors contain a basic transducer structure as shown in Figure 7-25. Figure 7-25 illustrates the top view of the pressure sensor silicon chip, showing the strain-gauge resistor diagonally placed on the edge of the diaphragm. Voltage is applied across pins 1 and 3, while the taps that sense the voltage differential transversely across the pressure-sensitive resistor are connected to terminals 2 and 4. An external series resistor is used to provide temperature compensation while reducing the voltage impressed on the sensor to within its rated value.

The recommended voltage drive is 3 V DC and should not exceed 6 V under any operating condition. The differential voltage output of the sensor, appearing between terminals 2 and 4, will be positive when the pressure applied to the "pressure" side of the sensor is greater than the pressure applied to the "vacuum" side. Nominal full-scale span of the transducer is 60 mV when driven by a 3 V constant voltage source.

When no pressure is applied to the sensor there will be some output voltage, called *zero pressure offset*. For the MPX700 sensor this voltage is guaranteed to be within the range of 0–35 mV. The zero pressure offset output voltage easily is nulled out by a suitable instrumentation amplifier. The output voltage of the sensor will vary in a linear manner with applied pressure. Figure 7-26 illustrates output voltage vs. pressure differential applied to the sensor, when driven by a 3 V source.

TABLE 7-3 Uncompensated pressure sensors. (Reproduced by permission of Motorola Inc.)

Device Series	Pressure Range kPa/psi (Max)	Over-pres-sure (kPa)	Offset mV (Typ)	Full-Scale Span mV (Typ)	Sensitivity (mV/kPa) (Typ)	Linearity % of FSS[1] (Min)	(Max)	Temperature Coefficient of Span °C (Typ)	Input Imped-ance Ohms (Typ)
MPX10D	1.45	100	20	35	3.5	−1	1	−0.19	475
MPX12D	1.45	100	20	55	5.5	0	5	−0.19	475
MPX50D	50/7.3	200	20	60	1.2	−0.1	0.1	−0.19	475
MPX100D/A	100/14.5	200	20	60	0.6	−0.1	0.1	−0.19	475
MPX200D/A	200/29	400	20	60	0.3	−0.25	0.25	−0.19	475
MPX201D/A	200/29	400	20	60	0.3	−0.35	0.35	−0.19	475
MPX700D	700/100	2100	20	60	0.086	0.5	0.50	−0.18	475

[1] Based on end point straight-line fit method. Best-fit straight-line linearity error is approximately half the listed value.

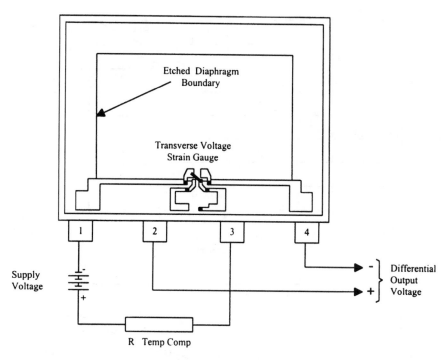

FIGURE 7-25 Sensor construction as applied to Motorola MPX series showing electrical connections. (Reproduced by permission of Motorola Inc.)

7.5.5.1.1 Temperature Compensation

Because this strain gauge is an integral part of the silicon diaphragm, there are no temperature effects due to differences in the thermal expansion of the strain gauge and the diaphragm, as often are encountered in bonded strain gauge pressure sensors. However, the properties of the strain gauge itself are temperature dependent, requiring that the device be temperature compensated if it is to be used over an extensive temperature range. Temperature compensation and offset calibration can be achieved rather simply with additional resistive components. Several approaches to external temperature compensation over both −40 to +125°C and 0 to +80°C ranges are presented in Motorola Applications Note AN 840 (Schwartz, Derrington, and Gragg, ●●●). Figure 7-27 shows a practical circuit for a digital pressure gauge.

The simplest method of temperature compensation, placing a resistance (R19 and R20) in series with the sensor driving voltage, is utilized in Figure 7-27. This provides good results over a temperature span of 0–80°C, yielding a 0.5% full-scale span-compensated device. Since the desired bridge driving voltage is about 3 V, placing the temperature compensating resistor in series with the bridge circuit has the additional advantage of reducing the power supply voltage, 15 V, to the desired 3 V level. Note that the 15 V power source must be held to within

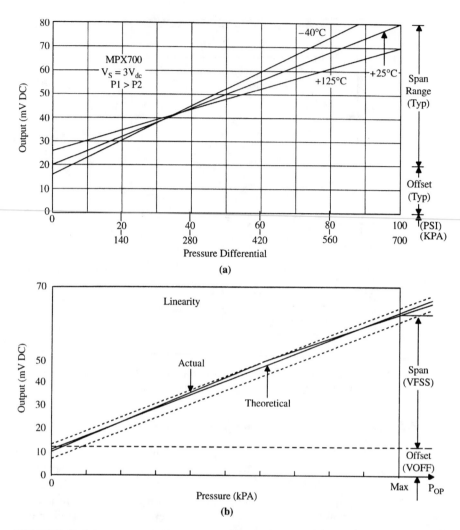

FIGURE 7-26 Characteristics of the MPX700 series devices: (a) output vs. pressure differential, (b) linearity specification comparison. (Reproduced by permission of Motorola Inc.)

a tight tolerance, since the output voltage of the transducer is ratiometric with the supply voltage. In most applications, an ordinary fixed 15 V regulator chip can be used to provide the required stable supply voltage.

The series method of compensation requires a series resistor which is equal to 3.577 times the bridge input resistance at 25°C. The range of transducer resistance is between 400 and 550 Ω, so the compensating network will be 1431–1967 Ω. If a temperature compensated span of greater than $\pm 0.5\%$ is satisfactory or the operating temperature range of the circuit is less than 80°C,

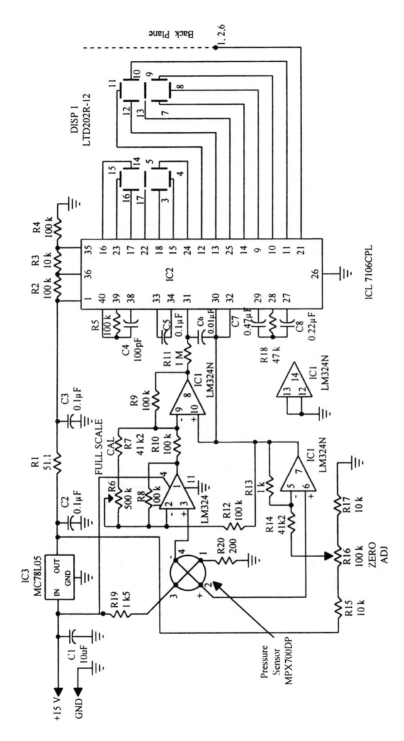

FIGURE 7-27 Schematic diagram of a digital pressure gauge using the MPX700 series. (Reproduced by permission of Motorola Inc.)

one value of compensating resistance can be used for any sensor resistance over the range 400–550 Ω. In the circuit of Figure 7-27, the temperature compensating network is composed of two resistors to allow the quiescent voltage of the sensor at pins 2 and 4 to be near the center level (2.5 V) of the analog and digital circuit that follows.

7.5.5.1.2 Signal Amplification

To amplify the transducer output (60 mV at 100 psi) to a useful level that can drive subsequent circuitry, common op amps such as LM324 could be used. The circuit in Figure 7-27 shows the application, which allows means to null out the DC offset output voltage of the transducer when no pressure is applied. The high input impedance of the IC1 ensures that the circuit does not load the basic transducer. In the practical circuit of Figure 7-27, the differential output of the instrumentation amplifier is fed to the ADC (IC2), to provide a digital readout of the pressure difference impressed on the transducer. For further details, see Caristi (Motorola Application Note AN-1105).

7.5.5.1.3 Signal Conditioning for Uncompensated Pressure Sensors

Today's unamplified solid-state sensors typically have an output voltage of tens of millivolts (Motorola's basic 10 kPa pressure sensor, MPX10, has a typical full-scale output of 58 mV, when powered with a 5 V supply). Therefore, a gain stage is needed to obtain a signal large enough for additional processing. This additional processing may include digitization by a microcontroller's analog-to-digital converter, input to a comparator, and the like.

An instrumentation amplifier for pressure sensors should have a high input impedance, a low output impedance, differential to single-ended conversion of the pressure-related voltage output, and high gain capability. In addition, it will be useful to have the gain adjustment without compromising common mode rejection and both positive and negative DC-level shifts of the zero pressure offset.

Varying the gain and offset is desirable since full-scale span and zero pressure offset voltages of pressure sensors will vary somewhat from unit to unit. Therefore, a variable gain is desirable to fine tune the sensor's full-scale span, and a positive or negative DC-level shift (offset adjustment) of the pressure sensor signal is needed to translate the pressure sensor's signal-conditioned output span to a specific level (e.g., with the high and low reference voltages of an ADC).

Pressure sensor interface circuits may require either a positive or a negative DC-level shift to adjust the zero pressure offset voltage. As described previously, if the signal-conditioned pressure sensor voltage is an input to an ADC, the sensor's output dynamic range must be positioned within the high and low reference voltages of the ADC; that is, the zero pressure offset voltage must be greater than (or equal to) the low reference voltage and the full-scale pressure voltage must be less than (or equal to) the high reference voltage (see

Figure 7-28(a)). Otherwise, voltages above the high reference will be digitally converted as 255 decimal (for an 8-bit ADC), and voltages below the low reference will be converted as 0. This creates nonlinearity in the analog-to-digital conversion.

A similar requirement that warrants the use of a DC-level shift is to prevent the pressure sensor's voltage from extending into the saturation regions of the operational amplifiers. This also would cause nonlinearity in the sensor output

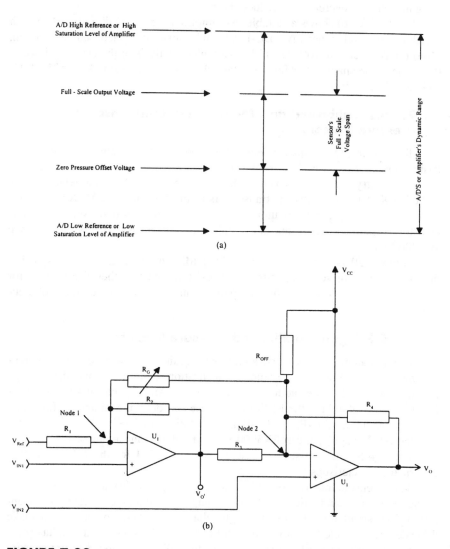

(a)

(b)

FIGURE 7-28 Sensor signal conditioning: (a) positioning the sensor's full-scale span within the ADC's or amplifier's dynamic range, (b) a suitable two amp signal conditioning stage

measurements. For example, if an op amp powered with a single-ended 5 V supply saturates near the low rail of the supply at 0.2 V, a positive DC-level shift may be required to position the zero pressure offset voltage at or above 0.2 V. Likewise, if the same op amp saturates near the high rail of the supply at 4.8 V, a negative DC-level shift may be required to position the full-scale pressure voltage at or below 4.8 V. It should be obvious that, if the gain of the amplifiers is too large, the span may be too large to be positioned within the 4.6 V window (regardless of ability to level shift the DC offset). In such a case, the gain must be decreased to reduce the span.

Figure 7-28(b) shows a suitable two-amplifier signal conditioning state with variable gain and a negative DC-level shift capability (Jacobsen and Baum, Motorola Application Note AN-1525). Complete analysis of the circuit is beyond the scope of the chapter. For further details, Jacobsen and Baum (AN-1525, 1995) and Jacobsen (1996).

7.5.5.2 Calibrated and Temperature Compensated Pressure Sensors

To provide precise span, offset calibration, and temperature compensation, basic sensor elements such as Motorola's X-ducer could be supplemented with special circuitry within the sensor package. An example of such a device family is the MPX2000 series pressure transducers from Motorola. The MPX2000 series sensors are available both as unported elements and as ported assemblies suitable for pressure, vacuum, and differential pressure measurements in the range 10–200 kPa.

Figure 7-29 is a block diagram of the MPX2000 series sensors, showing the arrangement of seven laser-trimmed resistors and two thermistors used for calibration of the sensor for offset, span, symmetry, and temperature compensation.

7.5.5.3 Signal-Conditioned Pressure Sensors

In this category of sensors, additional circuitry is added for signal conditioning (amplification), temperature compensation, calibration, and the like, so that the user needs fewer additional components. An example of such a sensor family from Motorola is the MPX5000. These sensors are available in full-scale pressure ranges of 50 kPa (7.3 psi) and 100 kPa (14.7 psi). With the recommended 5.0 V supply, the MPX5000 series produces an output of 0.5 V at no pressure to 4.5 V at full-scale pressure. (See Table 7-4 for the MPX5100DP's electrical characteristics.)

These sensors integrate on-chip bipolar op-amp circuitry and thin-film resistor networks to provide high-level analog output signal and temperature compensation. The small form factor and high reliability of on-chip integration make these devices suitable for automotive applications such as manifold absolute pressure sensing. Figure 7-30 is a schematic of the fully integrated pressure sensor.

To explain the advantage of signal conditioning on chip, refer to Figure 7-31. Figure 7-31(a) is a schematic of the circuitry to be coupled with an MPX2000

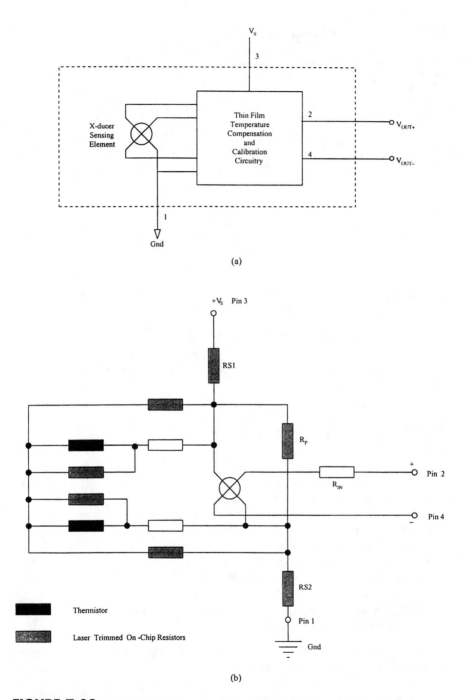

FIGURE 7-29 The MPX2000 series: (a) sensor block diagram, (b) arrangement of thermistors and laser-trimmed resistor sensors. (Reproduced by permission of Motorola Inc.)

TABLE 7-4 MPX5100DP electrical characteristics

Characteristics	Symbol	Minimum	Typical	Maximum
Pressure range (kPa)	P_{op}	0	—	100
Supply voltage (V)	V_S	—	5.0	6.0
Full-scale span (V)	V_{FSS}	3.9	4.0	4.1
Zero pressure offset (V)	V_{off}	0.4	0.5	0.6
Sensitivity (mV/kPa)	S	—	40	—
Linearity (%FSS)		−0.5	—	0.5
Temperature effect on span (%FSS)		−1.0	—	1.0
Temperature effect on offset (mV)		−50	0.2	50

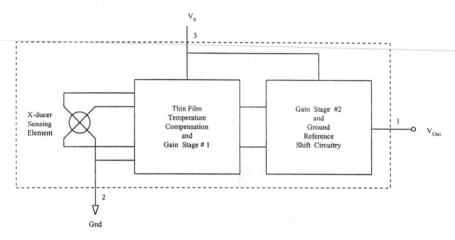

FIGURE 7-30 A fully integrated pressure sensor. (Reproduced by permission of Motorola Inc.)

series (which is compensated for temperature and calibrated for offset) to achieve ground referenced output with amplification.

Some devices similar to the MPX5100 go one step further by adding the differential-to-ground referenced conversion and the amplification circuitry on chip. This reduces the 18-component circuit in Figure 7-31(a) to a 1-signal conditioned sensor, as shown in Figure 7-31(b). Figure 7-32 is a schematic of a fully integrated pressure sensor such as the MPX5100.

7.5.5.4 Interface Between Pressure Sensors and Microprocessors or ADCs

In many practical situations, designers face the need to provide an interface between pressure sensors and microprocessors or microcontroller-based systems. In such cases, the designer should consider the level of on-chip signal conditioning or on-chip temperature compensation/calibration available in designing the system. In sensors with on-chip calibration and temperature compensation, the basic block diagram of a system could be depicted as in Figure 7-33.

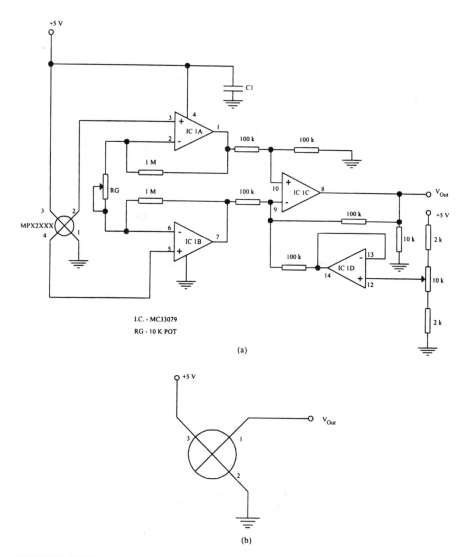

FIGURE 7-31 Simplification of external circuitry by internal signal conditioning: (a) high-level ground referenced output using an MPX2000 series transducer, (b) similar output-integrated device. (Reproduced by permission of Motorola Inc.)

When on-chip calibration and temperature compensation are not available, the gain stages shown need be designed to take care of such needs. While processor techniques are similar to other applications, such as temperature sensors, there are many advanced techniques for higher resolution or compensating for the offset and temperature. For such examples, see Schultz (Motorola AN-1318), Burri (Motorola AN-1097), Lucus (Motorola AN-1305), and Winkler (Motorola AN-1326). When configuring silicon pressure sensors with ADCs or

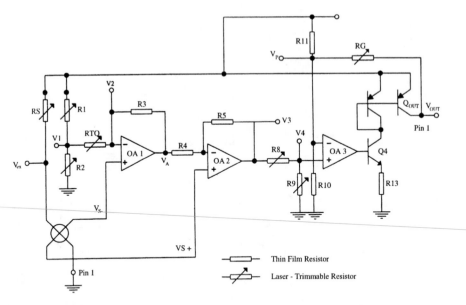

FIGURE 7-32 A fully integrated pressure sensor. (Reproduced by permission of Motorola Inc.)

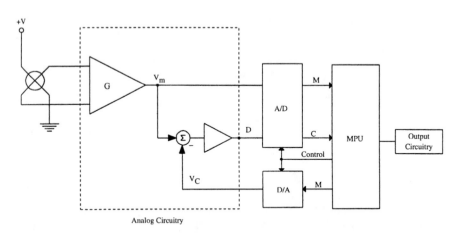

FIGURE 7-33 A basic block diagram for the interface between a compensated sensor and an ADC and microprocessor unit (MPU). (Reproduced by permission of Motorola Inc.)

microcontrollers with built-in ADCs, the ratiometric function of both the ADC and sensor could be used to minimize the need for additional components such as voltage reference sources. The ratiometric function of these elements makes all voltage variations from power supply rejected by the system.

The many advance techniques of using microcontroller-based sensor systems are beyond the scope of this chapter. Four Motorola application notes above describe such practical and useful techniques.

7.6 Silicon Accelerometers

With the demand from automotive and other industries in the latter 1980s, several sensor manufacturers developed micromachined silicon ICs for sensing acceleration. Today, several component manufacturers, such as IC Sensors, Analog Devices, and Motorola, have families of silicon accelerometers. Basic sensor elements as well as signal conditioned versions are available.

7.6.1 Basic Principles of Sensing Acceleration

The principles of acceleration sensing were simulated using a weight and spring connected to a frame to develop silicon accelerometers using the piezoresistive properties of silicon and building capacitive structures with variation of effective capacitance between plates attached to a seismic mass of silicon. To simulate the basic mechanical analogy of accelerometers and minimize secondary effects that complicate the measuring process (IC Sensors, •••; Quinnell, 1992) required several improvements in silicon-processing technology.

One such improvement was the advent of silicon fusion bonding (Quinnell, 1992). Fusion bonding, which bonds two wafers while preserving the crystalline structure of silicon, permits the creation of complex 3-D structures without introducing mechanical discontinuities or thermal-dependent stress. This structuring ability lets accelerometer manufacturers capture the seismic mass with a sealed cavity by bonding a cap and a base plate to the frame. By controlling the space between the mass and cavity, vendors can use the air sealed inside the cavity as a viscous damping fluid for the system's motion.

Silicon fusion bonding also provides an answer to another limitation: shock resistance. Simply falling off of a desk can produce a $200g$ shock when the sensor hits the floor. Despite silicon's toughness and flexibility, that kind of shock could break the springs in an accelerometer unless the seismic mass's motion is limited. Silicon fusion bonding allows the placement of bumpers and other mechanical stops to make the accelerometer much more shock resistant. Devices now routinely handle shocks as great as $2000g$.

In commercially available sensors, a single- or double-cantilevered or a membrane-supported mass is coupled with a piezoresistive or capacitive element.

7.6.1.1 Piezoresistive Sensors

Figure 7-34(a) is a diagram of a single-cantilevered design of an accelerometer using piezoresistive elements. Thin beams support one edge of a seismic mass, which is free to move within a cavity created by fusion bonding two additional wafers to the one containing the mass. Piezoelectric resistors fabricated

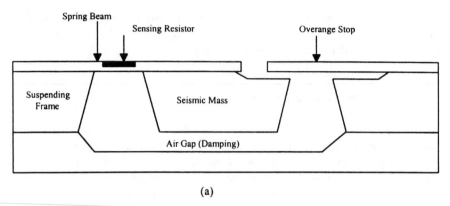

(a)

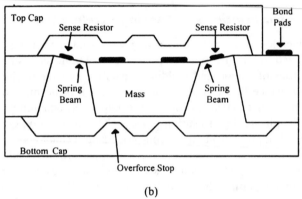

(b)

FIGURE 7-34 Silicon accelerometer simulating the mass and spring: (a) single cantilevered, (b) double cantilevered

at the beams measure the displacement by changing the resistance as the beams bend. The double-cantilevered approach shown in Figure 7-34(b) supports the mass from two sides.

While single-cantilevered types are simplest and sensitive, they have drawbacks such as transverse sensitivity (Quinnell, 1992). Double-cantilevered types can be designed with self-compensating effects for transverse forces. IC Sensors and Lucas Nova Sensor manufacture piezoresistive type sensors.

7.6.1.2 Capacitive Sensors

Figure 7-35 shows a typical capacitive sensing device. In these devices, the mass is supported on all four sides and transverse sensitivity is very much reduced. Capacitive sensors use top and bottom plates to form a capacitor divider with a seismic mass that is temperature insensitive. Sensing the change in capacitance requires relatively complex circuitry, however.

In some capacitive accelerometers, such as the ADXL series from Analog Devices, the seismic mass is not a single block but a series of interdigitated

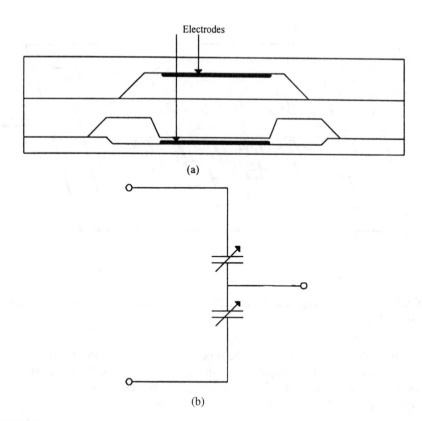

FIGURE 7-35 The basic arrangement of a capacitive sensor: (a) structure, (b) equivalent circuit

fingers, as shown in Figure 7-36. This allows the sensing of acceleration in the plane of the chip than in other types of sensors, where sensing is normal to the surface.

The drawback of complicated interface circuitry in a capacitive sensor is compensated for by the additional ability inherent in the capacitor structure. The presence of charge-carrying plates in the sensors provides a built-in means for applying an electrostatic force on the seismic mass. This capability lets the sensor be used in a closed-loop configuration.

Instead of letting the seismic mass move freely during acceleration, a closed-loop system applies a restoring force to the mass, keeping it relatively motionless. Restricting the movement of the mass has two advantages. First, it improves sensor linearity by confining the motion to the linear region of the spring's restoring force. Second, it extends the range of a sensor beyond the limits imposed by its housing on the seismic mass's movement. In such force-feedback systems, the restoring force, not the actual movement, serves as the measure of acceleration.

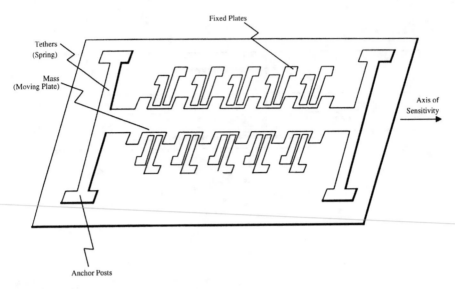

Fixed Plates

Tethers (Spring)

Mass (Moving Plate)

Axis of Sensitivity

Anchor Posts

FIGURE 7-36 Arrangement in the ADXL50. (Reproduced by permission of Analog Devices, Inc.)

The ability to apply a force to the proof mass has an additional advantage: It gives the sensor a self-test capability. This capability is particularly important in systems such as automotive airbags, where you cannot test the system by actually accelerating it yet testing is necessary for safety or reliability.

7.6.2 Practical Components

7.6.2.1 Piezoresistive Example: Model 3255 from IC Sensors

The Model 3255 is a fully signal conditioned accelerometer containing two chips: the silicon sensing element and a custom integrated circuit (ASIC) for signal conditioning. The Model 3255 accelerometer is available in various measurement ranges. With a supply voltage of 5 V, the output voltage is 2.5 V at no applied acceleration and the output range is 0.5–4.5 V for the full acceleration range. The output voltage is ratiometric with the supply voltage and will track the supply voltage in the range 5.0 ± 0.5 V. Only three connections need to be made to use the accelerometer: 5 V supply, ground, and signal output. Figure 7-37(a) is a photograph of the device showing the two chips and the sealed unit. Figure 7-37(b) shows the arrangement of the sensor element.

7.6.2.1.1 Sensor Element

The silicon sensor element is shown in Figure 7-37(b). A seismic mass and four flexures are formed using bulk micromachining processes. Each of the four beams contains two implanted resistors that are interconnected to form a Whetstone bridge. When the device undergoes acceleration, the mass moves up

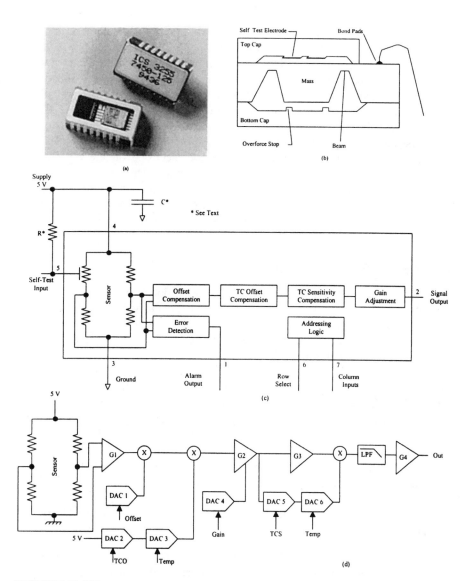

FIGURE 7-37 Model 3255 accelerometer: (a) photograph, (b) cross-section of accelerometer die, (c) functional block diagram, (d) simplified schematic diagram. (Reproduced by permission of IC Sensors.)

or down, causing four of the resistors to increase and the other four to decrease in value. This results in an output voltage change proportional to the applied acceleration. The eight resistors are interconnected such that the effect of any off-axis acceleration is canceled.

Silicon top and bottom caps are attached to the section containing the seismic mass and the beams. The silicon caps serve several purposes. Precision gaps are etched into the caps to provide air damping to suppress the resonant peak of the structure. Because the part is critically damped, the frequency response is flat up to several kilohertz with little dependence on temperature. Small elevated stops on the top and bottom caps limit the motion of the mass to a fraction of the deflection at which fracture occurs. The caps also form a chamber around the seismic mass to provide protection during the later stages of manufacturing and its operating lifetime.

Last, the top cap allows testing the accelerometer in the absence of acceleration. When a voltage is applied to a metal electrode on the top cap, an electrostatic force moves the mass toward the top cap. This results in a change in output voltage proportional to the sensitivity and to the square of the applied voltage. It thus is possible to generate an "acceleration" using an external voltage and check the functionality of the mechanical structure as well as the electronics.

7.6.2.1.2 Signal Conditioning

The signal conditioning circuitry amplifies the output of the sensor element and corrects the sensitivity and offset changes that occur with overheating. As a result, the output signal is accurate and no trimming is required by the user. The data used to set the performance of the accelerometer is stored in fused registers within the signal conditioning IC.

The signal conditioning IC converts the differential signal from the sensor element (nominally ±5 mV) into a single-ended signal in the 0.5–4.5 V range while correcting for the temperature-related signal variations. Signals are processed by differential amplifiers throughout most of the circuit to minimize common mode effects and noise. Switched-capacitor circuitry is used to save space and because high-accuracy gain stages can be made easily. As a result, the compensated accelerometers are interchangeable with a very small total error. The signal conditioning IC is made in 1.5 μm CMOS technology and intended for 5 V operation.

The signal path is shown in the block diagram in Figure 7-37(c). The output signal of the accelerometer die is processed by the following stages:

- The first stage provides a high impedance load for the sensor and amplifies the signal to maximize the dynamic range during subsequent processing.
- The offset of the sensor die is reduced to less than 0.5% of full scale at room temperature by adding a voltage generated by DAC 1. This DAC is controlled by a digital word representing the programmed offset value.
- The temperature coefficient of offset (TCO) of the sensor is compensated for by adding a voltage generated by DAC 2 and DAC 3. This voltage is controlled by digital words representing the temperature and the programmed TCO value. Both the offset and TCO voltages are derived from the supply to ensure that the signal remains ratiometric with the supply voltage.
- The signal gain is set by the value in DAC 4. The gain can be varied in a 5:1 range to allow for different full-scale specifications.

- The temperature coefficient of sensitivity (TCS) of the sensor is compensated in the next stage, built around a feed-forward loop using two DACs controlled by digital words representing the temperature and the programmed TCS value. The sensitivity decrease over temperature is compensated for by increasing the signal gain linearly with temperature.
- The output bias voltage can be set to either 0.5 or 2.5 V by connecting an input pad on the chip to ground during assembly of the part. This allows signals to be processed with either a bipolar or unipolar range.
- A two-pole passive filter removes signals generated by the internal oscillator and switched capacitor networks. Switching noise is further minimized by having separate digital and analog internal supply lines and the differential signal processing.
- The final stage provides a low-impedance output for driving resistive and capacitive loads without influencing the signal. The output will go in a impedance "tristate" mode if the part is not addressed.

The temperature word that controls DACs 3 and 6 is generated by an ADC that digitizes the output of a temperature PTAT source driven by a bandgap reference. The temperature word therefore is linearly proportional to the temperature but does not depend on the supply voltage. For application and performance details, see IC Sensors (1995).

7.6.2.2 Capacitive Example: The ADXL50 from Analog Devices

The ADXL50 is a complete acceleration measurement system on a single monolithic IC. Three external capacitors and a +5 V power supply are all that is required to measure accelerations up to $\pm 50g$. Device sensitivity is factory trimmed to 19 mV/g, resulting in a full-scale output swing of ± 0.95 V for a $\pm 50g$ applied acceleration. Its $0g$ output level is +1.8 V. A TTL compatible self-test function can electrostatically deflect the sensor beam at any time to verify device functionality. A functional block diagram of ADXL50 is shown in Figure 7-38.

The ADXL50 is a complete acceleration measurement system on a single monolithic IC. It contains a polysilicon surface-micromachined sensor and signal conditioning circuitry. The ADXL50 is capable of measuring both positive and negative acceleration to a maximum level of $\pm 50g$.

Figure 7-39(a) is a simplified view of the ADXL50's acceleration sensor at rest. The actual structure of the sensor consists of 42 unit cells and a common beam. The differential capacitor sensor consists of independent fixed plates and a movable "floating" central plate that deflects in response to changes in relative motion. The two capacitors are series connected, forming a capacitive divider with a common movable central plate. A force balance technique counters any impending deflection due to acceleration and drives the sensor back to its $0g$ position.

Figure 7-39(b) shows the sensor responding to applied acceleration. When this occurs, the common central plate or "beam" moves closer to one of the

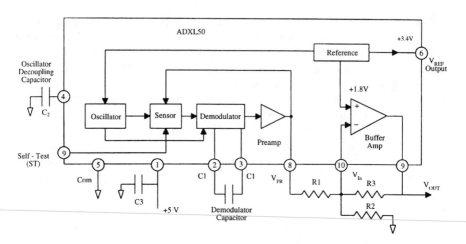

FIGURE 7-38 Functional block diagram of the ADXL50. (Reproduced by permission of Analog Devices, Inc.)

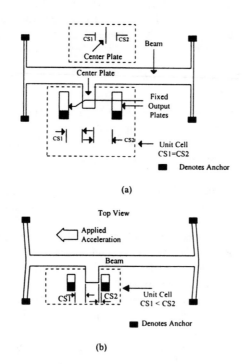

FIGURE 7-39 ADXL50 operation: (a) sensor element at rest, (b) sensor momentarily responding to acceleration, (c) functional block diagram. (Reproduced by permission of Analog Devices, Inc.)

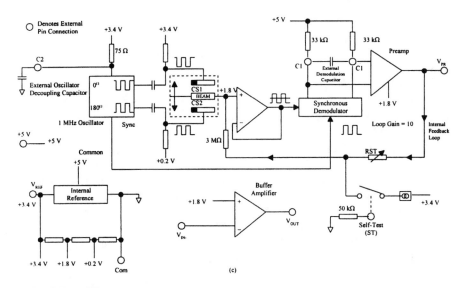

FIGURE 7-39 Continued

fixed plates and farther from the other. The sensor's fixed-capacitor plates are driven deferentially by a 1 MHz square wave; the two square wave amplitudes are equal but 180° out of phase with one another. When at rest, the values of the two capacitors are the same, and therefore, the voltage output at their electrical center (i.e., at the center plate) is 0.

When the sensor begins to move, a mismatch in the value of their capacitance is created, producing an output signal at the central plate. The output amplitude will increase with the amount of acceleration experienced by the sensor. Information concerning the direction of beam motion is contained in the phase of the signal with synchronous demodulation being used to extract this information. Note that the sensor needs to be positioned so that the measured acceleration is along its sensitive axis.

Figure 7-39(c) shows a block diagram of the ADXL50. The voltage output from the central plate of the sensor is buffered and applied to a synchronous demodulator. The demodulator also is supplied with a (nominal) 1 MHz signal from the same oscillator that drives the fixed plates of the sensor. The demodulator will rectify any voltage in sync with its clock signal. If the applied voltage is in sync and in phase with the clock, a positive output will result. If the applied voltage is in sync but 180° out of phase with the clock, the demodulator's output will be negative. All other signals will be rejected. An external capacitor, C1, sets the bandwidth of the demodulator.

The output of the synchronous demodulator drives the preamp, an instrumentation amplifier buffer that is referenced to +1.8 V. The output of the preamp is fed back to the sensor through a 3 MΩ isolation resistor. The correction voltage required to hold the sensor's center plate in the 0g position is a direct measure of the applied acceleration and appears at the V_{PR} pin. When the ADXL50 is subjected to acceleration, its capacitive sensor begins to move, creating a momentary output signal. This is signal conditioned and amplified by the demodulator and preamp circuits. The DC voltage appearing at the preamp output then is fed back to the sensor and electrostatically forces the center plate back to its original center position.

At 0g, the ADXL50 is calibrated to provide +1.8 V at the V_{PR} pin. With applied acceleration, the V_{PR} voltage changes to the voltage required to hold the sensor stationary for the duration of the acceleration and provides an output that varies directly with the applied acceleration. The loop bandwidth corresponds to the time required to apply feedback to the sensor and is set by external capacitor C1. The loop response is fast enough to follow changes in gravitational level up to that exceeding 1 kHz. The ADXL50's ability to maintain a flat response over this bandwidth keeps the sensor virtually motionless. This eliminates any nonlinearity or aging effects due to the sensor beam's mechanical spring constant, as compared to an open-loop sensor.

An uncommitted buffer amplifier provides the capability to adjust the scale factor and 0g offset level over a wide range. An internal reference supplies the necessary regulated voltages for powering the chip and +3.4 V for external use.

Applications and further details on the ADXL series devices can be found in Analog Devices (Application Note G2112A).

7.7 Hall Effect Devices

The basic Hall sensor is simply a small sheet of semiconductor material. A constant voltage source forces a constant bias current to flow in the semiconductor sheet. The output, a voltage measured across the width of the sheet, reads near 0 if a magnetic field is not present. If the biased Hall sensor is placed in a magnetic field oriented at right angles to the Hall current, the voltage output is in direct proportion to the strength of the magnetic field. This is the Hall effect, discovered by E. H. Hall in 1879 (Figure 7-40). When a magnetic field, B, is applied to a specimen (metal or semiconductor) carrying a current, I_c, in the direction perpendicular to I_c, a potential difference, V_H, proportional to the magnitude of the applied magnetic field B appears in the direction perpendicular to both I_c and B. The relationship is expressed in the form

$$V_H = K \times I_c \times B$$

where K represents a constant, the product sensitivity, which depends on the physical properties and dimensions of the material used for the Hall effect device.

The basic Hall sensor essentially is a transducer that will respond with an output voltage if the applied magnetic field changes in any manner. Differences in the response of devices generally are related to tolerances and specifications, such

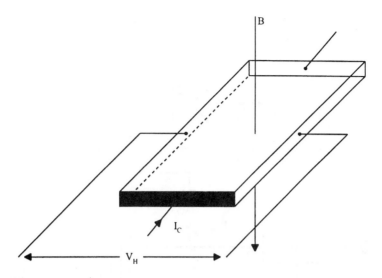

FIGURE 7-40 The Hall effect

as operate (turn on) and release (turn off) thresholds, as well as the temperature ranges and temperature coefficients of these parameters. Also available are linear output sensors that differ in sensitivity or respond per gauss change.

7.7.1 Linear Output Devices

Linear output Hall effect devices are the simplest Hall sensor devices. Practical devices such as Allegro Microsystems' UGN-3605 give an output voltage response to applied magnetic field changes. Electrical connections for the UGN-3605 are given in Figure 7-41(a). Applications of Hall devices are discussed in Swager (1989).

The output voltage of the devices such as the UGN-3605 is quite small, which can present problems, especially in an electrically noisy environment. Addition of a suitable DC amplifier and a voltage regulator to the circuit improves the transducer's output and allows it to operate over a wide range of supply voltages. Such combined devices are available and an example is the UGN-3501 from Allegro Microsystems.

7.7.2 Digital Output Devices

The addition of a Schmitt trigger threshold detector with built-in hysteresis, as shown in Figure 7-42, gives the Hall effect circuit digital output capabilities. When the applied magnetic flux density exceeds a certain limit, the trigger provides a clean transition from off to on with no contact bounce. Built-in hysteresis eliminates oscillation (spurious switching of the output) by introducing a magnetic dead zone in which switch action is disabled after the threshold value is passed.

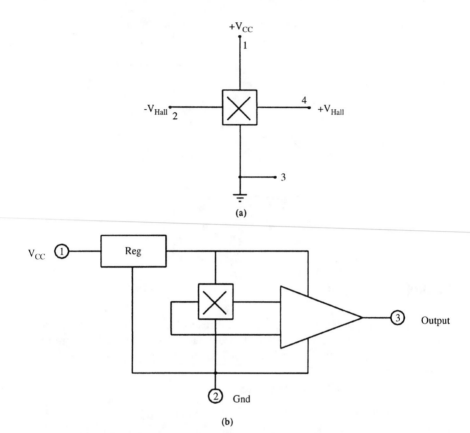

FIGURE 7-41 Linear output Hall effect device: (a) device connections for UGN-3605, (b) amplified version. (Reproduced by permission of Allegro Microsystems.)

An open-collector NPN output transistor added to the circuit gives the switch digital logic compatibility. The transistor is a saturated switch that shorts the output terminal to ground wherever the applied flux density is higher than the on trip point of the device. The switch is compatible with all digital families. The output transistor can sink enough current to directly drive many loads, including relays, triacs, SCRs, LEDs, and lamps.

7.8 Humidity and Chemical Sensors

7.8.1 Humidity Sensors

Humidity, usually understood to refer to the water content of the air, also can be sensed using silicon-based sensor elements. Relative humidity (RH), which is the ratio of absolute humidity to saturation humidity, has a value between 0 and 1 (0% and 100%). Several techniques are used to measure the relative humidity

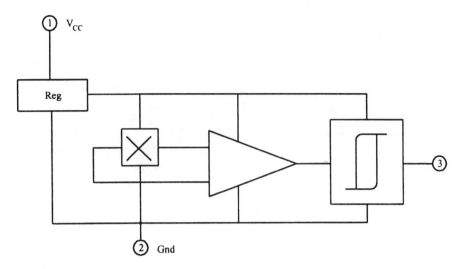

FIGURE 7-42 Digital output Hall effect switch. (Reproduced by permission of Allegro Microsystems.)

using capacitance, resistance, conductivity, and temperature measurements. Thermoset polymer or thermoplastic polymer-based materials are used on silicon or ceramic-based substrates to measure the RH. A comparison of RH sensors is available in Microswitch (1997b).

Capacitive RH sensors dominate both atmospheric and process measurements and are the only type of full-range RH measuring devices capable of operating accurately down to 0% RH. Because of their low temperature effect, they often are used over wide temperature ranges without active temperature compensation.

Thermoset polymer-based (as opposed to thermoplastic-based) capacitive sensors (see Figure 7-43) allow higher operating temperatures and provide better resistivity against chemical liquids; vapors such as isopropyl, benzene, toluene, and formaldehyde; oils; common cleaning agents; and ammonia vapor in concentrations common to chicken coops and pig barns. In addition, thermoset polymer RH sensors provide the longest operating life in ethylene oxide-based sterilization processes.

An example of a thermoset polymer-based capacitance RH device family is the Hycal IH36XX series from Microswitch. These devices come with on-chip signal conditioning and provide a fairly linear ratiometric output based on the DC supply.

In operation, water vapor in the active capacitor's dielectric layer equilibrates with the surrounding gas. The porous platinum layer shields the dielectric response from external influences while the protective polymer overlayer provides mechanical protection to the platinum layer from contaminants such as dirt, dust, and oil. A heavy contaminant layer of dirt, however, will slow down the sensor's response time, because it will take longer for water vapor to equilibrate in the sensor.

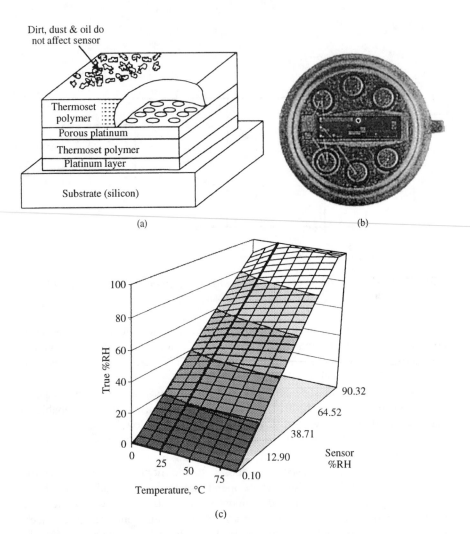

FIGURE 7-43 Thermoset polymer-based RH sensors: (a) basic construction, (b) relative humidity IC, (c) output voltage of IH-3602 vs. relative humidity. (Reproduced by permission of Microswitch.)

7.8.2 Temperature and Humidity Effects

The output of all absorption-based humidity sensors (capacitive, bulk resistive, conductive film, etc.) are affected by both temperature and %RH. Because of this, temperature compensation is used in applications that call for either higher accuracy or wider operating temperature ranges. When temperature compensating a humidity sensor, it is best to make the temperature measurement as close as possible to the humidity sensor's active area; that is, within the same moisture microenvironment. This is especially true when combining RH and temperature as a method of measuring the dew point.

HyCal's industrial-grade humidity and dew point instruments incorporate a HyCal 1000 Ω platinum resistance temperature detector on the back of the ceramic sensor substrate for unmatched temperature compensation measurement integrity. No on-chip signal conditioning is provided in these high temperature sensors (Figure 7-43(b)).

7.8.3 Chemical Sensors

Numerous technologies are utilized in the chemical sensing industry (see Walters, 1996). Silicon as a basic structure for chemical sensors has been investigated in numerous laboratories over the last 20 years. Based on metal oxide semiconductor gas sensors, some sensor manufacturers such as FiS Sensors (Japan) offer a wide range of products covering many applications, including carbon monoxide sensing, flammable gas detection, toxic gas detection, indoor air quality controls, and combustion monitoring and control.

The sensing element used in these devices is a mini-bead-type semiconductor, composed mainly of tin dioxide (SnO_2). A heater coil and an electrode wire are embedded in the element (Figure 7-44(a)). The element is installed in a metal housing, which uses double stainless steel mesh in the path of gas flow and provides an antiexplosion feature (Figure 7-44(b)). The sensor has three pins for output signal and heater power supply. The SB-50 uses an active charcoal filter as shown in Figure 7-44(c). The conductivity of tin dioxide-based metal oxide semiconductor material changes according to gas concentration changes. This is caused by adsorption and desorption of oxygen and the reaction between surface oxygen and gases. These reactions cause a dynamic change of electric potential on the SnO_2 crystal and results in a decrease in sensor resistance under the presence of reducing gases such as carbon monoxide, methane, and hydrogen.

Figure 7-44(d) and (e) show the pin layout and the equivalent circuit. Figure 7-44(f) shows the standard circuit of the SB series. The applied heater voltage regulates the sensing element temperature to obtain the specific performance of sensors. A change in the sensor resistance generally is obtained as a change in the output voltage across the fixed or variable load resistor (R_L) in series with the sensor resistance (R_S).

The sensitivity characteristics of semiconductor gas sensors are shown by the relationship between the sensors resistance (R_S) and concentration of gases. The sensor resistance decreases with an increase of the gas concentration based on a logarithmic function. The standard test conditions of each model are calibrated to meet a typical target gas and concentration; for example, methane 1000 ppm for flammable gas detection, hydrogen 100 ppm for hydrogen detection, or ethanol 300 ppm for solvent detection. Figure 7-45 shows the typical sensitivity characteristics of the SB series. In these diagrams, the sensor resistance change is normalized by the R_S at specific conditions. For further details, see FiS Inc. (1996).

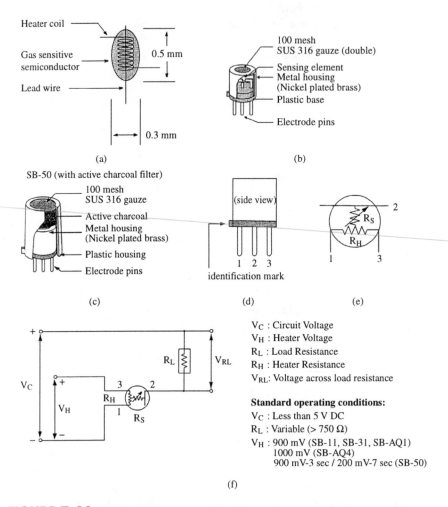

FIGURE 7-44 The SB series gas sensors: (a) sensing element, (b) structure for standard housing, (c) SB-50 housing, (d) pin layout, (e) equivalent circuits, (f) standard circuit. (Reproduced by permission of FiS Inc.)

7.9 IEEE P1451 Standard for Smart Sensors and Actuators

Sensors are used in a wide range of applications from industrial automation to patient-condition monitoring in hospitals. With advancement of silicon and Micro Electro Mechanical Systems (MEMS) technologies, more "smarts" are integrated into sensors. The emergence of the control networks and smart devices in the marketplace may provide economical solutions for connecting transducers (hereafter specified as sensors or actuators) in distributed measurement and control applications; therefore, networking small transducers is seriously considered by transducer manufacturers and users.

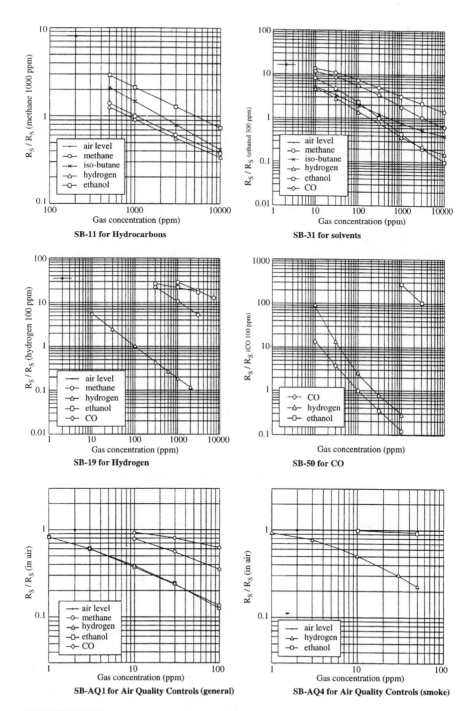

FIGURE 7-45 Typical sensitivity characteristics of the SB series gas sensors. (Reproduced by permission of FiS Inc.)

Control networks provide many benefits for transducers:

- Significant reduction in installation costs by eliminating many and long analog wires.
- Acceleration of control loop design cycles, reduction of commissioning time, and reduction of downtime.
- Dynamic configuration of measurement and control loops via software.
- Addition of intelligence by leveraging the microprocessors used for digital communication.

For anyone attempting to choose a sensor-interface or networking standard, the range of choices is overwhelming. Some standards are open, and some are proprietary to a company's control products. To remedy the situation, the IEEE Sensor Technology Committee TC-9 is developing the IEEE P1451, Standard for Smart Transducer Interface for Sensors and Actuators. The sensor market comprises widely disparate sensor types. Designers consume relatively large amounts of all types of sensors. However, the lack of a universal interface standard impedes the incorporation of "smart" features, such as an onboard electronic data sheet, onboard A/D conversion, signal conditioning, device-type identification, and communications hand-shaking circuitry, into the sensors. In response to the industry's need for a communication interface for sensors, the IEEE with cooperation from the National Institute of Standards and Technology (NIST), decided to develop a hardware-independent communication standard for low-cost smart sensors that includes smart transducer object models for control networks (Travis, 1995).

The IEEE P1451 standards effort, currently under development, will provide many benefits to the industry. P1451, "Draft Standard for Smart Transducer Interface for Sensors and Actuators," consists of four parts, namely:

(i) IEEE 1451.1 — Network Capable Application Processor (NCAP) information model,

(ii) IEEE 1451.2 — Transducer to Microprocessor Communications Protocols and Transducer Electronic Data Sheet (TEDS) formats,

(iii) IEEE P1451.3 — Digital Communication and Transducer Electronic Data Sheet (TEDS) formats for distributed multidrop systems, and

(iv) IEEE 1451.4 — Mixed-mode Communication Protocols and Transducer Electronic Data Sheet (TEDS) formats.

In the process of writing the draft document, the working group has defined the smart transducer interface module (STIM), transducer electronic data sheet (TEDS), transducer-independent interface (TII), and a set of communication protocols between the STIM and the network capable application processor (NCAP).

A system block diagram depicting the interface is shown in Figure 7-46. A STIM is specified to include up to 255 transducers, a signal converter or conditioning, a TEDS, and the necessary logic circuitry to support digital communication

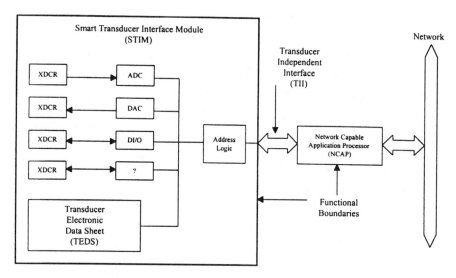

FIGURE 7-46 System block diagram depicting the transducer interface (Courtesy of NIST)

with NCAP. The TEDS is a small physical memory containing manufacturer's information and data for the transducer in a standardized data format. The TII, a 10-wire digital interface with provision for hot-swapping a sensor to a network, is used to access the TEDS, read sensors, and set actuators.

Figure 7-47(a) depicts a STIM and the associated digital interface as described in the P1451.2-1997 hot swap. The STIM shown here is under the control of a network-node microprocessor. In addition to their use in control networks, STIMs can be used with microprocessors in a variety of applications, such as portable instruments and data acquisition cards, as shown in Figure 7-47(b).

TABLE 7-5 The ten lines that make up the transducer-independent interface (Courtesy of NIST)

Line	Driven by	Function
DIN	NCAP	Address and data transport from NCAP to STIM
DOUT	STIM	Data transport from STIM to NCAP
DCLK	NCAP	Positive-going edge latches data on both DIN and DOUT
NIOE	NCAP	Signals that the data transport is active and delimits data transport framing
NTRIG	NCAP	Performs triggering function
NACK	STIM	Serves two functions: trigger acknowledge and data transport acknowledge
NINT	STIM	Used by the STIM to request service from the NCAP
NSDET	STIM	Grounded in the STIM and used by the NCAP to detect the presence of a STIM
POWER	NCAP	Nominal 5 V power supply
COMMON	NCAP	Signal common or ground

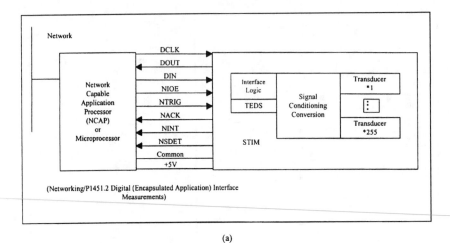

(a)

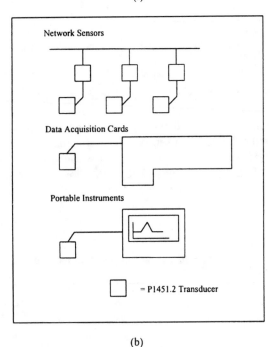

(b)

FIGURE 7-47 (a) Hardware partition proposed by P1451.2 and (b) possible use for the interface. (Source: Woods, 1996.)

The origin and function of each signal line of the ten-wire interface is listed in Table 7-5.

1451.2 was adopted by the IEEE as a full use standard, designated as IEEE Std. 1451.2-1997. The IEEE Std. 1451.2-1997 can be applied standalone, or it can be used with P1451.1. The two documents together will define a standard interface

for networked smart sensors and actuators. Likewise, the P1451.1 infor.
be implemented in a sensor control or field network without 1451.2.

The IEEE Std. 1451.2-1997 standard and IEEE P1451.1 D2.2 draft
ordered from the IEEE customer service department by calling 1-(800)-678-
(IEEE) in the United States and Canada, 1-(732)-981-0600 from outside
United States and Canada, or by faxing 1-(732)-981-9667.

7.10 P1451 and Practical Components

Existing microcontrollers fall short of fully implementing the standard in
silicon, either because of functionality or prohibitive cost. For example, the
standard transducer interface module (STIM) portion of the standard specifies
the sensor interface electronics, signal conditioning, data conversion, calibration,
linearization, basic communication capability, and a non-volatile 565-byte TEDS.
Some microcontrollers with integrated 8- or 10-bit ADCs or comparator-based
slope conversion can implement most of the STIM functionality, but are limited
in conversion speed and accuracy. Moreover, few available controllers have
economically integrated analog conversion together with high-density EEPROM
because of the additional process complexity requirements of both functions.

These limitations are overcome by recently introduced components such as
the AduC812 MicroConverter™ (Leonard, 1998) from Analog Devices, which
integrates key STIM elements with 12-bit, 5 µs data conversion on a single chip
for high-accuracy, fast-conversion-time applications such as battery monitoring,
pressure and temperature management, gas monitoring, and leak detection. In a
typical application, the AduC812 conditions and converts signals from various
types of sensors, sends signals to actuators and display devices, and communications with the host microprocessor over signal and control lines.

The AduC812 MicroConverter™ is supported by a development system
that includes documentation, applications board, power supply, serial port cable,
and software. Provided on a 3.5-inch floppy disk, the software consists of an
assembler, simulator, debugger, serial downloader, and example code.

References

Ajluni, C. "Pressure Sensors Strive to Stay on Top." *Electronic Design* (October 3, 1994), pp. 67–74.
Analog Devices, Inc. *Design*. In *Reference Manual*. Analog Devices Inc., USA, 1994.
Analog Devices, Inc. "Data Sheet: Serial Output Thermometers — TMP03/04," rev. 0, Analog
    Devices, Inc., MA, USA 1995a.
Analog Devices, Inc. "Data Sheet: Airflow and Temperature Sensor — TMP 12," rev. 0. Analog
    Devices, Inc., MA, USA 1995b.
Analog Devices, Inc. "Data Sheet: Low Voltage Temperature Sensors — TMP 35/TMP 36/TMP 37,"
    rev. 0, MA, USA 1996a.
Analog Devices, Inc. "Data Sheet: ADXL 50," rev. B. Analog Devices, Inc., MA, USA 1996b.
Analog Devices, Inc. "Accuracies of AD 590." Application note AN-272. Analog Devices, Inc.,
    Applications Reference Manual 1993, pp. 19-3 to 19-6.
Analog Devices, Inc. "Practical Design Techniques for Sensor Signal Conditioning," Analog Devices
    Inc., USA 1999.

Analog Devices, Inc. "Accelerometer Application Guide" (G 2112A). Analog Devices, Inc.

Bry Zek, J. "Characterization of MEMS Industry in Silicon Valley and Its Impact on Sensor Technology." Proceedings of Sensors Expo, October 1996, pp. 1–13.

Burri, M. "Calibration Free Pressure Sensor System." Application note AN-1097. Motorola Inc., Sensor Device Data, DL200/D, Rev. 2, 1995, pp. 4-19 to 4-23.

Caristi, A. J. "A Digital Pressure Gauge Using the Motorola MPX700 Series Differential Pressure Sensors." Application note AN-1105. Motorola Inc., Sensor Device Data, DL200/D, Rev. 2, pp. 4-27 to 4-31, 1995.

Dallas Semiconductor. *System Extension Data Book*. Dallas Semiconductor, TX, 1994–1995.

Demington, C. "Compensating for Non-Linearity in the MPX10 Series Pressure Transducer." Application note AN 935. Motorola Inc., Sensor Device Data, DL200/D, Rev. 2, 1995, pp. 4-4 to 4-10.

FiS Inc. *SB Products Review: Sensors and Systems Technology*. Japan: FiS Inc., 1996.

Freeman, W. "Solid State Temperature Sensors Protect and Control." PCIM (November 1993), pp. 39–45.

Guckel, H. "Micromechanisms Fabrication: Challenge in Micromechanisms and Microelectricity." ISSCC Proceedings, 1992 (IEEE Catalog Number 92CH3128-6), pp. 14–17.

Hauptmann, P. *Sensors—Principals and Applications*. Englewood Cliffs, NJ: Prentice-Hall, 1991.

IC Sensors. "Temperature Compensation—IC Pressure Sensors." Application note TN-002. IC Sensors, March 1985.

IC Sensors. "Signal Conditioning for IC Pressure Sensors." Application note TN-001. IC Sensors, April 1988.

IC Sensors. "Model 3255 Accelerometer." Application note TN-010. IC Sensors, April 1995.

IC Sensors. "Understanding Accelerometer Technology." Application note UATRO-9111. IC Sensors, Milpitas, CA, USA.

Jacobson, E. "Designing Amplifiers for Sensor Applications: A Cookbook Approach." *EDN* (January 4, 1996), pp. 119–128.

Jacobson, E., and J. Baum. "Optimize Sensor Systems Using Fixed Components." *EDN* (July 6, 1995), pp. 85–95.

Jacobson, E., and J. Baum. "The A-B-C's of Signal Conditioning Amplifier Design for Sensor Applications." Application note AN 1525. Motorola Inc., Sensor Device Data, DL200/D, Rev. 2, pp. 4-152 to 4-158, 1995.

Klonowski, P. "Use of the AD 590 Temperature Transducer in a Remote Sensing Application." Application note AN-273. Analog Devices, Inc., MA, USA.

Leonard, M. "Self-Programming Microcontroller Networks Sensors & Transducers," *Electronic Design* (February 9, 1998), pp. 92–94.

Le Fort, B., and B. Ries. "Taking the Uncertainty out of Thermocouple Temperature Measurement (with AD594/AD595)." Application note AN-274, Analog Devices, Inc., Norwood, MA, USA.

Lucus, B. "An Evaluation System for Direct Interface of MPX5100 Pressure Sensor with a Microprocessor." Application note AN 1305. Motorola Inc., Sensor Device Data, DL200/D, Rev. 2, pp. 4-42 to 4-57, 1995.

Marcin, J. "Thermocouple Signal Conditioning Using the AD594/AD595." Application note AN-369. Analog Devices, Inc., Norwood, MA, USA.

Microswitch. "Pressure, Force and Airflow Sensors." In *Catalog 15*. Microswitch, August 1996, pp. 74–91.

Microswitch. "Temperature and Moisture Sensors." Microswitch, May 1997.

Microswitch. "Moisture Tutorial-Relative Humidity Sensors." In *Temperature and Moisture Sensors Catalog*. Microswitch, May 1997, pp. B1–B7.

Motorola Inc. "Analog to Digital Converter Resolution Extension Using a Motorola Pressure Sensor." Application note AN 1100. Motorola Inc., Sensor Device Data, DL200/D, Rev. 2, pp. 4-24 to 4-26, 1995.

Motorola Inc. "Sensor Device Data Book." DL 200/D, Rev. 2, Q2/95. Motorola Inc., 1995.

Quinnell, R. A. "Silicon Accelerometers Tackle Cost-Sensitive Applications." *EDN* (September 3, 1992), pp. 69–76.

Schultz, W. "Interfacing Semiconductor Pressure Sensors to Microcomputers." Application note AN-1318. Motorola Inc., Sensor Device Data, DL200/D, Rev. 2, pp. 4-99 to 4-108, 1995.

Steele, J. "Get Maximum Accuracy from Temperature Sensors." *Electronic Design* (August 19, 1996), pp. 99–110.

Swager, A. W. "Hall Effect Sensors-Improved ICs Find Broad Application." *EDN* (May 11, 1989), pp. 75–92.

Swartz, C., C. Derrington, and J. Gragg. "Temperature Compensation Methods for the Motorola X-ducer Pressure Sensor Element." Motorola Application note AN 840. Motorola Inc., Phoenix, AZ, USA.

Travis, B. "Smart-Sensor Standard Will Ease Networking Woes." *EDN* (June 22, 1995), pp. 49–52.

Travis, B. "Temperature-Management ICs Combat System Meltdown." *EDN* (August 15, 1996), pp. 38–48.

Walters, D. "Chemical Sensing: An Emergent MEMS Technology." Proceedings of Sensors Expo, October 1996, pp. 173–186.

Winkler, C., and J. Baum. "Barometric Pressure Measurement Using Semiconductor Pressure Sensors." Application note AN-1326. Motorola Inc.

Wood, T. "The Hall Effect Sensor." *Sensors* (March 1986).

Woods, S. P. "IEEE-P 1451.2 Smart Interface Module." Proceedings of Sensors Expo, October 1996, Helmers Publishing Inc., pp. 25–46.

CHAPTER 8

Nonlinear Devices

8.1 Introduction

By exploiting the basic physics of semiconductor devices it is possible to design circuits that perform a wide variety of mathematical operations, including addition, subtraction, multiplication, and division as well as trigonometric, logarithmic, and exponential functions. Such circuits perform in the analog domain and frequently offer real advantages over more conventional digital computation. Operations where analog computation is preferable to digital include those where both the input and output signals must be analog, limited amounts of processing are required and no digital circuitry is present, the signal is differentiated to produce a rate signal, fast signals must be processed in real time, large dynamic ranges are involved, and complex or transcendental functions must be evaluated.

In an electronic design world, where a digital approach to design is preferred in many instances, much room remains for analog computation techniques, particularly in situations where a wide, dynamic range of signals or fast signals can be processed. To explain this situation, we take the case of a simple AC power meter. A simple AC power meter may be constructed very easily with a single analog multiplier as per Figure 8-1(a), where the moving coil meter can act as the integrator. A digital power meter would require conversion of both voltage and current to digital form, with considerable attention to the timing of the conversions, since the relative phase of the two signals is of critical importance. However, if with the advantage of a CPU (see Figure 8-1(b)) that has a display driven by it and a multiplexed ADC with spare capacity, the power metering facility could be added to the base system at minimal additional cost.

As another example, the first derivative of a varying analog signal is complex to calculate using digital techniques, compared to a simple C-R network for analog differentiation (Analog Devices, 1987, Section 5). Digitizing a signal with wide dynamic range also is expensive (see Figure 8-2). If such a signal is digitized

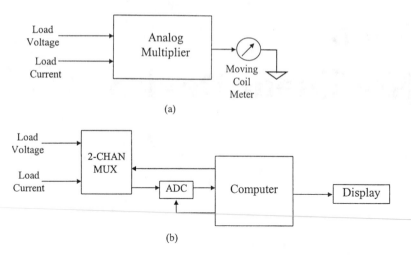

(a)

(b)

FIGURE 8-1 Power meter implementation: (a) analog approach, (b) digital approach

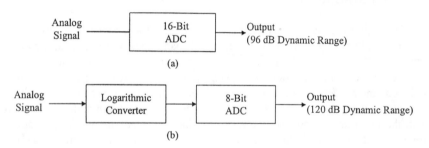

(a)

(b)

FIGURE 8-2 Analog processing advantage of signals with high dynamic range: (a) a 16-bit ADC, the expensive way, yielding a 96 dB dynamic range; (b) an inexpensive 8-bit ADC with a logarithmic converter, yielding a 120 dB dynamic range

with a 16-bit ADC (a comparatively expensive device), the ratio of an LSB to full scale is 96 dB; whereas if the signal were first applied to a logarithmic converter (frequently misnamed a *logarithmic amplifier*), then a dynamic range approaching 120 dB is practical with an 8-bit ADC.

Historically, analog computers have been slow devices. Even though high-frequency multipliers, modulators, logarithmic amplifiers, and other function generators have been available for many years, they generally have had relatively poor accuracy and stability and have not been considered analog computers. Within the last decade, a few classes of accurate nonlinear devices have entered the market: multipliers, modulators, and log amps (Analog Devices, 1990, Section 5). This chapter is an introduction to modern nonlinear devices, their design concepts, and special application areas.

8.2 A Basic Semiconductor Physics-Based Approach to Analog Computation Circuits

The operation of many analog computational circuits depends on the logarithmic properties of silicon junctions. An ideal logarithmic diode has the current voltage relationship

$$I = I_0(e^{qV/kT} - 1) \tag{8.1}$$

This could be rewritten as

$$V = \frac{kT}{q} \ln\left(\frac{I}{I_0} + 1\right) \approx \frac{kT}{q} \ln\left(\frac{I}{I_0}\right) \tag{8.2}$$

where

I = the current through the diode;

k = Boltzman's constant (1.38062×10^{-23});

V = the voltage across the diode;

q = a constant equal to unit charge, 1.60219×10^{-19} coulombs;

T = the absolute temperature in Kelvin;

I_0 = the extrapolated current for $E_0 = V = 0$ volts.

Referring to Figure 8-3, these equations clearly show that the current in a diode increases exponentially with voltage or, conversely, the voltage increases

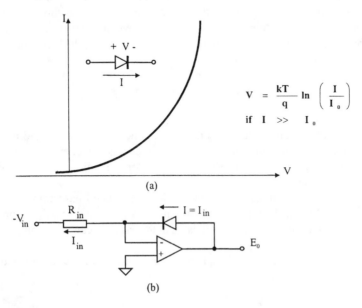

FIGURE 8-3 (a) Diode curve, (b) diode log converter

logarithmically with current. These equations are less clear in showing that I_0, the theoretical diode current at zero voltage, is temperature dependent and so the variation of a diode's behavior with temperature is by no means as simple as the equation would suggest; that is, the voltage is not proportional to absolute temperature at a fixed current. Several approximations concerning the logarithmic behavior of diodes are worth remembering:

$$\frac{kT}{q} = 26 \text{ mV (at } 28.58°C) \tag{8.3}$$

$$\frac{kT}{q} \ln_{10} = 60 \text{ mV (at } 29.25°C) \tag{8.4}$$

These approximations simplify the diode expression to

$$V = 60 \text{ mV} \log \frac{I}{I_0} \tag{8.5}$$

This simply says that V increases by 60 mV every time the current increases by a factor of 10 at 29.25°C.

If we were to place an ideal logarithmic diode in the feedback path (output to inverting input) of an operational amplifier and apply a current to the inverting input, the output voltage would be the logarithm of the input current times a temperature varying constant. For the circuit in Figure 8-3(b), if $I_{in} \gg I_0$,

$$E_0 = \frac{kT}{q} \ln \left(\frac{I_{in}}{I_0} \right) \approx 0.06 \log \frac{V_{in}}{R_{in} I_0} \tag{8.6}$$

It is unfortunate that real diodes are not ideal logarithmic diodes. In a real diode, the bulk resistivity, R_B, of the silicon limits the logarithmic accuracy at high currents and diffusion currents in surface inversion layers and generation-recombination effects in space-charge regions cause a scale factor error, m, at low currents. We therefore find that

$$E_0 = m\frac{kT}{q} \ln \left(\frac{I}{I_0} \right) + I_{RB} \tag{8.7}$$

where m varies with the current.

Even with similar diodes, m can vary (it is never less than 1 and may be as high as 4), as does the value of E_0 at which m changes. General purpose diodes therefore are impractical as logarithmic diodes for dynamic ranges of more than 100:1 (two decades).

Luckily, we can replace the diode with a grounded-base transistor as per Figure 8-4 and get a dynamic range of 1 million:1 (six decades) or more — the only disadvantage of such a circuit is that the signals can have only a single polarity.

From the Ebers and Moll equations (see Sheingold, 1976, for a detailed derivation), it may be shown that

$$E_0 = \frac{kT}{q} \ln \left(\frac{I_{in}}{I_{ES}} \right) - \frac{kT}{q} \ln \alpha_n \tag{8.8}$$

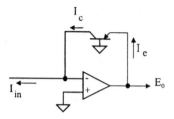

FIGURE 8-4   Transistor log converter

where $I_{in} \gg I_{ES}$; I_{ES} is the emitter saturation current; and α_n is the forward current-transfer ratio (α_n is not the grounded-base current gain).

Since I_{ES} is less than a picoampere and α_n is nearly unity over a wide range of currents, in the silicon planar transistors used to manufacture logarithmic converters, the effect of the second term generally may be disregarded; and the equation simplifies to

$$E_0 = \frac{kT}{q} \ln \left(\frac{I_{in}}{I_{ES}} \right) \tag{8.9}$$

Such logarithmic converters are temperature sensitive. In equation (8.9), kT/q has a temperature coefficient of 0.34%/°C around 25°C, and I_{ES} doubles for every 10°C temperature rise and varies with device size and geometry. Many of these basic concepts, in refined forms or in combination with other compensation circuits, are used in nonlinear circuits. (For a further discussion on these basic techniques, see Analog Devices, Inc., 1987; Sheingold 1976.)

8.3 Important Design Considerations in Nonlinear Devices

In the discussion of nonlinear devices two important design considerations are the dynamic range of a signal and the noise.

8.3.1 Dynamic Range

In many cases, a wide dynamic range is an essential aspect of a signal, something to be preserved at all costs. This is true, for example, in the high-quality reproduction of music and communication systems. However, often the signal must be compressed to a smaller range with no significant loss of information. Compression is used in magnetic recording, where the upper end of the dynamic range is limited by tape saturation and the lower end by the granularity of the medium. In professional noise-reduction systems, compression is "undone" by precisely matched nonlinear expansion during reproduction. Similar techniques are used in conveying speech over noisy channels, where the performance more likely is to be measured in terms of word intelligibility than audio fidelity. The reciprocal processes of compressing and expanding are implemented using "compandors," and many schemes have been devised to achieve this function. In terms of the signal voltage,

$$\text{Dynamic range (dB)} = 20\log_{10}\frac{\text{Largest signal voltage}}{\text{Smallest signal voltage}} \qquad (8.10)$$

Note that in a linear-impedance system, the power is proportional to the signal voltage (or current) squared. Accordingly,

$$\text{Dynamic range (dB)} = 10\log_{10}\frac{\text{Largest signal power}}{\text{Smallest signal power}} \qquad (8.11)$$

Also, it is useful to differentiate between the dynamic range of the signal and that of the processing system. The signal dynamic range is

$$\text{Signal dynamic range} = 20\log_{10}\frac{\text{Largest actual signal voltage}}{\text{Smallest actual signal voltage}} \qquad (8.12)$$

whereas the system dynamic range is

$$\text{System dynamic range} = 20\log_{10}\frac{\text{Largest permissible signal voltage}}{\text{Smallest detected signal voltage}} \qquad (8.13)$$

In system design, one should be concerned with the system's dynamic range, which should match or exceed the signal dynamic range.

8.3.2 Noise Limitations

The dynamic range of all signal-processing systems is limited by random noise, which sets a fundamental bound on the smallest signal that can be detected or otherwise utilized with an adequate signal-to-noise ratio (SNR). This noise may be generated by numerous mechanisms, including those associated with the source itself (e.g., antenna, photomultiplier, piezoelectric transducer) as well as by the active and passive devices in the amplifiers.

Noise cannot be discussed without reference to bandwidth, which will be unavoidably limited by the types of amplifier used. Deliberate filtering often is included in a signal-processing channel to reduce noise, as well as to improve the separation of wanted from unwanted signals. This may take the form of bandpass, low-pass, or high-pass functions or combinations of these, depending on the situation. Nonlinear filtering also may be used, for example, to minimize the disturbance of the signal path in the presence of impulsive noise.

The noise powers of uncorrelated sources add up, so noise voltages (or currents) must be added using a root-sum-of-squares (RSS) calculation. This leads to some rather startling consequences. Suppose a system has a major voltage noise source of magnitude E_a and several minor noise sources whose RSS sum to a magnitude of E_b. Then, E_a needs to be only twice E_b for the major source to contribute almost 90% of the total system noise. When $E_a/E_b = 5$, 98% of the noise is due to E_a.

It follows that the overall noise performance of a practical system can benefit greatly by (i) minimizing the input-referred noise of the first stage and (ii) using the highest possible gain in this stage. However, the second of these objectives frequently cannot be realized in systems that must handle signals of a large dynamic range, because the high gain would preclude distortion-free operation at maximum signal levels.

Noise frequently is specified in terms of a noise spectral density (NSD). The term reflects the fact that the total noise power is directly proportional to the system's noise bandwidth, B_N (in Hertz). The NSD therefore usually is of interest in specifying a channel's input-noise limitations. Note that, in general, the noise bandwidth is not equal to the -3dB bandwidth. B_N can be viewed as the bandwidth of an equivalent system with a "brick-wall" cessation of response at that frequency. A system with a single-pole low-pass corner at $f_0 = 1/2\pi T$ has a B_N value equal to $\pi f_0/2$, or $1.57 f_0$, while for two such real pole low-pass sections in cascade, B_N is $\pi f_0/4$ (see Figure 8-5). The total NSD will have both voltage and current components. Since the noise power is proportional to the square of either the voltage or current, these two noise components have the dimensions of volts/$\sqrt{Hz}$ and amps/$\sqrt{Hz}$.

Noise signals usually are small, and therefore nonlinear effects often are negligible. In such circumstances, it is permissible to use superposition methods to evaluate each contributing source independently, followed by an RSS calculation to calculate the total noise. A notable exception is the logarithmic amplifier, where even very small noise voltages at the input can cause heavy limiting in later stages in the amplifier. Special approaches to both noise analysis and noise specification are required in such cases.

For details on noise performance, see Analog Devices, Inc. (1992a) and Section 2.3.2 of Chapter 2. While noise limits the low end of a system's dynamic range, performance at the upper end of signal range is degraded by the increasing importance of nonlinear aspects of circuit behavior.

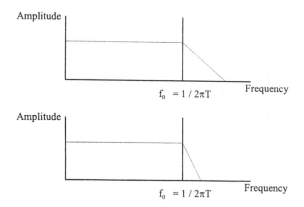

FIGURE 8-5 Filter noise bandwidths: (a) single-pole low-pass filter ($B_N = \pi f_0/2 = 1.57 f_0$), (b) two single-pole low-pass filters sections in cascade ($B_N = \pi f_0/4$)

8.4 Logarithmic Converters

The conversion of a signal to its equivalent logarithmic value involves a nonlinear operation, the consequences of which can be confusing if not fully understood. It is important to realize that many of the familiar concepts of linear circuits are irrelevant to log amps. For example, the incremental gain of an ideal log amp approaches infinity as the input tends to 0, and change of offset at the output of a log amp is equivalent to a change of amplitude at its input, not a change of input offset. The commonly used term *logarithmic amplifier* is something of a misnomer but is used lavishly.

If we consider the equation $y = \log(x)$, as shown in Figure 8-6(a), every time x is multiplied by a constant A, y increases by another constant A_1. Thus, if $\log(K) = K_1$, then $\log(AK) = K_1 + A_1$; $\log(A^2 K) = K_1 + 2A_1$; $\log(K/A) = K_1 - A_1$. As shown in figure 8.6(a) when x is 1, y approaches 0.

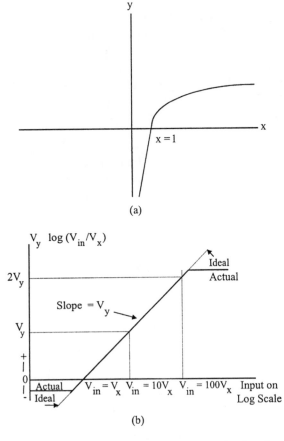

FIGURE 8-6 (a) Graph of $y = \log(x)$, (b) log amp transfer function

A practical log amp has the graph of transfer characteristics shown in Figure 8-6(b). Such a practical log amp has the transfer function

$$V_{out} = V_y \log_{10} \left(\frac{V_{in}}{V_x} \right) \tag{8.14}$$

This is valid over some range of input values, which may vary from 100:1 (40 dB) to over 1 million:1 (120 dB). The scale of the horizontal axis (the input) is logarithmic, and the ideal transfer characteristic is a straight line. When $V_{in} = V_x$, the logarithm is 0 ($\log 1 = 0$). V_x therefore is known as the *intercept voltage* of the log amp, because the graph crosses the horizontal axis at this value of V_{in}.

With inputs very close to 0, log amps cease to behave logarithmically and most then follow a linear V_{in}/V_{out} law. This behavior often is lost in device noise. Noise often limits the dynamic range of a log amp. The constant V_y has the dimensions of voltage, because the output is a voltage. The input, V_{in}, is divided by a voltage, V_x, because the argument of a logarithm must be a simple dimensionless ratio.

8.4.1 Practical Log Amps and Negative Values of x

The logarithm function is indeterminate for negative values of x. Log amps can respond to negative inputs in three different ways:

1. They can give a full-scale negative output as shown in Figure 8-7(a). This basic log amp saturates with negative inputs.
2. They can give an output proportional to the log of the absolute value of the input and disregard its sign, as shown in Figure 8-7(b). This type of log amp can be considered a full-wave detector with a logarithmic characteristic and often is referred to as a *detecting log amp*.
3. They can give an output proportional to the log of the absolute value of the input and have the same sign as the input, as shown in Figure 8-7(c). This type of log amp can be considered a video amp with a logarithmic characteristic and may be known as a *logarithmic video* (log video) *amplifier* or, sometimes, a *true log amp*.

8.4.2 Practical Implementation

Three basic architectures are used by manufacturers such as Analog Devices, Inc.: the basic diode log amp, the true log amp, and the successive detection log amp.

8.4.2.1 The Basic Diode Log Amp

As per the discussion in Section 8.2, a simple diode could be used for a log amp function (see Figure 8-3). In practice, the dynamic range of this configuration is limited to 40–60 dB because of nonideal characteristics of the diode. However, if the diode is replaced with a diode connected transistor, as shown in Figure 8-4, the dynamic range can be extended to 120 dB or more. This type of log amp has

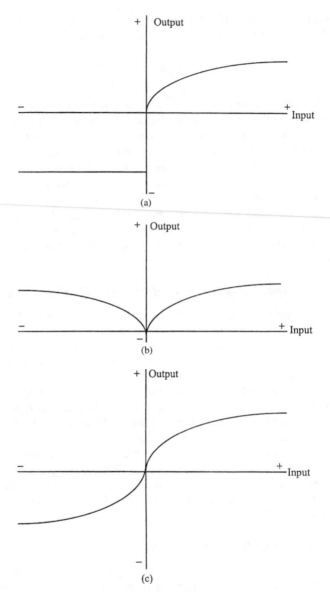

FIGURE 8-7 Log amps with negative values of x input; (a) basic log amp, (b) detecting log amp, (c) true log amp or log video amp

three disadvantages: both the slope and intercept are temperature dependent, it will handle only unipolar signals, and its bandwidth is both limited and dependent on the signal amplitude. Where several such log amps are used on a single chip to produce an analog computer that performs both log and antilog operations, the temperature variation in the log operations is unimportant, since it is compensated

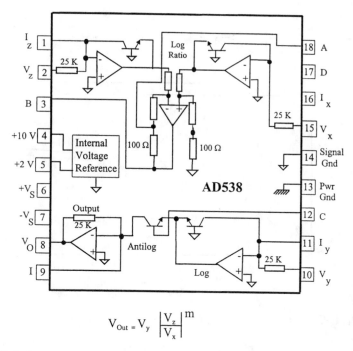

$$V_{Out} = V_y \left| \frac{V_z}{V_x} \right|^m$$

FIGURE 8-8 The AD-538 analog computational unit. (Reproduced by permission of Analog Devices, Inc.)

for by a similar variation in the antilogging. An example of a practical device that utilizes such techniques is the AD-538 analog computation unit (ACU) from Analog Devices, Inc. (Figure 8-8). This device, which has a transfer function of $V_{out} = V_y(V_z/V_x)$ can multiply, divide, and raise to powers. When actual logging is required, these types of devices require temperature compensation (Sheingold, 1976).

A major disadvantage of this type of log amp for high-frequency applications is its limited frequency response, limited by Miller capacitance or the residual feedback capacitance of these devices (Analog Devices, Inc., 1995). Practical limits are within a few hundred kHz. Therefore, for high-frequency applications detecting and true log architectures are used.

8.4.2.2 True and Detecting Log Amps

Although these two types differ in detail, the general principle behind their design is the same: Instead of one amplifier having a logarithmic characteristic, these designs use a number of similar, cascaded linear stages having well-defined large signal behavior.

As shown in Figure 8-9(a) consider N cascaded limiting amplifiers, the output of each driving a summing circuit as well as the next stage. If each amplifier has a gain of A dB, the small signal gain of the strip is NA dB. If

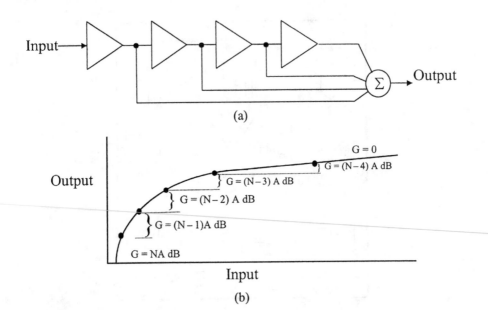

(a)

(b)

FIGURE 8-9 Multistage log amp: (a) architecture, (b) response for unipolar case

the input signal is small enough for the last stage not to limit, the output of the summing amplifier will be dominated by the output of the last stage. As the input signal increases, the last stage will limit. It now will make a fixed contribution to the output of the summing amplifier, but the incremental gain to the summing amplifier will drop to $(N - 1)A$ dB. As the input continues to increase, this stage in turn will limit and make a fixed contribution to the output, and the incremental gain will drop to $(N - 2)A$ dB, and so forth, until the first stage limits and the output ceases to change with increasing signal input.

The response curve therefore is a set of straight lines, as shown in Figure 8-9(b). The total of these lines, though, is a very good approximation to a logarithmic curve and, in practice, is an even better one, because few limiting amplifiers, especially high-frequency ones, limit quite as abruptly as this model assumes. Due to the compromise needed between log approximation and number of gain stages, gains of 10–12 dB are chosen in practical devices (Analog Devices, 1995).

This general model, which is ideal, becomes difficult to implement at high frequencies due to delays associated with each stage. If each stage has a delay of t ns, a signal that passes through all stages will have a delay of Nt ns compared to a signal that passes only one stage, which is delayed by only t ns. Some solutions for such difficulties are discussed in Analog Devices (1995). Multistage architectures such as these or their variations are video log amps or true log amps. However, the most common types of high-frequency log amps are the devices based on successive detection architecture.

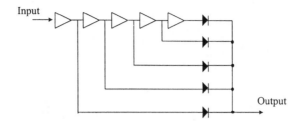

Input

Output

FIGURE 8-10 Successive detection logarithmic amplifier

8.4.2.3 Successive Detection Log Amps

The successive detection log amp consists of cascaded limiting stages as described previously, but instead of summing their outputs directly, these outputs are applied to detectors, and the detector outputs are summed as shown in Figure 8-10. If the detectors have current outputs, the summing process may involve no more than connecting all the detector outputs.

Log amps using this architecture have two types of output: the log output and a limiting output. In many applications, the limiting output is not used, but in some (FM receivers with "S" meters, for example) both are necessary. The log output of a successive detection log amplifier generally contains amplitude information, and the phase and frequency information is lost.

In the past, it has been necessary to construct high-performance, high-frequency successive detection log amps using a number of individual limiting amplifiers. These typically are assembled in complex and costly hybrids. Recent advances in IC processes have allowed this complete function to be integrated on a single chip.

The AD-640 log amp from Analog Devices is an example of successive detection type log amp for high-frequency use. The AD-640 log amp contains five limiting stages (10 dB per stage) and five full-wave detectors in a single IC package, and its logarithmic performance extends from DC to 145 MHz. A block diagram of AD-640 is shown in Figure 8-11.

With reference to Figure 8-11(b), the AD-640 has its log amp transfer function where V_x is calibrated to 1 mV exactly. The slope of the line is directly proportional to V_y. Base 10 logarithms are used in this context to simplify the relationship to decibel values. For $V_{in} = 10V_x$, the logarithm has a value of 1, so the output voltage is V_y. At $V_{in} = 100V_x$, the output is $2V_y$, and so on. V_y therefore can be viewed either as the slope voltage or as the volts per decade factor.

The AD-640 conforms to equation (8.1) except that its two outputs are in the form of current rather than voltage:

$$I_{out} = I_y \log(V_{in}/V_x) \tag{8.15}$$

Each of the five stages in the AD-640 has a gain of 10 dB and a full-wave detected output. The transfer function of the device is shown in Figure 8-11(b)

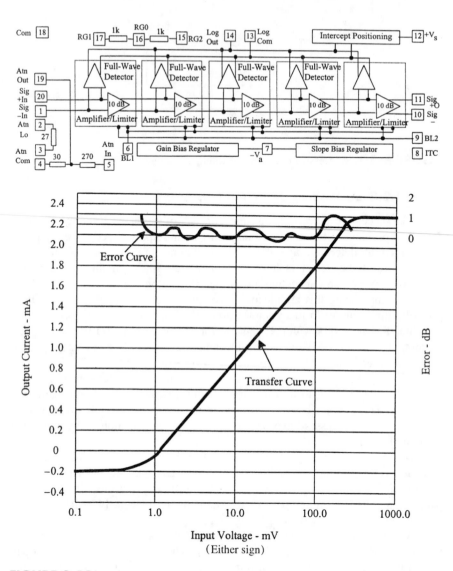

FIGURE 8-11 The AD-640 log amp: (a) block diagram, (b) transfer function and error curve. (Reproduced by permission of Analog Devices, Inc.)

along with the error curve. Note the excellent log linearity over an input range of 1–100 mV (40 dB). Although well suited to RF applications, the AD-640 is DC coupled throughout. This allows it to be used in low-frequency and very low-frequency systems, including audio measurements, sonar, and other instruments requiring operation to low frequencies or even DC. Unlike many other log amps, the AD-640 is laser trimmed to a high absolute accuracy of both slope and intercept and is fully temperature compensated. Some key features of the AD-640 are

1. 45 dB dynamic range, two can cascade to 95 dB.
2. Bandwidth DC to 145 MHz (120 MHz when cascaded).
3. Slope of 1 mA/decade, temperature stable.
4. Less than 1 dB log nonlinearity.
5. Balanced circuitry for stability and minimal external components.

For further details on applications, see Analog Devices (1992a, 1992c, 1995).

8.4.3 Key Parameters of Log Amps and Classifications

In selecting log amps for a given application, key parameters to be considered are listed in Table 8-1.

Over the years, logarithmic amplifiers have accumulated a confusing assortment of terms, some quite misleading. Here (Table 8-2), Analog Devices, Inc. attempts to classify log amps into three broad groups according to structure and application domain and try to be consistent in matters of terminology and nomenclature. For details, see Analog Devices (1992a, Section 9).

TABLE 8-1   Key parameters of log amps

Parameter	Description
Noise	Noise referred to the input (RTI) of the log amp, which may be expressed as a noise figure, as noise spectral density (voltage, current, or both), or as noise voltage (or noise current or both)
Dynamic range	Range of signal over which the amplifier behaves in a logarithmic manner (expressed in decibels)
Frequency response	Range of frequencies over which the log amp functions correctly
Slope	Gradient of transfer characteristic in V/dB or mA/dB
Intercept point	Value of input signal at which output is 0
Log linearity	Deviation of transfer characteristic (plotted on log/line axes) from a straight line (expressed in decibels)

TABLE 8-2   Types of log amps and their behavior. (Reproduced by permission of Analog Devices, Inc.)

Type	Performance
Translinear log amps	Based on logarithmic (or translinear) properties of bipolar transistors; wide dynamic range, poor AC performance
Baseband log amps	Respond to instantaneous value of rapidly changing input, "progressive compression technique" often used, sometimes called *video log amps*, "true log amp" accepts bipolar inputs sign of output following input
Demodulating log amps	AC input signal is rectified, output is the modulated envelope of the input, often called *successive detection log amp*

8.5 Multipliers and Dividers

A multiplier is a device having two input ports and an output port. The signal at the output is the product of the two input signals. If both input and output signals are voltages, the transfer characteristic is the product of the two voltages divided by a scaling factor, K, which has the dimension of voltage (see Figure 8-12(a)). From a mathematical point of view, multiplication is a four-quadrant operation; that is to say, both inputs may be either positive or negative, as may be the output. Some of the circuits used to produce electronic multipliers, however, are limited to signals of one polarity. If both signals must be unipolar, we have a single-quadrant multiplier and the output also is unipolar. If one of the signals is unipolar, but the other may have either polarity, the multiplier is a two-quadrant multiplier and the output may have either polarity (and is bipolar). The circuitry used to produce one- and two-quadrant multipliers may be simpler than that required for four quadrant multipliers; and since in many applications full four-quadrant multiplication is not required, it is common to find accurate devices that work only in one or two quadrants.

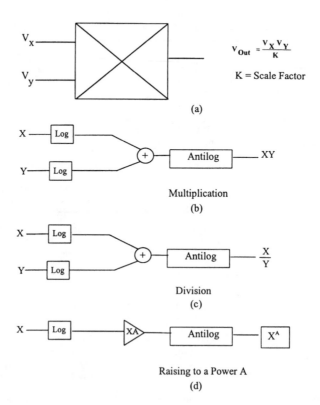

$$V_{Out} = \frac{V_X V_Y}{K}$$

K = Scale Factor

(a)

Multiplication

(b)

Division

(c)

Raising to a Power A

(d)

FIGURE 8-12 Analog multiplier: (a) basic block diagram, (b) multiplication, (c) division, (d) raising to a power A

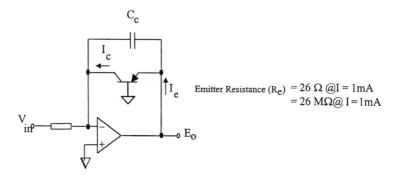

FIGURE 8-13 Log converter compensation problem

Many techniques can be used for analog multipliers, and some of these are discussed in Sheingold (1976). Most common forms are the use of log/antilog circuits or the use of a Gilbert cell (Gilbert, 1968a, 1968b). The AD-538 from Analog Devices is such a monolithic integrated circuit (Figure 8-8).

Some disadvantages of such circuits are its unipolar inputs and the variation of its bandwidth with the signal amplitude. The problem of bandwidth variation arises from the variation of emitter resistance, R_E, with current in the grounded-base transistor. With reference to Figure 8-13, R_E is inversely proportional to the emitter current, being approximately 26 Ω at 1 mA. The bandwidth of the circuit is inversely proportional to the product of $R_E C_C$ (C_C may be an external compensation capacitor or merely stray capacitance) and thus is proportional to the transistor current. Therefore, if the logarithmic converter works over a 120 dB dynamic range, its bandwidth will vary by 1 million:1, which can be inconvenient. For details, see Analog Devices (1987).

In addition to adopting other improved logarithmic conversion techniques, a popular technique used in commercial analog multiplier ICs is the Gilbert cell. There is a linear relationship between the collector current of a silicon junction transistor and its transconductance (gain) that is given by the following equation:

$$\frac{dI_c}{dV_{Be}} = \frac{q}{kT} I_c \tag{8.16}$$

where

I_c = collector current;

V_{BE} = base-emitter voltage;

q = electron charge (1.60219×10^{-19});

k = Boltzmann's constant (1.38062×10^{-23});

q/kT = $1/(25.69$ mV) at 25°C.

This relationship may be exploited to construct a multiplier with a long-tailed pair of silicon transistors (Analog Devices, 1987), as shown in Figure 8-14.

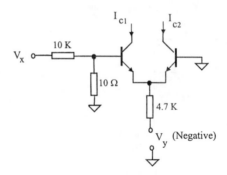

FIGURE 8-14 Basic transconductance multiplier circuit

At 25°C, this circuit provides an output of $q/kT[(V_y + V_{BE})/(4.7 \times 10^3)]$ $(10/10,010)V_x$ for the case of $I_{c1} - I_{c2} = \Delta I_c$.

However, this is a rather poor multiplier because

1. The y input is offset by the V_{BE}, which changes nonlinearly with V_y.
2. The x input is nonlinear as a result of the exponential relationship between I_c and V_{BE}.
3. The scale factor varies with temperature.

Gilbert realized that this circuit could be linearized and made temperature stable by working with currents rather than voltages and by exploiting the logarithmic properties of transistors, as per the case shown in Figure 8-15(a).

The x input to the Gilbert cell takes the form of a differential current, and the y input is a unipolar current. The differential x currents flow in two diode-connected transistors, and the logarithmic voltages compensate for the exponential V_{BE}/I_c relationship. Furthermore, the q/kT scale factors cancel each other. This gives the Gilbert cell the linear transfer function for $I_c = (I_{c1} - I_{c2})$:

$$\Delta I_c = \frac{\Delta I_x I_y}{I_x} \tag{8.17}$$

As it stands, the basic Gilbert cell shown in Figure 8-15(a) has three inconvenient features:

1. Its x input is a differential current.
2. Its output is a differential current.
3. Its y input is a unipolar current.

This makes the cell a two-quadrant multiplier.

By cross-coupling two such cells and using two voltage-to-current converters (as shown in Figure 8-15(b)), we can convert the basic architecture to a four-quadrant device with voltage inputs. A practical example of this type of a multiplier is the AD-534 from Analog Devices. In Figure 8-15(b), Q1A and Q1B

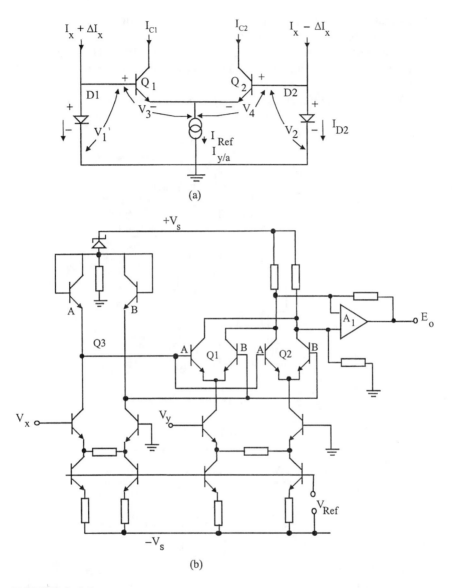

FIGURE 8-15 (a) Two-quadrant and (b) four-quadrant multipliers based on Gilbert cell. (Reproduced by permission of Analog Devices, Inc.)

and Q2A and Q2B form the two core long-tailed pairs of the two Gilbert cells, while Q3A and Q3B are the linearizing transistors for both cells. Figure 8-15(b) shows an operational amplifier acting as a differential current to single-ended voltage converter; but for higher speed applications, the cross-coupled collectors of Q1 and Q2 form a differential open collector current output (as in the

AD-834 multiplier with a 500 MHz range). Motorola's MC 1494 and 1495 are other examples of four-quadrant multipliers.

A basic wideband multiplier using the AD-834 (for 500 MHz bandwidth) is shown in Figure 8-16(a). Figure 8-16(b) indicates the block diagram of a 10 MHz multiplier with direct-divide capability.

For details on multipliers, see Analog Devices (1995, Section 3).

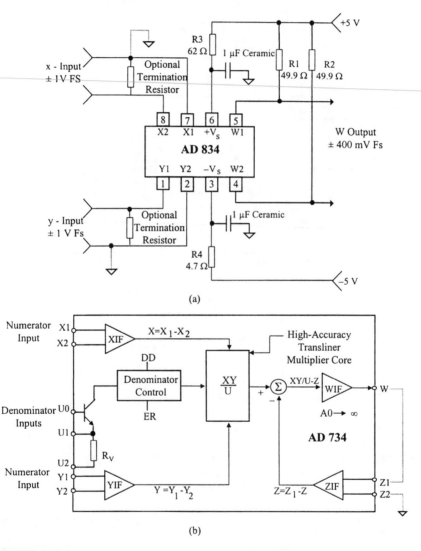

(a)

(b)

FIGURE 8-16 Practical multipliers: (a) A wideband application using the AD-834, (b) a 10 MHz multiplier with direct-divide capability using the AD-734. (Reproduced by permission of Analog Devices, Inc.)

8.6 RMS-to-DC Converters

The root mean square (RMS) is a fundamental measurement of the magnitude of an AC signal. Defined practically, the RMS value assigned to the AC signal is the amount of DC required to produce an equivalent amount of heat in the same load. Defined mathematically, the RMS value of a voltage is the value obtained by squaring the signal, taking the average, and then taking the square root. The averaging time must be sufficiently long to allow filtering at the lowest frequencies of operation desired. A complete discussion of RMS-to-DC converters can be found in Kitchen and Counts (1986) and application of these in multimeters are discussed in Kularatna (1996). Two basic techniques are used in RMS-to-DC converters: explicit and implicit.

8.6.1 Explicit Method

The explicit method is shown in Figure 8-17(a). The input signal is first squared by a multiplier. The average value is taken by using an appropriate filter, and the square root is taken using an op amp with a second squarer in the feedback loop. This circuit has limited dynamic range because the stages following the squarer must try to deal with a signal that varies enormously in amplitude. This restricts the method to inputs with a maximum dynamic range of approximately 10:1 (20 dB). However, excellent bandwidth (greater than 100 MHz) can be achieved with high accuracy if a multiplier such as the AD-834 is used as a building block (see Figure 8-17(b)).

8.6.2 Implicit Method

Figure 8-18 shows the circuit for computing the RMS value of a signal using the implicit method. Here, the output is fed back to the direct-divide input of a multiplier such as the AD-734. In this circuit, the output of the multiplier varies linearly (instead of as the square) with the RMS value of the input. This considerably increases the dynamic range of the implicit circuit as compared to the explicit circuit. The disadvantage of this approach is that it generally has less bandwidth than the explicit computation.

8.6.3 Monolithic RMS/DC Converters

While it is possible to construct such an RMS circuit from an AD-734, it is far simpler to design a dedicated RMS circuit. The V_{in}^2/V_z circuit may be current driven and only one quadrant if the input first passes through an absolute value circuit. Figure 8-19(a) shows a block diagram of a typical monolithic RMS/DC converter such as the AD-536. It is subdivided into four major sections: absolute value circuit (active rectifier), squarer/divider, current mirror, and buffer amplifier. The input voltage, V_{in}, which can be AC or DC, is converted to a unipolar current, I_{in}, by an absolute value circuit. I_{in} drives one input of the one-quadrant squarer/divider, which has the transfer function I_{in}^2/I_f. The output current, I_{in}^2/I_f,

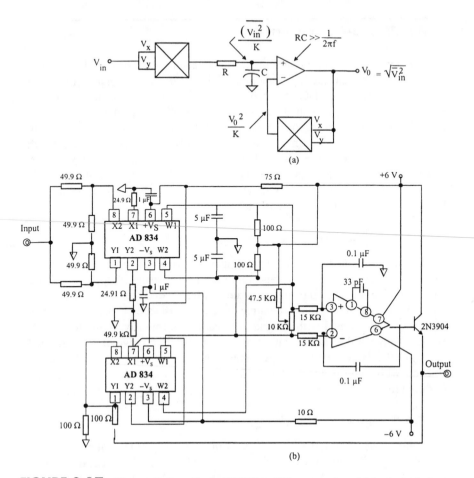

FIGURE 8-17 The explicit method of RMS-to-DC conversion: (a) basic technique, (b) wideband RMS measurement using the AD-834. (Reproduced by permission of Analog Devices, Inc.)

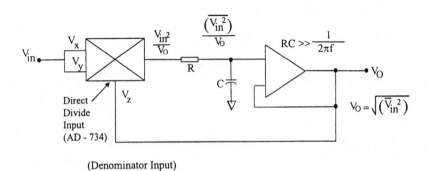

(Denominator Input)

FIGURE 8-18 Implicit RMS computation

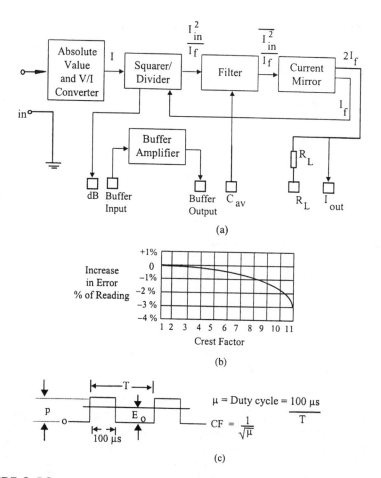

FIGURE 8-19 The AD-536 RMS/DC converter: (a) block diagram, (b) error against crest factor, (c) input waveform used for (b). (Reproduced by permission of Analog Devices, Inc.)

of the squarer/divider drives the current mirror through a low-pass filter formed by R_1 and an externally connected capacitor, C_{AV}. If the $R_1 C_{AV}$ time constant is much greater than the longest period of the input signal, then I_{in}^2/I_f is effectively averaged. The current mirror returns a current, I_f, that equals the average value of I_{in}^2/I_f back to the squarer/divider to complete the implicit RMS computation. Therefore,

$$I_f = \left(\frac{I_{in}^2}{I_f}\right) = I_{in}(\text{RMS}) \qquad (8.18)$$

The current mirror also produces the output current, I_{out}, which equals $2I_f$. The circuit provides a decibel output also, which has a temperature coefficient of approximately 3300 ppm/°C and must be temperature compensated.

TABLE 8-3 A representative set of RMS/DC converters from Analog Devices, Inc.

Part No.	Bandwidth	Full-Scale Input Voltage Range	Remarks
AD-536	450 kHz	>100 mV input	±15 V rails
	2 MHz	>1 V	
AD-636	1 MHz	Up to 200 mV	Low-power (±5 V) rails
	8 MHz		
AD-637	8 MHz	>1 V	Chip select/power down function available (±3 to ±18 V rails)
	600 kHz	200 mV	
AD-736	350 kHz	100 mV	Low-power precision converter
	460 kHz	200 mV	±5 to ±16 V power rails
AD-737	170 to 350 kHz	100 mV	Low-cost, low-power true RMS
	190 to 460 kHz	200 mV	±5 to ±16 V rails

There are a number of RMS/DC converters in monolithic form. A representative list from Analog Devices is shown in Table 8-3. For practical applications, design details, and selection of RMS/DC converters, see Analog Devices (1992c, 1995), Kitchen and Counts (1986), and Kularatna (1996).

8.7 Function Generators

Some interesting applications of nonlinear devices involve function generation using AD-538-type devices and AD-639 trigonometric function generators. To fully appreciate these devices, keep in mind that some digital techniques such as direct digital synthesis (DDS) still have restrictions at high frequencies, beyond several megahertz. An example of these could be shown using the AD-538. The arc-tangent circuit shown in Figure 8-20(a) is typical of AD-538 applications where Y is 1 so that $V_o = (Z/X)^M$ for $M < 1$.

In an approximation to the arc-tangent function, the AD-538 may be made to compute the angle represented by two rectangular coordinates, which, since they are applied to the X and Z inputs, we shall call X and Z rather than the more usual X and Y. If X and Z are within the range 100 μV to 10 V, the error in the computed angle is under 1° (the AD-639 can perform a similar computation with fewer components but cannot work over a wide dynamic range).

The circuit exploits the fact that

$$T = \frac{(\tan T)^{1.21}}{1 + (\tan T)^{1.21}} \tag{8.19}$$

where T is the angle normalized to 90°.

The AD-538 and the external amplifier calculate $\log(\tan T)$ from X and Z, amplifie it by a factor of 1.21 to raise to the 1.21 power, and perform an implicit calculation to calculate the angle (which is expressed in terms of the

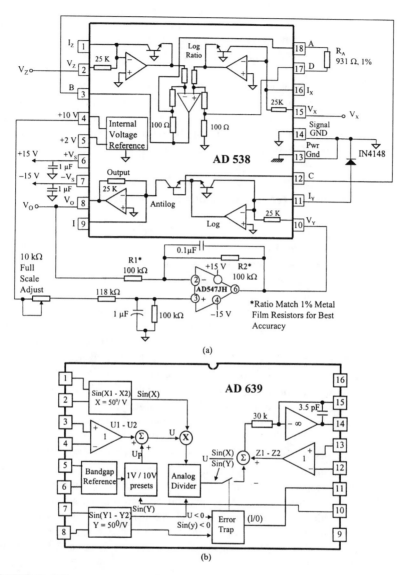

FIGURE 8-20 Nonlinear devices used as function generators: (a) arc-tangent function using the AD-538, (b) the AD-639, a universal trigonometric function generator. (Reproduced by permission of Analog Devices, Inc.)

reference voltage). Under these conditions, the output voltage tends to V_{REF} as the angle tends to 90° although, in fact, the circuit cannot be used much above 89.5° because at 90° tangents become infinite and before that the circuit becomes unstable. R_1 and R_2 must be matched for highest accuracy, and the circuit is stabilized by the 0.1 μF integrating capacitor in the amplifier feedback path. The circuit works in a single quadrant since both X and Z must be positive.

Trigonometric functions more normally are calculated by the AD-639. No external components (other than supply decoupling capacitors) are required to compute sines, cosines, tangents and cotangents, and secants and cosecants with the AD-639. Little more than an extra reference voltage (which may be generated from the internal reference with an operational amplifier and a couple of resistors) is required for versines and coversines. For details of the less common functions, consult the AD-639 data sheet (Analog Devices, 1999) and various application notes — the sine, cosine, and tangent will be described here. Some interesting applications of the AD-639 are discussed in Analog Devices (1987).

8.8 Benistor, a Newly Introduced Device

An interesting novel component surfaced recently (Bindra, 1998) from the Bensys Corporation, the Benistor™. (The name Benistor is a combination of the company's name and transistor.) The Benistor, which performs similar to multielectrode vacuum tubes, can independently control the voltage and current output from the device.

Introduced in 1998, the device block diagram, shown in Figure 8-21(a), comprises four blocks: the power controller (PC), the current separator (CS), the current controller (CC), and the voltage threshold controller (VTC). In the first commercial device, BEN 35100, the PC is a simple PNP transistor that acts as a switch or a variable resistor between the power source and the load. The CS block incorporates three NPN transistors to enable the voltage controller and current controller to work simultaneously or separately.

Comprising two open-collector op amps and two resistors, the current controller acts as a voltage/current converter for the power controller. There are two control inputs to this block: the noninverting input CC and the inverting input $\overline{CC}$. The amount of voltage input at the noninverting current control is directly proportional to the current output to the load, and the amount of voltage at the inverting electrode is inversely proportional to the output current. Functioning as a window comparator, the VTC controls the buffer's base current, in either a switching or self-switching mode. The two controls of the VTC are effective voltage control (EVC) and maximum voltage control (MVC), which determine the threshold voltages of the output. While the voltage at the EVC establishes the threshold for switching from off to on, the voltage at MVC pin determines the on/off states. In effect, the voltage settings on these two pins pre-establish the output voltage window.

In summary, the Benistor is a multielectrode device consisting of impedance command pins for precise control of output current and voltage. Consequently, eight electrodes completely define the Benistor. These are shown in Figure 8-21(b).

The SS electrode (not used in the BEN 35100 version) sets the initial state of the Benistor's self-switching mode of operation as either on or off at the beginning of an input power pulse wave. Having only two states, it is either grounded (on) or floating (off). Likewise, the CE provides the reference voltage

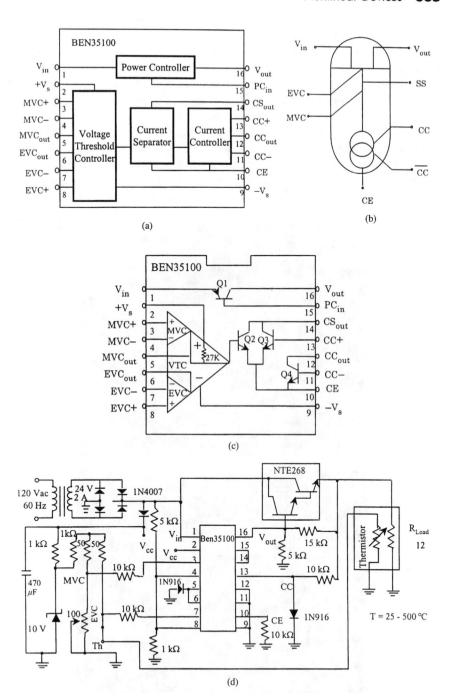

FIGURE 8-21 The Benistor: (a) block diagram, (b) symbol, (c) BEN 35100 equivalent circuit, (d) a soldering iron application for precise temperature control. (Reproduced by permission of the Bensys Corporation.)

for the device, while the CC determines the window of output current, and the VTC pre-establishes the window of output voltage. Based on these settings, the power controller will deliver to the load only that part of the input power signal, with respect to the amount of current and voltage, within the two pre-established ranges. In effect, together, the voltage and current-control electrodes can provide the system designer virtually any output possibility.

Because the device can accept AC, DC, and pulse input and provide output in any of these modes or any combination (based on conditions at the control electrodes), the Benistor inspires a new way of thinking in power control and power conversion. Combining the capability of all three previous values, it provides designers a unique method of controlling power parameters. Figure 8-22 indicates

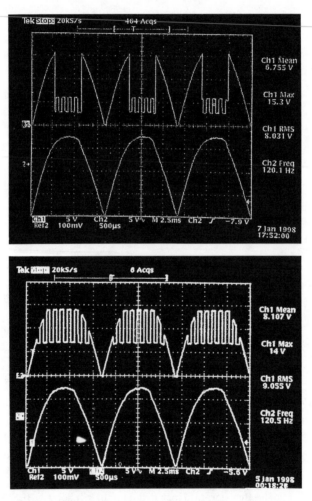

FIGURE 8-22 Waveform control example. (Reproduced by permission of Bensys Corporation.)

complex waveforms that can be obtained from the same input (bottom traces). For further details, see Bindra (1998), and Bensys Corporation (●●●, ●●●, ●●●, ●●●) and U.S. Patent No. 5,598,093.

References

Analog Devices. Linear Design Seminar. Analog Devices, Inc., 1987.

Analog Devices. High Speed Design Seminar. Analog Devices, Inc., 1990.

Analog Devices. *Amplifier Applications Guide*. Analog Devices, Inc. (Norwood, MA, USA) 1992a.

Analog Devices. *High Speed Design Techniques*. Analog Devices, Inc. (Norwood, MA, USA) 1992b.

Analog Devices. "AD640: DC Coupled Demodulating 120 MHz Logarithmic Amplifier." In *Special Linear Reference Manual*, rev. A. Analog Devices, Inc. (Norwood, MA, USA) 1992c, pp. 3-31 to 3-46.

Analog Devices. Linear Design Seminar. Analog Devices, Inc. (Norwood, MA, USA) 1995.

Analog Devices. "AD639 — Universal Trigonometric Function Converter — Data Sheet." Analog Devices, Inc. (Norwood, MA, USA), Designer's Reference Manual (CD ROM), 1999.

Bensys Corporation. "Preliminary Data Sheet, BEN 35100." Bensys Corporation, Sunnyvale, CA, USA September 11, 1998.

Bensys Corporation. "Incandescent Lamp Flasher." Application note. Bensys Corporation, Sunnyvale, CA, USA ●●●.

Bensys Corporation. "DC to DC Converters." Application note. Bensys Corporation, Sunnyvale, CA, USA ●●●.

Bensys Corporation. "Lithium Ion Battery Charger." Application note. Bensys Corporation, Sunnyvale, CA, USA ●●●.

Bensys Corporation. *User Manual — EB 16 Evaluation Board*. Bensys Corporation, Sunnyvale, CA, USA ●●●.

US Patent Number 5,598,093, January 28, 1997: "Low Dissipation Controllable Electron Value for Controlling Energy Delivered to a Load and Method Therefor."

Bindra, Ashok. "Unique Valve Independently Controls Power Parameters." *Electronic Design* (July 6, 1998), pp. 35–40.

Gilbert, Barrie. *ISSCC Digest of Technical Papers* (February 16, 1968), pp. 114–115.

Gilbert, Barrie. *Journal of Solid State Circuits* SC-3 (December 1968), pp. 353–372.

Hughes, Richard Smith. *Logarithmic Amplifiers*. Dedham, MA: Artech House, 1986.

Kitchen, Charles, and Lew Counts. *RMS to DC Conversion Application Guide*, 2nd ed. Analog Devices, Inc., 1986.

Kularatna, N. *Modern Electronic Test and Measuring Instruments*. London: IEEE, 1996.

Sheingold, Daniel H. *Nonlinear Circuits Handbook*. Analog Devices, Inc., 1976.

CHAPTER 9

Rechargeable Batteries and Their Management

9.1 Introduction

The insatiable demand for smaller, lightweight portable electronic equipment has dramatically increased the need for research on rechargeable (or secondary) battery chemistries. In addition to achieving improved performance on lead acid and nickel cadmium (NiCd) batteries, during the last decade, many new chemistries such as nickel metal hydride (NiMH), lithium ion (Li-ion), rechargeable alkaline, silver-zinc, zinc-air, lithium polymer, and the like have been introduced.

Higher energy density, superior cycle life, environmental friendliness, and safe operation are among the general design targets of battery manufacturers. To complement these developments many semiconductor manufacturers have introduced new integrated circuit families to achieve the best charge/discharge performance and longest possible lifetime from battery packs.

This chapter describes the characteristics of battery families such as sealed lead acid, NiCd, NiMH, Li-ion, rechargeable alkaline, and zinc-air together with modern techniques used in battery management ICs, without elaborating on the battery chemistries. Concepts and applications related to systems management bus and smart battery system also are introduced.

9.2 Battery Terminology

9.2.1 Capacity

Battery or cell capacity is measured as an integral of current (i) over a defined period of time (t):

$$\text{Capacity} = \int_0^t i \, dt$$

This equation applies to either the charge or discharge; that is, the capacity added or capacity removed from a battery or cell. The capacity of a battery or cell is measured in milliampere-hours (mAh) or ampere-hours (Ah).

Although the basic definition is simple, many different forms of capacity are used in the battery industry. The distinctions among them reflect differences in the conditions under which the capacity is measured.

Standard capacity measures the total capacity that a relatively new but stabilized production cell or battery can store and discharge under a defined standard set of application conditions. It assumes that the cell or battery is fully formed, charged at the standard temperature at the specification rate, and discharged at the same standard temperature at a specified standard discharge rate to a standard end-of-discharge voltage (EODV). The standard EODV itself is subject to variation depending on discharge rate as discussed.

When any of the application conditions differ from standard, the capacity of the cell or battery changes. The term **actual capacity** includes all nonstandard conditions that alter the amount of capacity the fully charged new cell or battery is capable of delivering when fully discharged to a standard EODV. Examples of such situations might include subjecting the cell or battery to a cold discharge or a high-rate discharge.

That portion of actual capacity which can be delivered by the fully charged new cell or battery to some nonstandard EODV is called **available capacity**. Therefore, if the standard EODV is 1.6 volts per cell, the available capacity to an EODV of 1.8 volts per cell would be less than the actual capacity.

Rated capacity is the minimum expected capacity when a new but fully formed cell is measured under standard conditions. This is the basis for C rate and depends on the standard conditions used, which may vary depending on the manufacturers and the battery types.

If a battery is stored for a period of time following a full charge, some of its charge will dissipate. The capacity that remains and can be discharged is called the **retained capacity**.

9.2.2 C Rate

The C rate is the rate in amperes or milliamperes numerically equal to the capacity rating of the cell given in ampere-hours or milliampere-hours. For example, a cell with a 1.2 Ah capacity has a C rate of 1.2 amps. The C concept simplifies the discussion of charging for a broad range of cell sizes, since the cells' responses to charging are similar if the C rate is the same. Normally, a 4 Ah cell will respond to a 0.4 amp (0.1 C) charge rate in the same manner that a 1.4 Ah cell will respond to a 0.14 amp (also 0.1 C) charge rate.

The rate at which current is drawn from a battery affects the amount of energy that can be obtained. At low discharge rates the actual capacity of a battery is greater than at high discharge rates. This relationship is shown in Figure 9-1.

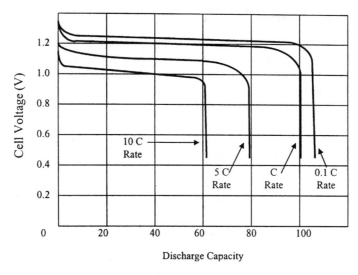

FIGURE 9-1 Capacity vs. discharge rate of a typical cell

9.2.3 Energy Density

The energy density of a cell is its energy divided by its weight or volume. This is called the **gravimetric energy density** when weight is used, and the **volumetric energy density** when volume is used. (The terms **energy density** and **specific energy** sometimes are used for volumetric and gravimetric measures, respectively; Bennett and Brown, 1997.)

9.2.4 Cycle Life

Cycle life is a measure of a battery's ability to withstand repetitive deep discharging and recharging using the manufacturer's cyclic charging recommendations and still provide the minimum required capacity for the application. Cyclic discharge testing can be done at any of various rates and depths of discharge to simulate conditions in the application. It must be recognized, however, that cycle life has an inverse logarithmic relationship to depth of discharge.

9.2.5 Cyclic Energy Density

For comparison, a better measure of rechargeable battery characteristics is a composite characteristic that considers energy density over the service life of the battery. This composite characteristic, cyclic energy density, is the product of energy density and cycle life at that energy density and measured in the dimensional units watt-hour-cycles/kilogram (gravimetric) or watt-hour-cycles/liter (volumetric).

9.2.6 Self-Discharge Rate

The self-discharge rate is a measure of how long a battery can be stored and still provide the minimum required capacity and be rechargeable to the rated capacity. This commonly is measured by placing batteries on shelf at room (or elevated) temperature and monitoring open circuit voltage over time. Samples are discharged at periodic intervals to determine remaining capacity and recharged to determine rechargeability.

9.2.7 Charge Acceptance

Charge acceptance is the willingness of a battery or cell to accept a charge. This is affected by cell temperature, charge rate, and the state of charge.

9.2.8 Depth of Discharge

The depth of discharge is the capacity removed from a battery divided by its actual capacity, expressed as a percentage.

9.2.9 Voltage Plateau

The voltage plateau is the protracted period of very slowly declining voltage that extends from the initial voltage drop at the start of a discharge to the knee of the discharge curve (Figure 9-2).

9.2.10 Midpoint Voltage

The midpoint voltage is the battery voltage when 50% of the actual capacity has been delivered (see Figure 9-2).

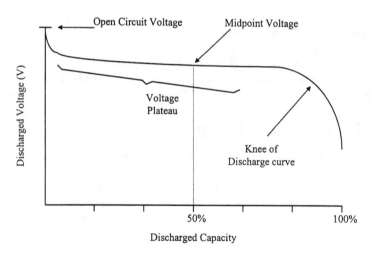

FIGURE 9-2 Nominal discharge performance for sealed lead-acid cells

9.2.11 Overcharge

Overcharge is the continued charging of a cell after it has become fully charged. When a cell is not yet fully charged, the electrical energy of the charge current is converted to chemical energy in the cell by the charging reactions. But, when all the available active material has been converted to a charged state, the energy available in the charging current goes to produce gases from the cell or to activate other nonuseful chemical reactions.

9.3 Battery Technology: An Overview

Many types of rechargeable chemistry are used in electronic systems. Today, most batteries are NiCd, NiMH, and sealed lead acid, with lithium-ion making headway into portable systems. The choice of a particular battery technology for a given system is typically limited by size, weight, cycle life, and cost. A comparison of some basic characteristics of the five major types of battery chemistry is depicted in Table 9-1.

NiCd batteries presently power most rechargeable consumer appliances. The technology is mature and well understood. However, cadmium is coming under increasing regulatory scrutiny (including mandatory recycling in some jurisdictions) and the maturity of NiCd technology also means most of the capacity and life cycle improvements already have been made.

NiMH offers incremental improvements in energy density both by weight and volume over NiCd. Li-ion is better still, offering over twice the watts per liter and per kilogram of NiCd batteries. As always, this higher performance comes at a higher price. NiMH and Li-ion are increasing in popularity as upgrade options or in applications that support a higher price/performance point. The advantages of NiMH and Li-ion chemistry, however, also come at the cost of greater electrical

TABLE 9-1 Battery chemistry characteristics

Parameter	Units/Conditions	Sealed Lead Acid	NiCd	NiMH	Li-ion	Rechargeable Alkaline
Cell voltage	Volts	2.0	1.2	1.2	3.6	1.5
Relative cost	NiCd = 1	0.6	1	1.6	2	0.5
Self-discharge	%/month	2–4%	15–30%	18–20%	6–10%	0.3%
Cycle life	Cycles to reach 80% of rated capacity	500–2000	500–1000	500–800	1000–1200	<25
Overcharge tolerance	—	High	Medium	Low	Very low	Medium
Energy by volume	Watt hour/liter	70–110	120–150	250–300	280–320	220
Energy by weight	Watt hour/kg	30–45	40–50	70–80	110–130	80

fragility. Li-ion particularly is more easily and extensively damaged by less than optimal battery management, so much so that fail-safe circuits to disconnect the cells from the load under overcurrent or overtemperature conditions usually are built into the battery pack.

Rechargeable alkaline batteries mimic the form and replace the function of disposable household batteries. While initially more expensive, they cost less over their lifetime than the equivalent in disposable batteries. They are the least expensive form of rechargeable chemistry for low current applications and have the lowest self-discharge rate. However, they have the shortest cycle life in deep discharge applications.

Lead-acid batteries are most familiar in automobiles because they are the most economical chemistry for delivering large currents. Lead-acid batteries also have a long trickle life and therefore serve well for classic "floating" applications. Although flooded lead-acid technology is popular for automobile and similar applications, sealed lead-acid batteries serve the electronic engineering environment. On the downside, lead-acid has the least capacity by volume and weight.

Table 9-1 lists the major advantages and disadvantages of the five chemistries. Chemistry selection involves trade-offs driven by the technical requirements and economics of the application.

9.4 Lead-Acid Batteries

9.4.1 Flooded Lead-Acid Batteries

The flooded lead-acid battery of today basically uses the design developed by Faure in 1881. It consists of a container with multiple plates immersed in a pool of dilute sulfuric acid. Recombination is minimal, so water is consumed throughout the battery life and the batteries can emit corrosive and explosive gases when experiencing overcharge. So-called maintenance-free forms of flooded batteries provide excess electrolytes to accommodate water loss throughout a normal life cycle. Most industrial applications for flooded batteries are found in motive power, engine starting, and large system power backup. Today, other forms of battery have largely supplanted flooded batteries in small- and medium-capacity applications, but in larger sizes flooded lead-acid batteries continue to dominate. By far, the biggest application for flooded batteries is starting, lighting, and ignition service on automobiles and trucks. Large flooded lead-acid batteries also provide motive power for equipment ranging from forklifts to submarines and provide emergency power backup for many electrical applications, most notably the telecommunication network.

9.4.2 Sealed Lead-Acid Batteries

Sealed lead-acid batteries first appeared in commercial use in the early 1970s. Although the governing reactions of the sealed cell are the same as other forms of lead-acid batteries, the key difference is the recombination process that occurs in the sealed cell as it reaches full charge. In conventional flooded lead-acid systems, the excess energy from overcharging goes into the electrolysis of water in the electrolyte with the resulting gases being vented. This occurs because the excess electrolyte prevents the gases from diffusing to the opposite plate and possibly recombining. Thus, electrolyte is lost on the overcharge and must be replenished. The sealed-lead cell, like the sealed nickel-cadmium cell, uses recombination to reduce or eliminate this electrolyte loss.

Sealed lead-acid batteries for electronics applications are somewhat different from the type commonly found in the automobile. There are two types of sealed lead-acid batteries: the original gelled electrolyte and retained (or absorbed) system. The gelled electrolyte system is obtained by blending silica gel with an electrolyte, causing it to set up in gelatin form. The retained system employs a fine glass fiber separator to absorb and retain liquid electrolyte. Sometimes the retained system is called an absorbed glass mat (AGM). The AGM also is known as a *starved design. Starved* means the absorption limits of the glass separator create a limitation to the AGM design relating to diffusion properties of the separator. In certain cases, the AGM battery must be racked and trayed in a specific position for optimum performance. Both types, gelled and AGM, are called *valve regulated lead acid* (VRLA) systems. Today, sealed-lead cells are operating effectively in many markets previously closed to lead-acid batteries. For a detailed account of lead-acid cells, see Gates Energy Products Inc. (1992), Hirai (1990), and Moore (1993). Meanwhile, some manufacturers have introduced special versions of sealed lead-acid batteries with higher volumetric energy density; for example, the Portable Energy Products, Inc., Thinline™ series and the Bolder Technologies Corporation Thin Metal Film (TMF™), both with comparatively higher energy densities (Moneypenny and Wehmeyer, 1994; Nelson, 1997).

9.4.2.1 Discharge Performance of Sealed-Lead Acid Cells

The general shape of the discharge curve, voltage as a function of capacity (if the current is uniform) is shown in Figure 9-2. The discharge voltage of the starved-electrolyte sealed-lead acid battery typically remains relatively constant until most of its capacity is discharged. It then drops off sharply. The flatness and the length of the voltage plateau relative to the length of the discharge are major features of sealed-lead cells and batteries. The point at which the voltage leaves the plateau and begins to decline rapidly often is identified as the knee of the curve. Starved-electrolyte sealed-lead acid batteries may be discharged over a wide range of temperatures. They maintain adequate performance in cold environments and may produce actual capacities higher than their standard capacity when used in hot environments. Figure 9-3 illustrates the relationships between capacity and cell temperature. Actual capacity is expressed as a percentage of rated capacity as measured at 23°C.

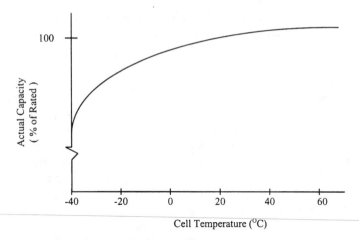

FIGURE 9-3 Typical discharge capacity as a function of cell temperature

9.4.2.2 Capacity During Battery Life

The initial actual capacity of sealed-lead acid batteries is almost always lower than the battery's rated or standard capacity. However, during the battery's early life, the actual capacity increases until it reaches a stabilized value, which usually is above the rated capacity. The number of charge/discharge cycles or length of time on float charge required to develop a battery's capacity depends on the specific regime employed. Alternatively, a battery that is on charge at 0.1 C usually is stabilized after receiving a 300% (of rated capacity) overcharge. The process may be accelerated by charging and discharging at low rates.

Under normal operating conditions, the battery's capacity will remain at or near its stabilized value for most of its useful life. Batteries then will begin to suffer some capacity degradation due to their age and the duty to which they have been subjected. This permanent loss usually increases slowly with age until the capacity drops below 80% of its rated capacity, which often is defined as the end of useful battery life. Figure 9-4 shows the capacity variation with cycle life that can be expected from sealed-lead acid batteries.

9.4.2.3 Effect of Pulse Discharge on Capacity

In some applications, the battery is not called on to deliver a current continuously. Rather, energy is drawn from the battery in pulses. By allowing the battery to "rest" between these pulses, the total capacity available from the battery is increased. Figure 9-5 shows typical curves representing the voltage delivered as a function of discharged capacity for pulsed and constant discharge at the same rate.

For the pulsed curve, the upper line represents the open-circuit voltage and the lower line represents the voltages during the periods when the load

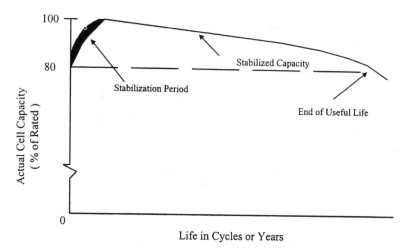

FIGURE 9-4 Typical cell's capacity during its life

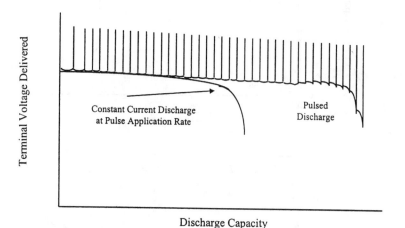

FIGURE 9-5 Typical pulsed and constant discharge curves

is connected. The use of discharged capacity as the abscissa eliminates the rest periods and shows only the periods of useful discharge.

9.4.3 Charging

In general, experience with sealed-lead acid cells and batteries indicates that application problems are more likely to be caused by undercharging than overcharging. Since the starved-electrolyte cell is relatively resistant to damage from overcharging, designers may want to ensure that the batteries are fully charged, even at the expense of some degree of overcharge. Obviously, excessive overcharging, either in magnitude or duration, still should be avoided.

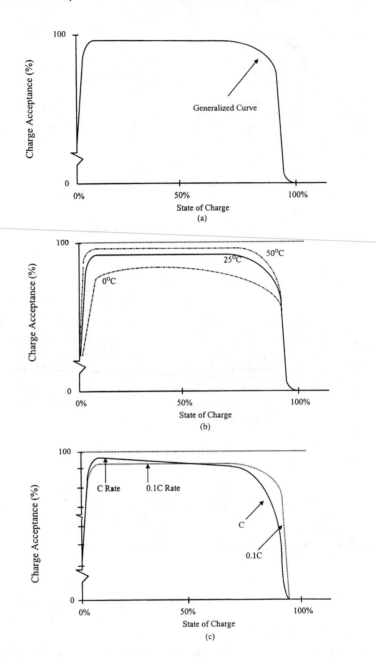

FIGURE 9-6 Charge acceptance: (a) effect of state of charge on charge acceptance, (b) charge acceptance at various temperatures, (c) charge acceptance at various charge rates

The charge acceptance of sealed-lead acid batteries in most situations is quite high, typically greater than 90%. A 90% charge acceptance means that for every ampere-hour of charge introduced into the cell, the cell will be able to deliver 0.9 Ah to a load. Charge acceptance is affected by a number of factors including cell temperature, charge rate, state of charge, age, and the method of charging (Figure 9-6).

The state of charge of the cell, to some extent, will dictate the efficiency with which the cell will accept a charge. When the cell is fully discharged, the charge acceptance initially is quite low. As the cell becomes only slightly charged, it accepts current more readily and the charge acceptance jumps quickly, approaching 98% in some situations. The charge acceptance stays at a high level until the cell approaches full charge.

As mentioned earlier, as the cell becomes fully charged, some of the electrical energy begins generating gas, which represents a loss in charge acceptance. When the cell is fully charged, essentially, all the charging energy generates gas, except for the very small current that makes up for internal losses that otherwise would be manifested as self-discharge. A generalized curve representing these phenomena is shown in Figure 9-6(a).

As with most chemical reactions, temperature has a positive effect on the charging reactions in the sealed-lead acid cell. Charging is more efficient at higher temperatures than at lower temperatures, all other parameters being equal, as shown in Figure 9-6(b).

The starved-electrolyte sealed-lead cell charges very efficiently at most charging rates. The cell can accept a charge at accelerated rates (up to the C rate) as long as the state of charge is not so high that excessive gas is generated. And the cell can be charged at low rates with excellent charge acceptance.

Figure 9-6(c) shows the generalized curve of charge acceptance now further defined by charging rates. When examining these curves, note that, at high states of charge, low charge rates provide better charge acceptance.

In the starved-electrolyte sealed-lead acid cell at typical charging rates, the bulk of the gases is recombined and there is virtually no venting of gases from the cell if it is overcharged.

9.5 Nickel-Cadmium Batteries

Sealed NiCd batteries are especially well suited to applications where a self-contained power source increases the versatility or reliability of the product. Among the significant advantages of NiCd families are higher energy density and discharge rates, fast recharge capability, and long operating and storage life. This places NiCd families at the top of usage in the portable products. In addition, the NiCd batteries are capable of operating over a wide temperature range and in any orientation with reasonable continuous overcharge capability.

9.5.1 Construction

NiCd secondary batteries operate at 1.2 V, using nickel oxyhydroxide for the active material in the positive electrode. The active material of the negative electrode consists of cadmium, and an alkali solution acts as the electrolyte. In NiCd batteries, a reaction at the negative electrode consumes the oxygen gas that generates at the positive electrode during overcharging. The design prevents the negative electrode from generating hydrogen gas, permitting a sealed structure. NiCd batteries mainly adopt cylindrical or prismatic configurations.

9.5.2 Discharge Characteristics

The discharge voltage of a sealed NiCd cell typically remains relatively constant until most of its capacity is discharged. It then drops off rather sharply. The flatness and length of the voltage plateau relative to the length of discharge are major features of sealed NiCd cells and batteries. The discharge curve, when scaled by considering the effects of all the application variables, provides a complete description of the output of a battery. Differences in design, internal construction, and conditions of actual use of cell affect the performance characteristics. As an example, Figure 9-7 illustrates the typical effect of discharge rate.

9.5.3 Charge Characteristics

Nickel-cadmium batteries are charged by applying a current of proper polarity to the terminals of the battery. The charging current can be pure direct current (DC) or it may contain a significant ripple component such as half-wave or full-wave rectified current.

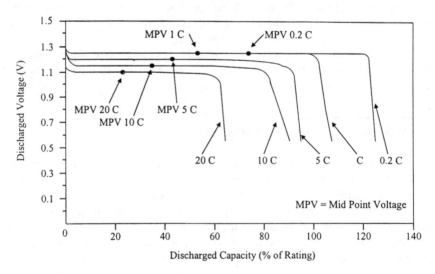

FIGURE 9-7 Discharge curves for NiCd cells: (a) typical curves at 23°C, (b) voltage depression effect

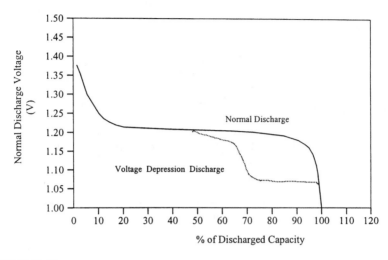

FIGURE 9-7 Continued

This section on charging sealed nickel-cadmium batteries refers to charging rates as multiples (or fractions) of the C rate. These C rate charging currents can be categorized into descriptive terms, such as standard charge, quick charge, fast charge, or trickle charge as shown in Table 9-2.

When a nickel-cadmium battery is charged, not all of the energy input goes to converting the active material to a usable (chargeable) form. Charge energy also goes to converting active material into an unusable form, generates gas, or is lost in parasitic side reactions.

Figure 9-8 shows the charge acceptance of NiCd cells. The ideal cell, with no charge acceptance loss, would be 100% efficient. All the charge delivered to the cell could be retrieved on discharge. But nickel-cadmium cells typically accept charge at different levels of efficiency, depending on the state of charge of the cell, as shown by the bottom curve of Figure 9-8.

Figure 9-8 describes this performance for successive types of charging behavior (zones 1, 2, 3, and 4). Each zone reflects a distinct set of chemical mechanisms responsible for loss of charge input energy.

TABLE 9-2 Categories of rates for charging NiCd cell

Method of Charging	Charge rate (multiples of C rate)	Recharge Time (hours)	Charge Control
Standard	0.05	36–48	Not required
	0.1	16–20	
Quick	0.2	7–9	Not required
	0.25	5–7	
	0.33	4–5	
Fast	1	1.2	
	2	0.6	
	4	0.3	Required
Trickle	0.02–0.1	Used for maintaining a fully charged battery	

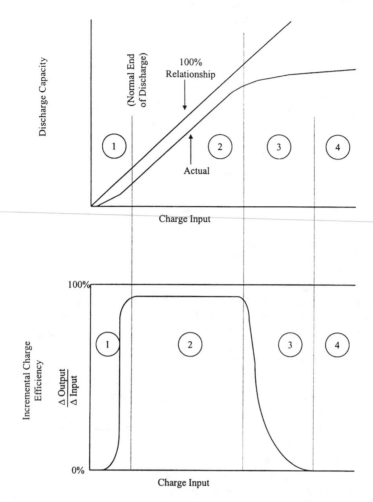

FIGURE 9-8 Charge acceptance of a sealed NiCd cell at 0.1 C and 23°C

In zone 1, a significant portion of the charge input converts some of the active material mass to an unusable form; that is, charged material not readily accessible during medium- or high-rate discharges, particularly in the first few cycles. In zone 2, the charging efficiency is only slightly less than 100%; small amounts of internal gas and parasitic side reactions are all that prevent the charge from being totally efficient. Zone 3 is transition region.

As the cell approaches a full charge, the current input shifts from charging positive active material to generating oxygen gas. In the overcharge region, zone 4, all the current coming into the cell generates gas. In this zone, the charging efficiency is practically none.

The boundaries between zones 1, 2, 3, and 4 are indistinct and vary depending on cell temperature, construction, and charge rate. The level of charge acceptance in zones 1, 2, and 3 also is influenced by cell temperature and charge rate. For details, see Gates Energy Products (1992).

9.5.4 Voltage Depression Effect

When some NiCd batteries are subjected to numerous partial discharge cycles and overcharging, cell voltage decreases below, 1.05 V/cell before 80% of the capacity is consumed. This is called the **voltage depression effect**; and the resulting lower voltage may be below the minimum voltage required for proper system operation, giving the impression that the battery has worn out (see Figure 9-7(b)). If cells are exposed to overcharging, particularly at higher temperatures, and this is quite common, the voltage may be about 150 mV lower than the normal cell voltage. Voltage depression is an electrically reversible condition that disappears when the cell is completely discharged and recharged. The voltage depression effect sometimes is called erroneously the **memory effect**, and clearing the same by charging and discharging is called conditioning.

9.6 Nickel-Metal Hydride Batteries

For those battery users who need high power in a small package and are willing to pay a higher price, an option is the nickel-metal hydride families, which offer a significant increase in cell power density. These extensions of NiCd cell technology to a new chemistry have become popular with product applications such as notebook computers and cellular phones. The first practical NiMH batteries entered the market in 1990.

In many ways, nickel-metal hydride batteries are the same as NiCd types: They use nickel for the positive electrode but a recently developed material, a hydrogen-absorbing alloy, for the negative electrode. With an operating voltage of 1.2 V, they provide high capacity, more energy density than NiCd models.

The NiCd cell is more tolerant of fast recharging and overcharging than NiMH cells. NiCd cells hold their charge longer than NiMH cells. NiCd cells will withstand 500–2000 charge/discharge cycles, compared to about 500 cycles for NiMH cells. Further, NiCd cells withstand a wider temperature range than NiMH cells.

On the other hand, NiMH cells seldom exhibit the notorious "memory effect" that NiCd cells sometimes do. As with any new technology, the prices for NiMH applications are higher than those of NiCds (Small, 1992; Eager, 1991; Furukawa, 1993; Briggs, 1994).

The voltage profile of NiMH cells during discharge is very similar to that of the NiCd cells. NiMH cells' open-circuit voltage is 1.3–1.4 V. At moderate discharge rates, NiMH cells' output voltage is 1.2 V. Both NiCd and NiMH cells have relatively constant output voltage during their useful service. Figure 9-9 is a typical graph from a battery company, comparing the output voltage of 700 mAh NiCd and 1100 mAh NiMH AA cells under load. Note that the NiMH cell's greater capacity results in approximately 50% longer service life.

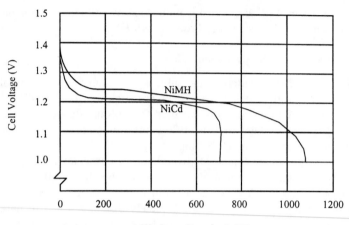

FIGURE 9-9 Discharge characteristics of NiCd and NiMH batteries

Figure 9-10 is another typical battery company graph showing that NiCd and NiMH batteries and cells charge in a similar fashion as well. However, the little bumps at the end of the two cell's charge curves bare closer examination. You always will see these negative excursions, even though absolute cell voltages vary significantly with temperature.

The negative excursions signal a fully charged cell more or less independent of temperature, a useful quirk that sophisticated battery chargers exploit. Note that the NiCd cell's negative-going voltage excursion after reaching a full charge is more pronounced than that of the NiMH cell.

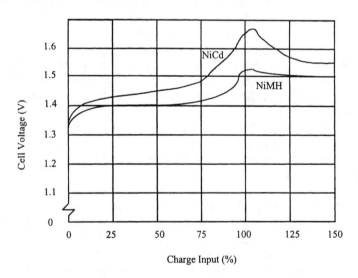

FIGURE 9-10 Battery voltage at the achievement of 100% charge

There are several reasons for replacing NiCd batteries with NiMH types:

- NiCd batteries contain cadmium, which is harmful to the environment.
- NiMH has nearly the same operating voltage as NiCd, making them interchangeable.
- NiMH batteries have 30–40% higher capacity at the same physical size as NiCd batteries.
- NiMH batteries have a 90-minute charging capability.

9.7 Lithium-Ion Batteries

The demand for portable systems is increasing at a dramatic rate. To remain competitive, companies are offering lighter-weight and longer-run-time systems. Meeting these goals requires improvements in battery technology beyond traditional NiCd and NiMH systems.

Li-ion is a promising technology that can improve capacity for a given size and weight of a battery pack. With an energy density by weight about twice that of nickel-based chemistry (see Table 9-1), Li-ion batteries can deliver lighter weight packs of acceptable capacity. Li-ion also has about three times the cell voltage of NiCd and NiMH batteries; therefore, fewer cells are needed for a required voltage. The Li-ion batteries are becoming widely used in notebook PCs and many other portable systems because of their high energy density, declining costs, and readily available management circuits.

The first noticeable difference between Li-ion and nickel-based batteries is the higher internal impedance of the lithium-based batteries. Figure 9-11

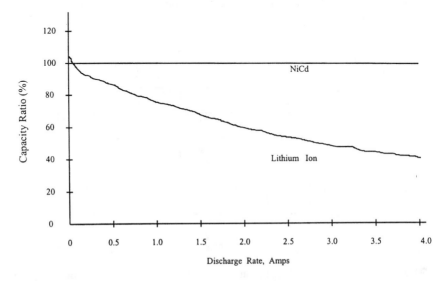

FIGURE 9-11 Li-ion and NiCd capacity vs. discharge current

shows this by graphing the actual discharge capacity of a Li-ion cell at different discharge currents compared to a NiCd cell.

At a 2 amp discharge rate (2 C), less than 80% of the rated capacity is available from the Li-ion cell, compared to nearly 95% of the rated capacity for NiCd. For systems with discharge currents greater than 1 amp, the capacity realized from the Li-ion battery may be less than expected. Parallel battery stack configurations often are used in Li-ion battery packs to help reduce the severity of this problem. Li-ion technology requires re-examination of the charge and discharge characteristics of portable systems. Due to the nature of the chemistry, Li-ion batteries cannot tolerate overcharging and overdischarging.

The most important factor in using a Li-ion battery in a portable consumer product is safety. Fortunately, battery safety is comprehensively addressed by battery manufacturers in concert with the manufacturers of protection circuits and charge control circuits. For details, see Bennett and Brown (1997). Commercially available packs have an internal protection circuit that limits the cell voltage during charge to between 4.1 and 4.3 V per cell, depending on the manufacturer. Voltages higher than this rating could permanently damage the cell. A discharge limit of between 2.0 and 3.0 V (depending on the manufacturer) is necessary to avoid reducing the cycle life of the battery and damaging the battery.

The anode, or negative electrode, in a Li-ion cell is composed of a material capable of acting as a reversible Li-ion reservoir. The material usually is a form of carbon, such as coke, graphite, or pyrolytic carbon. The cathode, or positive electrode, also is a material that can act as a reversible lithium-ion reservoir.

Currently, the preferred cathode materials are $LiCoO_2$, $LiNiO_2$, or $LiMn_2O_4$ because of their high oxidation potentials of about 4 V versus lithium metal. Commercially available Li-ion cells use a liquid electrolyte made up of mixtures that are predominately organic carbonates containing one or more dissolved lithium salts (Levy, 1995).

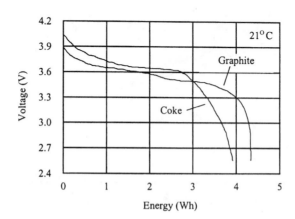

FIGURE 9-12 Li-ion discharge profile for different electrodes. (Reproduced by permission of Moli Energy Ltd.)

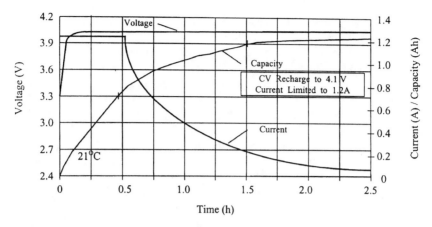

FIGURE 9-13 Li-ion charge profile at constant potential charging at 4.1 V and current limited to 1.2 amps. (Reproduced by permission of Moli Energy Ltd.)

Today, the predominant Li-ion technologies use coke or graphite as an anode material. Figure 9-12 illustrates the differences in the two types of cells during discharge. The graphite anode discharge voltage is relatively flat during a majority of the discharge cycle, while the coke anode discharge voltage is more sloped (Juzcow and St. Louis, 1996).

The energy available from the graphite anode cell is higher for a given capacity due to the higher average discharge voltage. This may be useful in systems that need the maximum watt-hour capacity for a given battery size. Also, the charge and discharge cutoff voltages between the two Li-ion systems vary among manufacturers.

Figure 9-13 shows the typical charge profile for Li-ion batteries. The charge cycle begins with a constant current limit and makes the transition to a constant-voltage limit, typically specified between 4.1 and 4.3 V ±1%. This allows maximum charge capacity with no cell damage.

Charging to a lower voltage limit does not damage the cell, but the discharge capacity will be reduced. A 100 mV difference could change the discharge capacity by more than 7%. Basically, the difficult aspect of this type of charger is the wide dynamic range required from the switching current regulator given the tight voltage tolerance. Some chargers provide single-cell monitoring, while others rely on the internal protection circuit to do this.

9.8 Reusable Alkaline Batteries

Alkaline technology has been used in primary batteries for several years. With the development of the reusable alkaline manganese technology, secondary alkaline cells quickly made their way into many consumer and industrial applications. In many applications, reusable alkaline cells can be recharged from 75 to over 500 times, unlike single-use alkaline batteries, and initially have three times the capacity of a fully charged NiCd battery. These cells do not compete with NiCds in high-powered applications, however.

Intensive research and development activities carried out at Battery Technologies Inc. (BTI), Canada, and at the Technical University in Graz, Austria, in the late 1980s and early 1990s resulted in the successful commercialization of the rechargeable alkaline manganese dioxide zinc (RAM™) system. BTI has chosen to sell licenses and production equipment, where necessary, for the manufacturing and worldwide marketing rights of its proprietary RAM technology. For example, Rayovac Corporation, one of the licensees, launched its line of reusable alkaline products under the name Renewal™ in the United States, Pure Energy Battery Corporation in Canada (Pure Energy™), and Young Poong Corporation in South Korea (Alcava™). For details, see Nossaman and Parvereshi (1995), Sengupta (1995), and Ivad and Kordesch (1997).

The chemistry behind the reusable approach depends on limiting the zinc anode to prevent overdischarge of the MnO_2 cathode. Additives are incorporated to control hydrogen generation and other adverse effects on charging. The rated cycle life is 25 cycles to 50% of initial capacity. A longer cycle life is possible, depending on drain rate and depth of discharge. To take advantage of the reusable alkaline cell and increase its life, a special "smart charger" is required.

Using reusable alkaline cells can drive down the total battery cost to the consumer. This cost saving can be determined by looking at the cumulative capacity of a reusable cell vs. the one-time use of a primary alkaline cell. Figure 9-14 illustrates the capacity of AA cells being discharged down to 0.9 V at 100 mA. It shows that, although the initial use of the reusable alkaline is

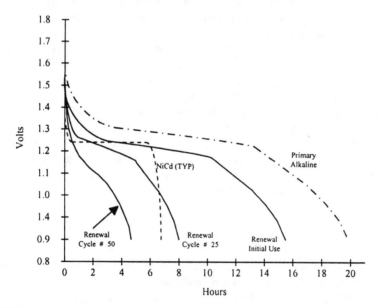

FIGURE 9-14 A 100 mA discharge curve comparison of NiCd, primary alkaline, and reusable alkaline batteries. (Source: Benchmarq Microelectronics.)

TABLE 9-3 Capacity of AA cells at various depths of discharge (values in mAh). (Source: Nossaman and Parvereshi, 1996.)

Condition	100% DOD	125 mA to 0.9 V 30% DOD	10% DOD
Cycle 1	1500	450	150
Cycle 50	400	—	—
Cumulative 50	33,000	22,000	7000
Cumulative 100	—	44,000	15,000
Cumulative 500	—	—	73,000

almost that of a primary alkaline battery, the reusable one can be recharged for continued use. Table 9-3 shows the increase in cumulative capacity by limiting the depth of discharge (DOD) and achieving more cycles.

Overcharging also affects the cycle life of reusable alkaline batteries. These batteries are not tolerant of overcharging and high continuous charge currents and may be damaged if high current is forced into them after they have reached a partially recharged state. Proper charging schemes should be used to prevent an overcharged condition.

9.9 Zinc-Air Batteries

Primary Zn-air batteries have been in existence for over 50 years in applications such as hearing aids and harbor buoys. The lightweight and high energy content in their technology has promoted research on Zn-air rechargeable chemistry by companies such as AER Energy Resources. The initial focus was on electric vehicles, but shifted toward portable appliances. Another application is solar powered rural telecom systems.

Rechargeable Zn-air technology is an air breathing technology in which the oxygen in the ambient air is used to convert zinc into zinc oxide in a reversible process. The cells use air breathing carbon cathode to introduce oxygen from air into potassium-hydroxide electrolyte. The cathode is multilayered with a hydrophilic layer, and the anode is composed of metallic zinc.

The characteristic voltage of the AER Energy rechargeable zinc-air system is a nominal 1 V. During discharge, the AER Energy cells operate at a voltage between 1.2 and 0.75 V. The current and power capability of the system is proportional to the surface area of the air breathing cathode. For more current and power, a larger surface area cell is required. For less current and power, a smaller cell may be used. Compared to other rechargeable chemistries, Zn-air needs an air manager for an intake and exhaust of air to allow the chemical process.

Figure 9-15 compares the gravimetric and volumetric energy densities of several rechargeable battery technologies. It clearly shows that Zn-air batteries require less weight and volume. Discharge and charge characteristics of Zn-air batteries are shown in Figure 9-16.

The cells exhibit a flat voltage profile over the discharge cycle. The typical charge voltage is 2 V per cell using a constant voltage/current taper approach.

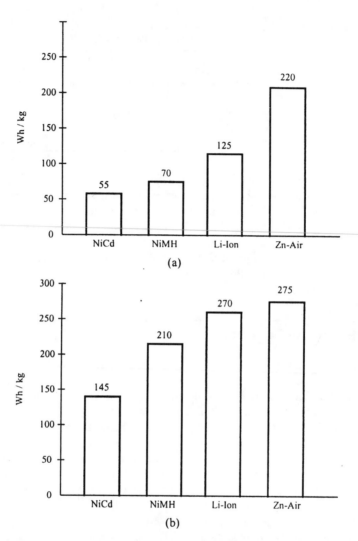

FIGURE 9-15 Energy density comparison of batteries: (a) gravimetric, (b) volumetric. (Reproduced by permission of AER Energy Resources.)

The battery's life cycle varies between 50 and 400, depending on the depth of discharge. The cost per watt-hour is apparently lower than in nickel- and lithium-based chemistries. For details, see Cutler (1997).

9.10 Battery Management

Two decades ago, battery management meant a reliable, fast, and safe charging methodology for a battery bank, together with the monitoring facilities

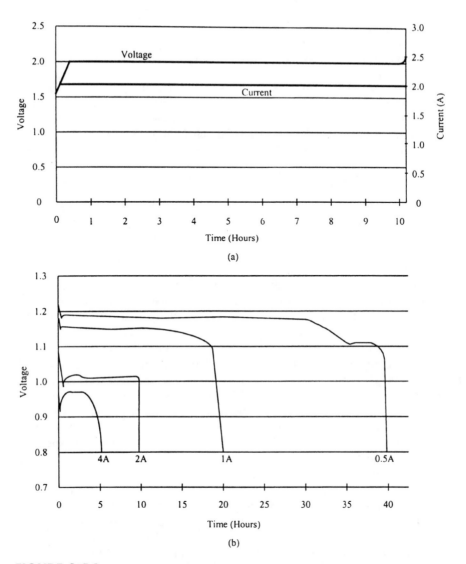

FIGURE 9-16 Charge/discharge characteristics of Zn-air chemistry: (a) charge, (b) discharge. (Reproduced by permission of AER Energy Technologies.)

to detect the discharge condition of the battery pack. With new battery technologies emerging, demands from the cost-sensitive portable product market, and the medium-power-range products such as uninterruptible power supplies (UPS) and telecom power units, the attributes of a modern battery management system must include

- Battery charging methods and charge control.
- End of discharge determination.

- Gas gauging.
- Monitoring battery health issues.
- Communication with the host system or power management subsystems.

The following sections discuss concepts and techniques related to managing nickel-based, sealed lead-acid, and Li-ion battery systems. A discussion on rechargeable alkaline batteries is beyond the depth of this chapter. Kuribayashi (1993), Levy (1995), and Ivad and Kordesch (1997) provide some details on managing rechargeable alkaline cells.

9.10.1 Charging Systems

While four major battery families can accept either a standard (16–24 hour) or fast (2–4 hour) charge, the discussion here is limited to fast charging methods. Slower charging schemes tend to be found in simpler, price-sensitive applications, which do not need (or cannot afford) much beyond a charger and a low battery indicator.

The objective of fast charging a battery is to cram as much energy as it takes to bring the battery back to fully charged state in the shortest possible time without damaging the battery or permanently affecting its long-term performance. Since current is proportional to energy divided by time, the charging current should be as high as the battery systems reasonably will allow. For the constant-current cells (NiCd and NiMH), a 1 C charge rate typically will return more than 90% of the battery's usable discharge capacity within the first hour of charging. The constant-potential cells (lead-acid and lithium-ion) are a bit slower to reach the 90% mark but generally can be completely recharged within 5 hours.

Fast charging has compelling benefits but places certain demands on the battery system. A properly performed fast charge, coordinated to the specifications of a battery rated for such charging, will deliver a long cycle life. The high charging rates involved, however, cause rapid electrochemical reactions within the cells of the battery. After the battery goes into overcharge, these reactions cause a sharp increase in internal cell pressure and temperature.

Uncontrolled high-rate overcharge quickly causes irreversible battery damage. Therefore, as the battery approaches a full charge, the charging current must be reduced to a lower "top-off" level, or curtailed entirely.

9.10.1.1 Charge Termination Methods

If a rapid charge is applied to a battery pack, it is necessary to select a reliable method to terminate charging at the fully charged position. Two practical approaches for charge termination are temperature termination and voltage termination.

9.10.1.1.1 Temperature Termination Method

Temperature is the main cause of failure in a rechargeable cell, so it makes sense to monitor the cell temperature to determine when to shut off the charge to a battery. Three methods of charge termination, based on temperature, are common: maximum temperature cutoff, temperature difference, and temperature slope. The maximum temperature cutoff system is the easiest and cheapest to implement but the least reliable. Using a bimetallic thermal switch or a positive temperature coefficient thermistor, a simple, low-cost circuit can shut down a charging current at an appropriate temperature.

The temperature difference (DT) method measures the difference in ambient and cell temperatures to compensate for a cool environment. The DT method requires monitoring two temperature sensors, one for the battery temperature and one for the ambient temperature. This method may be unsuitable if the difference between cell and ambient temperature is very large.

The DT method can become unreliable with a quickly changing ambient temperature unless an equal thermal mass is attached to the ambient sensor. This means that the DT method is suitable for a primary charge termination at lower charge rates, up to C/5, if the ambient temperature is not going to change often. The DT method also provides an excellent backup charge termination scheme.

The temperature slope (dT/dt) method, a more sophisticated temperature termination scheme, measures the change in temperature over time. This method uses the slope of the battery temperature curve and, therefore, is less dependent on changes in ambient temperatures or in large differences between ambient and battery temperatures. Accurately adjusted to a particular pack, and with careful attention paid to the type and placement of the temperature sensor, the dT/dt method works very well. This method is suitable for charge rates up to 1 C and provides an excellent backup method.

9.10.1.1.2 *Voltage Termination Methods*

Four commonly available voltage termination methods are maximum voltage, negative delta voltage, zero slope, and inflection point. The maximum voltage (V_{max}) method senses the increase in battery voltage as the battery approaches full charge. However, this is accurate only on a highly individualized basis. It is necessary to know the exact value at the voltage peak, otherwise the batteries may be over- or undercharged.

Temperature compensation also is required because of the negative temperature coefficient of battery voltage. The maximum voltage will increase if the batteries are cold, causing an undercharge because the charging voltage will reach the maximum voltage trip point early. If the batteries are hot, the maximum voltage may never be reached and the batteries will be cooked. Therefore, the V_{max} method generally is not recommended for fast charge rates.

The negative delta voltage ($-\Delta V$) is the most popular fast charge termination scheme. It relies on the characteristic drop in cell voltage that occurs when a battery enters overcharge, as shown in Figure 9-17. With most NiCd cells, the voltage drop is a very consistent indicator and the $-\Delta V$ method is fine for charge rates up to 1 C. An inherent problem with this method is that the batteries must be driven into the overcharge region to cause the voltage decrease. Pressures

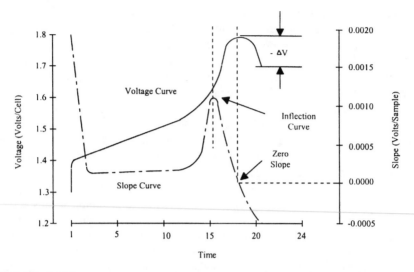

FIGURE 9-17 Termination methods based on changes in voltage and voltage slope

and temperatures rise very rapidly at fast charge rates beyond 1 C. In cyclic applications, the battery must be able to endure that continual abuse.

Another concern is that cells like the NiMH types do not always have the characteristic decrease in voltage, compared to NiCd cells, as shown in Figure 9-10. This creates a problem of forward compatibility when moving from NiCd to NiMH cells. Most manufacturers of NiMH cells do not advocate the $-\Delta V$ method of charge termination.

The zero slope method monitors the point where the slope of the battery voltage reaches 0. This method is reliable for rapid charge rates up to 4 C and is less susceptible to noise on the voltage sense lines. However, a few types of batteries, such as the NiCd button cells, may have a voltage slope that never quite reaches 0. Therefore, the zero slope method is better suited as a backup method.

In the inflection point (dV/dt) method the system monitors the change in voltage over time and is the most sensitive indicator for preventing overcharge. The inflection point method relies on the changes in the voltage slope shown in Figure 9-17, which occur during charging and are an excellent primary termination method for up to 4 C charge rates.

The change in the voltage slope is an extremely reliable and repeatable indication of charge. It does not rely on a decrease in voltage, which may not always occur. Instead, this method looks for the flattening of the voltage profile as the battery reaches full charge. By monitoring the relative change in the steepness of the voltage slope, this method avoids having to use absolute numbers.

9.10.1.2 NiCd and NiMH Fast Charge Methods

Nickel-based batteries, such as the NiCd and NiMH types, have become the most popular choice for portable products. Although it is not correct to consider the NiCd and NiMH electrochemistry or charging regimens as interchangeable, they are similar enough to discuss together.

There is no one best way to fast charge a NiCd or NiMH battery. Variables introduced by the allowable cost and size of the application, the choice of charge termination method(s), and the specific battery vendor's recommendations all influence the final choice of charging technique. Figure 9-18(a) shows the voltage, pressure, and temperature characteristics of a NiCd cell being charged at a 1 C rate. Figure 9-18(b) shows similar data for NiMH cell.

These curves illustrate the need for a reliable termination of the high-current portion of the charge cycle, and assist in understanding the various fast charge termination methods outlined in Table 9-5. For both types of electrochemistry, the ideal fast charge termination point is at 100–110% of the returned charge. The charging current is then reduced to the top-off value for 1–2 hours, to bring the cell to a state of slight overcharging.

This compensates for the inefficiencies of the charging process (e.g., heat generation). If the specific application will have the battery on standby for more

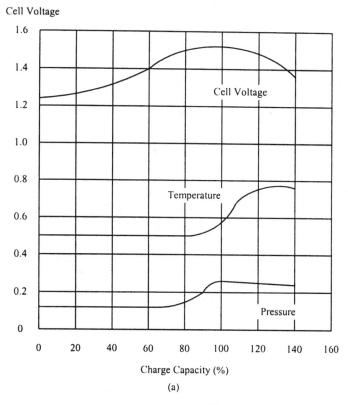

FIGURE 9-18 Charging indications for nickel-based batteries at a 1 C charge rate: (a) cell voltage, temperature, and pressure for a typical NiCd cell at a 1 C rate charging, (b) NiMH voltage and temperature characteristics

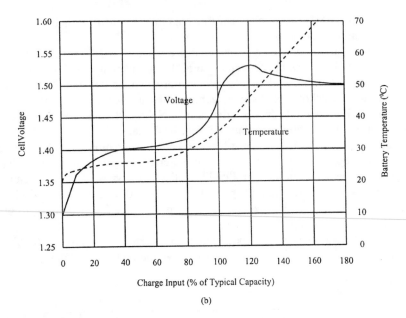

(b)

FIGURE 9-18 Continued

than several weeks or at high temperatures, the top-off charge is followed by a continuous, low-level "trickle" charge to counter the self-discharge characteristics of NiCd and NiMH cells.

Under certain conditions, particularly following intervals of storage, a NiMH battery may give an erroneous voltage peak as charging commences. For this reason, the charger should deliberately disable any voltage-based charge termination technique for the first 5 minutes of the charging interval.

See Table 9-4 for representative charging recommendations. The most appropriate method should be selected in consultation with the manufacturer. Table 9-5 summarizes the fast charge termination methods for NiCd and NiMH cells.

TABLE 9-4 Representative charging recommendations for different batteries

	NiCd or NiMH	Sealed Lead Acid	Li-Ion
Charging current	1 C	1.5 C	1 C
Voltage per cell (V)	1.80	2.5	4.20 ± 0.05
Charging time (hours)	~3	~3	2.5–5.0
Method for optimum fast charge termination point	See Table 9-5	Current cutoff	Typically a timer[1]
Backup charge termination method	See Table 9-5	Timer	—
Top-off rate	0.1 C	0.002 (trickle)	
Temperature range (°C)	10–40° (NiCd) 15–30° (NiMH)	0–30°	0–40°

[1] Depending on manufacturer's recommendation.

TABLE 9-5 Fast charge termination methods for NiCd and NiMH batteries

Charging Technique	Description
Negative delta voltage $(-\Delta V)$	Looks for the downward slope in cell voltage, which a cell exhibits ($\gg$30–50 mV for NiCd, 5–15 mV for NiMH) on entering overcharging. Very common in NiCd applications due to its simplicity and reliability
Zero ΔV	Waits for the time when the voltage of the cell under charge stops rising and is "at the top of the curve" prior to the downslope seen in overcharging. Sometimes preferred over $-\Delta V$ for NiMH, due to relatively small downward voltage slope of NiMH
Voltage slope (dV/dt)	Looks for an increasing slope in cell voltage (positive dV/dt), which occurs somewhat before the cell reaches 100% returned charge (prior to the zero ΔV point)
Inflection point cutoff (d^2V/dt^2)	As a cell approaches full charge, the rate of its voltage rise begins to level off. This method looks for a 0 or, more commonly, slightly negative value of the second derivative of cell voltage with respect to time
Absolute temperature cutoff	Uses the cell's case temperature (which will undergo a rapid rise as the cell enters high-rate overcharging) to determine when to terminate high-rate charging. A good backup method, it is too susceptible to variations in ambient temperature to make a reliable primary cutoff technique
Incremental temperature cutoff (Δ TCO)	Uses a specified increase in the cell's case temperature, relative to the ambient temperature, to determine when to terminate high-rate charging. It is a popular, relatively inexpensive and reliable cutoff method
Delta temperature/delta time ($\Delta T/\Delta t$)	Uses the rate of increase of a cell's case temperature to determine the point at which to terminate the high-rate charge. This technique is inexpensive and reliable once the cell and its housing have been properly characterized

9.10.1.3 Sealed Lead-Acid Batteries

Unlike nickel-based batteries, sealed lead-acid batteries are charged using a "constant-potential" (CP) regimen. CP charging employs a voltage source with a deliberately imposed current limit (a current-limited voltage regulator). A significantly discharged battery undergoing CP charging initially will attempt to draw a high current from the charger. The current-limiting function of the CP regimen keeps the peak charging current within the battery's ratings.

Following the current-limited phase of the charging profile, a sealed lead-acid battery exposed to a constant voltage will exhibit a tapering current profile, as shown in Figure 9-19. When the returned charge reaches 110–115% of the rated capacity, allowing a dischargeable capacity of 100% of nominal, the charge cycle is complete.

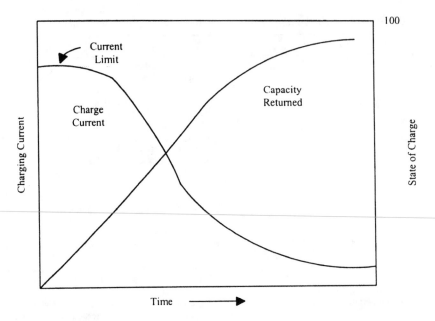

FIGURE 9-19 Typical current and capacity returned vs. charge time for CP charging

The specifics of fast charging sealed lead-acid batteries are more vendor dependent than those of NiCd or NiMH units. The information in Table 9-4 uses data from GS Battery Inc. The primary termination method, current cutoff, looks at the absolute value of the average charging current flowing into the battery.

When that current drops below 0.01 C, the battery is fully charged. If it will be in standby for a month or more, a trickle current of 0.002 C should be maintained. The backup termination method, according to the vendor's recommendations, should be a 180 minute time out on the charging cycle (Schwartz, 1995).

To satisfy more stringent charge control recommendations, where the battery temperature, voltage, and current need to be sampled, many dedicated charge controller ICs are available. The bq2031 lead-acid fast charge IC from Benchmarq Microelectronics Inc. and UC3906 (sealed lead-acid charger) from Unitrode Integrated Circuits are examples.

The UC3906 battery charger controller contains all the necessary circuitry to optimally control the charge and hold the cycle for sealed lead-acid batteries. These integrated circuits monitor and control both the output voltage and current of the charger through three separate charge states: a high-current bulk-charge state, a controlled overcharge, and a precision float charge or standby state. Figure 9-20 is a block diagram and one implementation of the UC3906 in a dual-step current charger. Sacariesen and Parvereshi (1995) and Unitrode Application Note U-104 provide details about charge control in sealed lead-acid batteries using the UC3906.

9.10.1.4 Lithium-Ion Chargers

Li-ion batteries require a constant potential charging regimen, very similar to that used for lead-acid batteries. Typical recommendations for Li-ion fast charging are listed in Table 9-4. As with lead-acid batteries, a Li-ion cell under charge will reduce its current draw as it approaches full charge.

If the cell vendor's recommendation for charging voltage (generally 4.20 V ±50 mV at 23°C) is followed, the cells will be able to completely recharge from any "normal" level of discharge within 5 hours. At the end of that time, the charging voltage should be removed. Trickle current is not recommended.

If the voltage on a Li-ion cell falls below 1.0 V, recharging should not be attempted. If the voltage is between 1.0 V and the manufacturer's nominal minimum voltage (typically, 2.5–2.7 V), it may be possible to salvage the cell by charging it with a 0.1 C current limit until the voltage across the cell reaches the nominal minimum, followed by a fast charge.

Due to the special characteristics of Li-ion batteries, most manufacturers incorporate custom circuits into their battery packs to monitor the voltage across each cell within the battery and provide protection against overcharging, battery reversal, and other major faults. These circuits are not to be confused with charging circuits. For example, the MC 33347 protection circuit is such a monolithic IC from Motorola (Alberkrack, 1996).

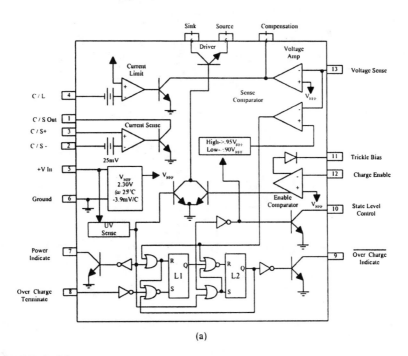

(a)

FIGURE 9-20 The UC3906 and its implementation: (a) block diagram, (b) implementation in a dual-step current charger. (Reproduced by permission of Unitrode Inc.)

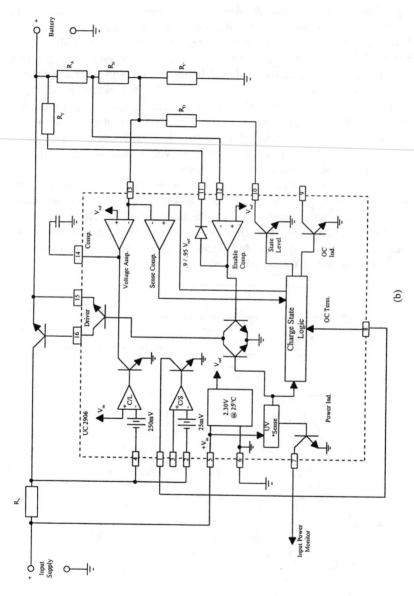

FIGURE 9-20 Continued

9.10.2 End-of-Discharge Determination

Determination of the point at which a battery has delivered all of its usefully dischargeable energy is important to the longevity of the cells that form that battery. Discharging a single cell too far often will cause irreversible physical damage inside the cell.

If multiple cells are placed in series, unavoidable imbalances in their capacities can cause the phenomenon known as *cell reversal*, in which the higher-capacity cells force a backward current through the lowest-capacity cell. Knowing the end-of-discharge (EOD) point provides a "zero capacity" reference for coulometric gas gauging.

The actual determination of the EOD point typically is done by monitoring cell voltage. For the most accurate determination of the EOD when the load is varying, correction factors for the load current and the battery's state of charge should be applied, especially to lead-acid and Li-ion batteries. The essentially flat discharge profiles of NiCd and NiMH make these corrections a matter of user discretion for most load profiles. Table 9-6 shows voltages commonly used to indicate the end-of-discharge point for the four battery types.

9.10.3 Gas Gauging

The gas gauging or fuel gauging concept discussed here does not refer to the gases that may evolve by the battery reactions but rather to the concept of the battery as a fuel tank powering the product. Gas gauging, therefore, involves a real-time determination of a battery's state of charge, relative to the battery's nominal capacity when fully charged.

It is possible to make an inexpensive and moderately useful state of charge measurement from a simple voltage reading, if the battery being used has a sloping voltage profile. Hence, Li-ion batteries and, to a lesser extent, sealed lead-acid batteries, should be amenable to such an approach. In practice, the results are less than optimal: Cell voltages depend on loading, internal impedance, cell temperature, and other variables.

This reduces the attractiveness of the battery voltage method of gas gauging; and that it is unsuitable for NiCd or NiMH batteries, due to their essentially

TABLE 9-6 Typical end-of-discharge voltages

Cell Type	EOD Voltage (V)	Comments
Lead acid	1.35 – 1.9 (1.8 typical)	Dependent on loading, state of charge, cell construction, and manufacturer
NiCd	0.9	Essentially constant
NiMH	0.9	Essentially constant within recommended range of discharge rates
Li-ion	2.50 – 2.70	Dependent on manufacturer, loading, and state of charge

flat voltage profile, makes it commercially untenable. A clever and effective alternative is the "coulometric" method.

Coulometric gas gauging, as its name implies, meters the actual charge (current × time) going into and out of the battery. By integrating the difference of current in and current out, it is possible to determine the charge status of the battery at any given time. There are real-world details, of course, that must be observed in the actual implementation of such a gas gauge; some of the most important of these are:

1. It is necessary to have an accurate starting point for the integrator, corresponding to a known state of charge in the battery. This problem often is resolved by zeroing the integrator when the battery reaches its EOD voltage.
2. It is necessary to compensate for the temperature. The actual capacity of lead-acid batteries increases with temperature, that of nickel-based batteries decreases as battery temperature rises.
3. Appropriate conversion factors should be applied for the particular charge regimen and discharge profile used. Under conditions of highly variable battery loading, dynamic compensation may be advisable.

9.10.4 Battery Health

With *battery health* defined as a battery's actual capacity relative to its rated capacity, the health of the battery can be determined and maintained in three steps:

1. Discharge the battery to the EOD point, preferably into a known load.
2. Execute a complete charge cycle, while gas gauging the battery.
3. Compare the battery's measured capacity to its rated capacity.

This sequence will simultaneously "condition" the battery (e.g., overcome the so-called memory effect of capacity of NiCd batteries) and indicate the capacity of the battery after conditioning. This information can be used to ascertain whether the battery is in good shape or approaching the end of its useful life.

9.11 The System Management Bus, Smart Battery Data Specifications, and Related Standards

In the mid 1990s, new industry standards were proposed to standardize the battery and power management subsystems within portable products. The following standards were proposed:

1. System Management Bus (SMB) Specification.
2. Smart Battery Data Specification.
3. Smart Battery Charger Specification.
4. Smart Battery Selector Specification.

These specifications form the Smart Battery Systems Specification, which presents a solution for many of the issues related to batteries used in portable equipment such as laptop computer systems, cellular telephones, and video cameras. Fundamental to the system is the concept that the battery contains all the necessary components to determine the battery's state of charge, predict the time to full and empty charging, specify the charging voltage and current, and determine when the battery is fully charged or fully discharged.

A typical smart battery system is shown in Figure 9-21(a). It consists of an AC/DC converter (unregulated), power switch, system power supply, smart battery charger, and smart battery selector, all of which communicate with the system host and the system elements themselves via the SMB. In this case, smart battery A powers the system while smart battery B is getting conditioned or charged.

The system management bus is a two-wire interface through which simple power-related chips can communicate with the rest of the system. It uses I^2C as its backbone (Paret & Fenger, 1997). A system using an SMB passes messages to and

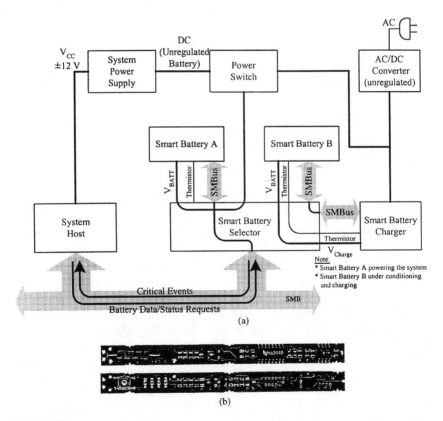

(a)

(b)

FIGURE 9-21 The smart battery system and an SMB module: (a) a typical multiple smart battery system, (b) the bq219XL module, (c) the bq219XL connections. (Reproduced by permission of Benchmarq Microelectronics, Inc.)

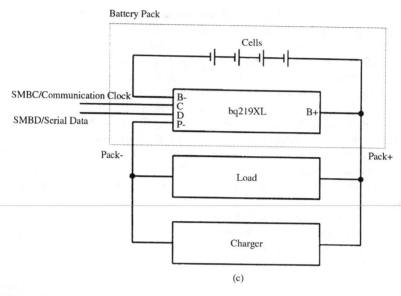

Battery Pack

Cells

SMBC/Communication Clock

SMBD/Serial Data

B-
C
D
P-

bq219XL

B+

Pack-

Pack+

Load

Charger

(c)

FIGURE 9-21 Continued

from devices instead of tripping individual control lines. Removing the individual control lines reduces the pin count. Accepting messages ensures future expandability. With a system management bus, a device can provide manufacturer information, tell the system its mode or part number, save its state in case operation is suspended, report different types of errors, accept control parameters, and return its status. The SMB may share the same host device and physical bus, as long as an appropriate electrical bridge is provided between the respective devices. Benchmarq et al. (1995a) provides details related to communication protocols available for use by devices on an SMB. Using gas gauge ICs such as the bq2092 or bq2040, Benchmarq provides a complete smart battery module such as its bq219XL (Figure 9-21(b)). Designed for battery pack integration, the bq219XL combines the bq2092 or bq2040 gas gauge IC with a serial EEPROM on a small printed circuit board. This smart battery module provides a complete solution for the design of intelligent battery packs using the SMB protocol and supports the smart battery data (SBD) commands in the SMB/SBD specifications. The board includes all the necessary components to accurately monitor battery capacity and communicate critical battery parameters to the host system or battery charger. The bq219XL also includes four LEDs coupled with a push-button switch to activate the LEDs to show remaining battery capacity in 25% increments.

Contacts are provided on the bq219XL for direct connection to the battery stack (B^+, B^-) and the two-wire interface (C, D). For further details on the operation of the gas gauge and communication interface refer to the bq2092 or bq2040 data sheets. A tutorial on the bq2040 for battery system designers is provided in Benchmarq (1998a).

For mobile communication products, the full-function SMB architecture is too costly and hence the SBS implementer's forum is attempting to develop a specification that meets the needs of the market by eliminating many advanced features (Heacock, 1998).

A complete description of SBS standards and implementation is beyond the scope of this introductory chapter; for further details, see Benchmarq et al. (1995a, 1995b, 1996a, 1996b), Dunstan (1995), and Heacock (1998).

9.12 Semiconductor Components for Battery Management

In practical systems, rechargeable batteries could be charged using a variety of components, from simple voltage regulator ICs to microprocessor-based systems (Kerridge, 1993). Specially designed battery-management ICs provide fine control and a useful indication of a battery's charge condition both on and off charge.

Battery voltage, charge/discharge current, and cell case temperature supply the clues to information about battery-management ICs. However, each clue is the result of a chain of events far removed from the chemical reaction necessary to monitor and control. Further complexities arise because the three properties interact. Further, instantaneous measurements of any of the three attributes are virtually useless; all battery-management ICs include some sort of time to add meaning to the data. Table 9-7 shows some representative battery-management ICs from Benchmarq, Unitrode, and Dallas Semiconductor.

There are many means of providing a complete universal battery management system. Years ago, most companies involved in the design of chargers for

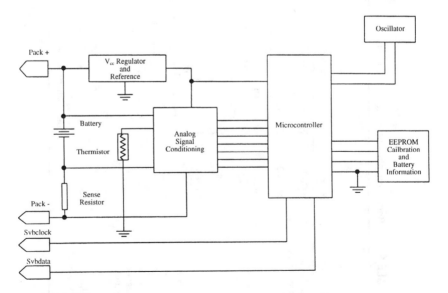

FIGURE 9-22 A microcontroller-based universal battery monitor

TABLE 9-7 Battery-management integrated circuits

Manufacturer	Part Number	Description	Cell Type					Availability of SMB
			NiCd	NiMH	Li-ion	Lead-Acid	Rechargeable Alkaline	
Benchmarq Microelectronics	bq2004	Fast charge IC	Yes	Yes	—	—	—	—
	bq2000	Programmable multichemistry fast charge management IC	Yes	Yes	Yes	—	—	—
	bq2010	Gas gauge IC	Yes	Yes	—	—	—	—
	bq2018	Gas gauge IC with single wire serial interface (Power Minder IC)	Yes	Yes	Yes	—	—	—
	bq2040	Gas gauge IC	Yes	Yes	Yes	—	—	Yes
	bq2031	Fast charge IC	—	—	—	Yes	—	—
	bq2058	Li-ion pack supervisor (3- and 4-cell packs)	—	—	Yes	—	—	—
	bq2060	Gas gauge IC	Yes	Yes	Yes	—	—	Yes
	bq2902	Charge/discharge controller	—	—	—	—	Yes	—
	UC3905	Charge controller	Yes	Yes	—	—	—	—
Unitrode Integrated Circuits	UC3906	Charger	—	—	—	Yes	—	—
	UC3909	Switchmode charger	—	—	—	Yes	—	—
Dallas Semiconductor	DS 1633	High-speed charger	Yes	Yes	Yes	Yes	Yes	—
	DS 1837	Quick battery recharge	Yes	Yes	Yes	Yes	Yes	—

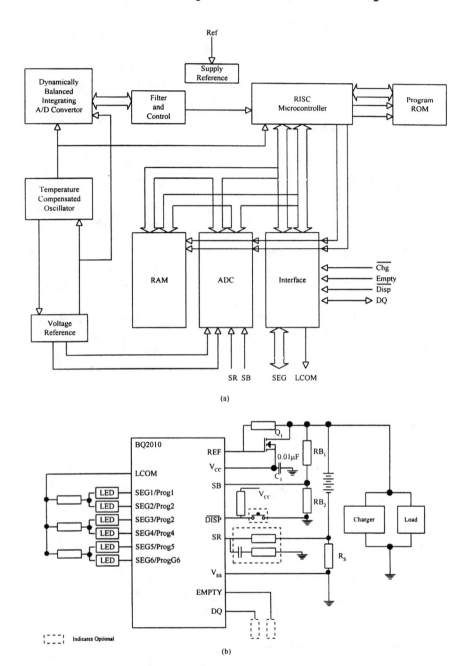

FIGURE 9-23 The bq2010: (block diagram, (b) typical implementation. (Reproduced by permission of Benchmarq Microelectronics Inc.)

battery banks had design resources dedicated internally to provide such solutions. Today, with the rapid changes in rechargeable battery chemistry and the associated charge/discharge management requirements, many companies prefer to use standardized semiconductor components dedicated to this function. Two possible approaches involve using a customized microcontroller or an ASIC dedicated to battery management. Figure 9-22 is the block diagram of a microcontroller-based universal battery monitor.

Using ASIC solutions, such as bq2010 from Benchmarq Microelectronics, the standby current necessary for battery monitoring circuits can be minimized while providing many gas gauge functions. A block diagram of the bq2010 forms Figure 9-23(a).

Specialized component manufacturers recently introduced many gas gauge ICs for multichemistry (NiCd/NiMH/Li-ion) environments (Freeman, 1995; Benchmarq, 1997).

VLSI components such as the bq2040 with an SMB interface are intended for battery-pack or in-system installation to maintain an accurate record of available battery charge. In addition to supporting the SMB protocol, it supports smart battery data specification, rev. 1.0 (Dunstan, 1995; see Figure 9-24). Another example is Linear Technology Corporations' LTC® 1325 battery management IC, which provides a complete system that can accommodate all plausible types of electrochemistry and their concomitant charging needs and charge termination algorithms, with few or no hardware changes for different battery types.

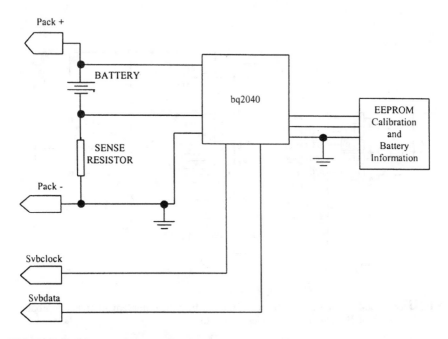

FIGURE 9-24 The implementation of a universal battery monitor using the bq2040. (Reproduced by permission of Benchmarq Microelectronics, Inc.)

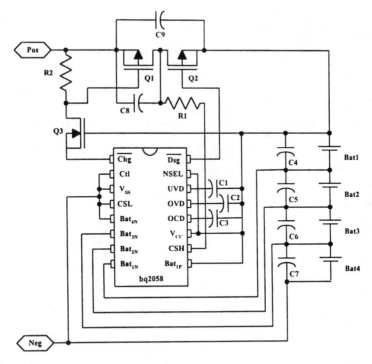

FIGURE 9-25 Li-ion pack safety. (Reproduced by permission of Benchmarq Micro-electronics, Inc.)

The two critical aspects of battery management for Li-ion cells are charge control and pack safety protection. The fundamental purpose of the safety circuitry is to protect the Li-ion cells from abusive conditions. Figure 9-25 illustrates the basic concept of a protection circuit using the bq2058. The control IC monitors each series potential in the pack for overvoltage (charge), undervoltage (discharge), and overcurrent conditions. The three thresholds are set to match the requirements of the manufacturer and Li-ion battery type. For further details on this circuit, the bq2058 data sheet is suggested. In designing Li-ion battery management systems, temperature effects must be carefully considered (Fundaro, 1998).

References

Alberkrack, J. "A Programmable In-Pack Rechargeable Lithium Cell Protection Circuit." HFPC Conference Proceedings, Las Vegas, Nevada, USA September 1996, pp. 230–237.

Benchmarq Microelectronics Inc. "Using the bq2040 Smart Battery System Gas Gauge IC," application note. Benchmarq Microelectronics Inc., Dallas, TX, USA January 1998.

Benchmarq Microelectronics Inc. *1997 Data Book*. Benchmarq Microelectronics Inc., 1998b.

Benchmarq Microelectronics Inc., Duracell Inc., Energizer Power Systems, Intel Corporation, et al. "Smart Battery System Specifications — System Management Bus Specifications," rev. 1.0. Benchmarq Microelectronics, Dallas, TX, USA February 15, 1995a.

Benchmarq Microelectronics Inc., Duracell Inc., Energizer Power Systems, Intel Corporation, et al. "Smart Battery System Specifications — Smart Battery Data Specifications," rev. 1.0. Benchmarq Microelectronics, Dallas, TX, USA February 15, 1995b.

Benchmarq Microelectronics Inc., Duracell Inc., Energizer Power Systems, Intel Corporation, et al. "Smart Battery System Specifications — Smart Battery Charger Specifications," rev. 1.0. Benchmarq Microelectronics, Dallas, TX, USA June 27, 1996a.

Benchmarq Microelectronics Inc., Duracell Inc., Energizer Power Systems, Intel Corporation, et al. "Smart Battery System Specifications — Smart Battery Selector Specifications," rev. 1.0. Benchmarq Microelectronics, Dallas, TX, USA September 5, 1996b.

Bennett, Phillip D., and George W. Brown. "Introduction to Applying Li-Ion Batteries." Proceedings of Portable by Design Conference, Santa Clara, CA, USA 1997, pp. 125–134.

Bowen, N. L. "System Considerations for Lithium-Ion Batteries." Portable by Design Conference Proceedings, Santa Clara, CA, USA 1996, pp. 179–191.

Briggs, A. "NiMH Technology Overview." Portable by Design Conference Proceedings, Santa Clara, CA, USA 1994, pp. BT-42–BT-45.

Cummings, G., D. Brotto, and J. Goodhart. "Charge Batteries Safely in 15 Minutes by Detecting Voltage Inflection Points." *EDN* (September 1, 1994), pp. 89–94.

Cutler Tim. "Rechargeable Zinc-Air Design Options for Portable Devices." Proceedings of Portable by Design Conference, Santa Clara, CA, USA 1997, pp. 112–118.

Duley, R., and H. David. "Battery Management Techniques for Portable Systems." Proceedings of Portable by Design Conference, Santa Clara, CA, USA 1994, pp. BT-1–BT-11.

Dunstan, R. A. "Standardized Battery Intelligence." Proceedings of Portable by Design Conference, Santa Clara, CA, USA 1995, pp. 239–244.

Eager, J. S. "The Nickel-Metal Hydride Battery." Proceedings of Power Conversion International Proceedings, September 1991, pp. 398–404.

Falcon, C. B. "Fast Charge Termination Methods for NiCd and NiMH Batteries." *PCIM* (March 1994), pp. 10–18.

Freeman, D. "Aspects of Universal Battery Management in Advanced Portable Systems." Proceedings of Portable by Design Conference, Santa Clara, CA, USA 1995, pp. 515–524.

Freeman, D., and D. Heacock. "Lithium-Ion Battery Capacity Monitoring Within Portable Systems." HFPC Conference Proceedings, San Jose, CA, USA 1995, pp. 1–8.

Fundaro, P. "Thermal Effects on Li-Ion Battery Management." Proceedings of Portable Technology Asia Seminar, Taiwan, May 1998.

Furukawa, N. "Developers Spur Efforts to Improve NiCd, NiMH Batteries." *JEE* (October 1993), pp. 46–50.

Gates Energy Products Inc. *Rechargeable Batteries Applications Handbook.* Boston: Butterworth–Heinemann, 1992.

Harold, P. "Batteries Function in High Temperature Environments." *EDN*, (July 9, 1987), p. 232.

Heacock, D. "Enabling Smart Batteries for Portable Devices." Proceedings of Global Forum on Mobile Handsets, London, June 1998.

Heacock, D., and D. Freeman. "Single Chip IC Gauges NiMH or NiCd Battery Charge." *PCIM* (March 1994), pp. 19–24.

Hirai, T. "Sealed Lead-Acid Batteries Find Electronic Applications." *PCIM* (January 1990), pp. 47–51.

Ivad, Josef Daniel, and Karl Kordesch. "In-application Use of Rechargeable Alkaline Manganese Dioxide/Zinc (RAM^{TM}) Batteries." Proceedings of Portable by Design Conference, Santa Clara, CA, USA 1997, pp. 119–124.

Juzkow, M. W., and C. St. Louis. "Designing Lithium-Ion Batteries into Today's Portable Products." Proceedings of Portable by Design Conference, Santa Clara, CA, USA 1996, pp. 13–22.

Kerridge, B. "Battery Management ICs." *EDN Asia* (August 1993), pp. 38–50.

Kovacevic, B. "Microprocessor-Based Nickel Metal Hydride Battery Charger Prevents Overcharging." *PCIM* (March 1993), pp. 17–21.

Kuribayashi, I. "Needs for Small Batteries Spur Progress in Lithium-Ion Models." *JEE* (October 1993), pp. 51–54.

Levy, S. C. "Recent Advances in Lithium Ion Technology." Portable by Design Conference Proceedings, Santa Clara, CA, USA 1995, pp. 316–323.

McClure, M. "Energy Gauges Add Intelligence to Rechargeable Batteries." *EDN* (May 26, 1994), p. 125.

Moneypenny, G. A., and F. Wehmeyer. "Thinline Battery Technology for Portable Electronics." HFPC Conference Proceedings, San Jose, CA, USA April 1994, pp. 263–269.

Moore, M. R. "Valve Regulated Lead Acid vs. Flooded Cell." Power Quality Proceedings, Irvine, CA, USA October 1993, pp. 825–827.

Nelson, Bob. "Pulse Discharge and Ultrafast Recharge Capabilities of Thin-Metal Film Technology." Proceedings of Portable by Design Conference, Santa Clara, CA, USA 1997, pp. 13–18.

Nossaman, P., and J. Parvereshi. "In Systems Charging of Reusable Alkaline Batteries." Proceedings of HFPC Conference, San Jose, CA, 1995.

Paret, D., and Fenger, C. "The I^2C Bus — From Theory to Practice." John Wiley & Sons, England, 1997.

Quinnell, R. A. "The Business of Finding the Best Battery." *EDN* (December 5, 1991), pp. 162–166.

Sacarisen, P. S., and J. Parvereshi. "Lead Acid Fast Charge Controller with Improved Battery Management Techniques." SouthCon/95 Conference, Ft. Lauderdale, FL, USA March 1995.

Schwartz, P. "Battery Management." Proceedings of Portable by Design Conference, Santa Clara, CA, USA 1995, pp. 525–547.

Sengupta, U. "Reusable Alkaline™ Battery Technology: Applications and System Design Issues for Portable Electronic Equipment." Proceedings of Portable by Design Conference, Santa Clara, CA, USA 1995, pp. 562–570.

Small, C. H. "Nickel-Hydride Cells Avert Environmental Headaches." *EDN* (December 10, 1992), pp. 156–161.

Swager, A. W. "Fast Charge Batteries." *EDN* (December 7, 1989), pp. 180–188.

Swager, A. W. "Smart Battery Technology — Power Management's Missing Link." *EDN* (March 2, 1995), pp. 47–64.

Unitrode Inc. "Improved Charging Methods for Lead-Acid Batteries Using the UC 3906." Application note U-104. Applications Handbook, Unitrode Inc., 1997, pp. 3-78–3-88.

CHAPTER **10**

Programmable Logic Devices

Coauthors: M. Kalyanapala and Nirupa Rubasingam

10.1 Introduction

The ever-rising demand for huge amounts of data to move faster and faster through processors and networks has resulted in an endless sea of logic devices. To accomplish these tasks, semiconductor technology is moving toward speedier, more powerful processors and denser, more complex "glue chips" (devices that surround processors performing control logic functions) to manipulate information. The breadth of product offerings from the ASIC and programmable logic device vendors allows designers to select the device that is most cost effective for a target application. However, confusing terminology used by manufacturers (Maxfield, 1996) and various other reasons make choosing among these devices a trying task. Understanding the basic architectures around which these glue chips evolved, their progress, and their design methodologies will ease the task of selecting the right device for an application. Until 1970, most designers were using standard digital components in designing electronic products and systems. However, these fixed-function devices posed certain limits in achieving lower printed circuit board (PCB) real estate as well as product reliability due to the use of many interconnections. The need to design more compact digital circuits than is possible with standard digital components, without increasing design time and cost hampered design flexibility and led to the development of programmable logic. Figure 10-1 shows logic devices classified into five general types.

The programmable logic device (PLD) market has expanded in so many directions that the term has become a generic term for several subclasses of products, including programmable logic arrays, programmable array logic, programmable multilevel devices, and field programmable gate arrays. Programmable memories such as PROM, EPROM, and EEPROM are the earliest and simplest forms of PLD.

413

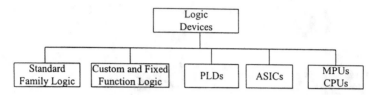

FIGURE 10-1 Logic device types

This chapter is an introduction to the designer who needs to use simple PLD applications and to identify the possibilities for using more complex forms of programmable devices such as field programmable gate arrays.

10.2 Basic Concepts

A PLD is an integrated circuit that is user configurable and capable of implementing digital logic functions. A typical PLD is composed of a programmable array of logic gates and interconnections with array inputs and outputs connected to the device pins through fixed logic elements such as inverters and flip-flops, based on the fact that all digital circuitry can be built using the basic elements such as NAND and NOR gates.

Any combinatorial logic function can be expressed as a sum of products: The variables are ANDed together to form products and the products are then ORed to form a sum. All PLDs are made up of combinations of AND gates, OR gates, inverters, and flip-flops. Programmable elements usually fuse connect the gates and flip-flops implement the desired logic functions. Each programmable logic device family requires a unique algorithm for fuse programming and verification on commercial programming equipment. The algorithm is a combination of voltage and timing required for addressing and programming fuses in the user array. The PLD's programming circuitry is enabled by pulsing one or more pins to a specific voltage level, generally different from normal logic levels. Once the programming circuitry is enabled, inputs become addressing nodes for the input and product lines within the PLD. The actual fuse link is at the intersection of the input product lines. Once addressed, the fuse can be programmed by pulsing the output associated with the location of the fuse link. A fuse can be verified to be programmed by enabling the programming circuitry, supplying the fuse address to the device's inputs, and reading the level of the device's output. In erasable PLDs, the fuses can be reset, effectively erasing the old design, so the PLD can be reprogrammed with a new design. For an average PLD user, the actual method by which the PLD "stores" data is not all that important. Most of the good PLD development tools available today handle code implementation automatically, leaving the developer free to concentrate on code development.

To keep the PLD easy to understand and use, a special convention has been adopted. Figure 10-2 is the representation of a three-input AND gate. Note that

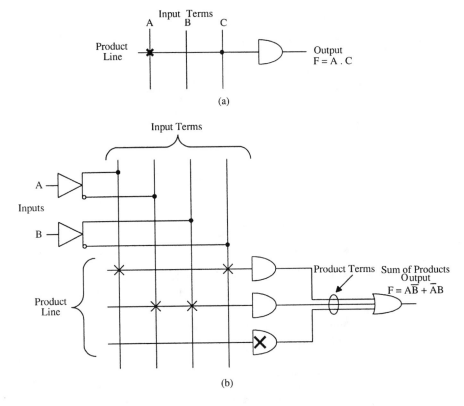

FIGURE 10-2 (a) basic symbols and (b) example

only one line is shown as the input to the AND gate. This line commonly is called the *product line*. The inputs are shown as vertical lines, and at the intersection of these lines are the programmable fuses. An x represents an intact fuse, which makes that input part of the product term. No x represents a blown fuse, which means that input will not be part of the product term. A dot at the intersection of any line represents a hardwired connection.

In Figure 10-2(b), the symbols are extended to develop a simple two-input programmable AND array feeding an OR gate. Note that buffers have been added to the inputs, which provide both true and complement outputs to the product lines. The intersection of the input terms forms a 4 × 3 programmable AND array. From these symbols, we can see that the output of the OR gate is programmed to the following equation, $F = A\overline{B} + \overline{A}B$. Note that the bottom AND gate has an x marked inside the gate symbol. This means that all fuses are left intact, which results in that product line having no effect on the sum term. In other words, the output of the AND gate will be a logic 0. When all the fuses are blown on a product line, the output of the AND gate always will be a logic 1. This has the effect of locking up the output of the OR gate to a logic level 1.

10.2.1 PROM as a Programmable Device

The PROM was the first widely used programmable logic family. Its basic architecture is an input decoder configured from AND gates, combined with a programmable OR matrix on the outputs. As shown in Figure 10-3, this allows every output to be programmed individually from every possible

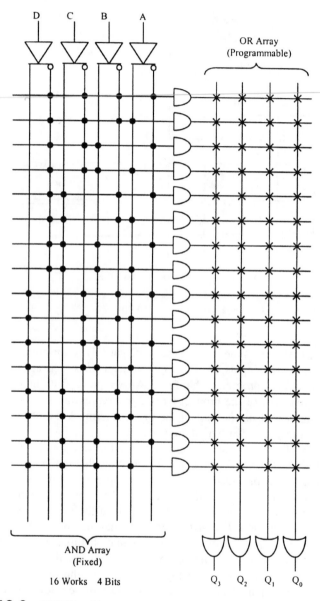

FIGURE 10-3 PROM architecture

input combination. In this example, a PROM with four inputs has 2^4, or 16, possible input combinations. With the output word width being 4 bits, each of the 16×4 bit words can be programmed individually. Applications such as data storage tables, character generators, and code converters are just a few design examples ideally suited for the PROM. In general, any application that requires every input combination to be programmable is a good candidate for a PROM. However, PROMs have difficulty accommodating large numbers of input variables. Eventually, the size of the fuse matrix will become prohibitive, because for each input variable added, the size of the fuse matrix doubles.

10.2.2 PLD Architecture

To overcome the limitation of a restricted number of inputs, the PLD utilizes a slightly different architecture, as shown in Figure 10-4. The same AND-OR implementation is used as with PROMs, but now the input AND array is programmable instead of the output OR array. This has the effect of restricting the output OR array to a fixed number of input AND terms. The trade-off is that, now, every output is not programmable from every input combination, but more inputs can be added without doubling the size of the fuse matrix. For example, if we were to expand the inputs on the PLD shown in Figure 10-4 and on the PROM in Figure 10-3 to ten, we would see that the fuse matrix required for the PLD would be 20×16 (320 fuses) vs. 4×1024 (4096 fuses for the PROM). It is important to realize that not every application requires every output to be programmable from every input combination. This makes the PLD a viable product family.

Simple PLDs such as programmable array logic (PAL^{TM} of Advanced Micro Devices) and generic array logic (GAL^{TM} of Lattice Semiconductor) derive their usefulness from an internal structure designed to allow the programming of either arbitrary products or sums or both. For example, PAL devices have a programmable AND-array and a fixed OR-array (Figure 10-4). PAL devices have two basic versions, with and without memory elements (flip-flops). Combinational PALs are devices that are based on a PAL structure but contain no memory elements. Combinational PALs are useful for a wide variety of random logic functions, including decoders, interface logic, and other applications that require a simple decoding of device inputs. The 16L8, a typical combinational PAL, is diagrammed in Figure 10-5. The diagram for the 16L8 shows that the device has ten dedicated inputs (pins 1 through 9 and 11) to the programmable AND array. Each input to the array is available in its true or a complemented form, allowing any combination of the inputs to be expressed on any row of the array. Each row of the 16L8 array corresponds to one product term of the device.

The 16L8 has eight output gates, each of which is fed by a seven-input OR gate. Each output of this device is capable of implementing a logic function composed of seven or fewer product terms. The eighth product term is used to control the three-state output buffer.

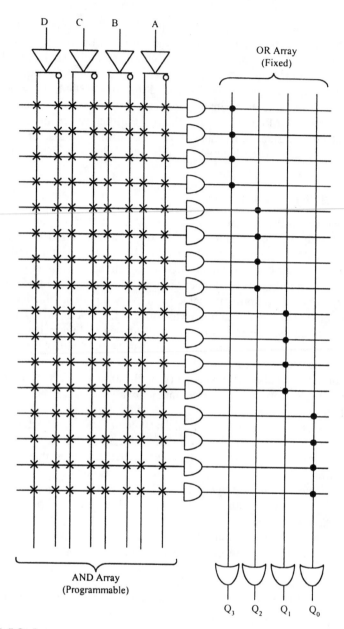

FIGURE 10-4 PLD architecture

Some devices, such as GAL family from Lattice Semiconductor, integrate an output logic macro cell on some or all of their output pins (Figure 10-6). The macro cell is configured by the designer on a pin-by-pin basis to implement the desired functions. This allows design of synchronous logic blocks.

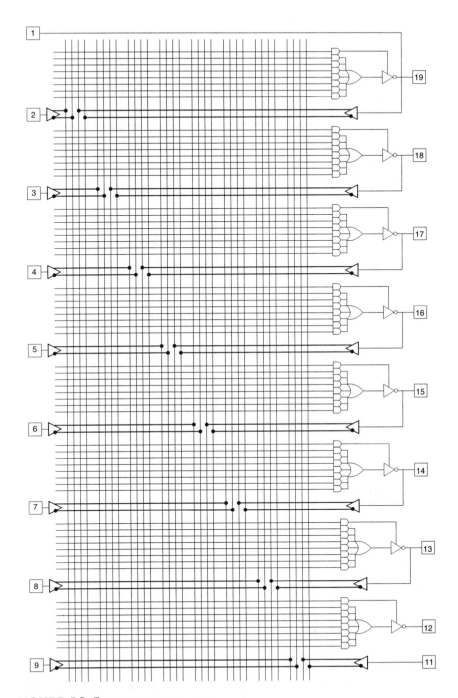

FIGURE 10-5 The 16L8 PAL device

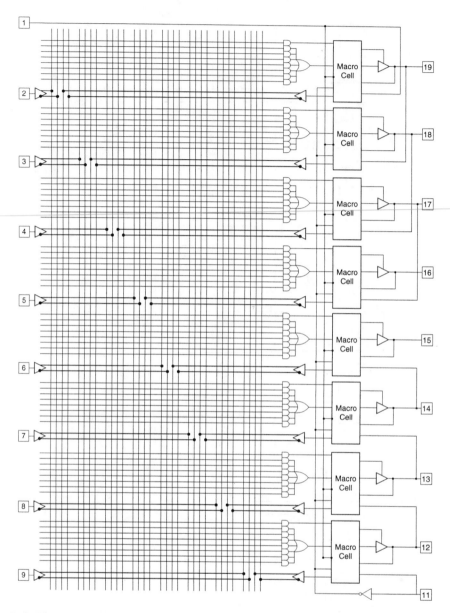

FIGURE 10-6 The 16V8- (GAL-)type PLD with output macro cells

PLDs equipped with output macro cells offer total output flexibility. Figure 10-7 shows examples of these features as implemented in the TIBPAL22V10 device. Fuses S0 and S1 allow selection between registered or combinational outputs as well as output polarity. Figure 10-8 illustrates the user options.

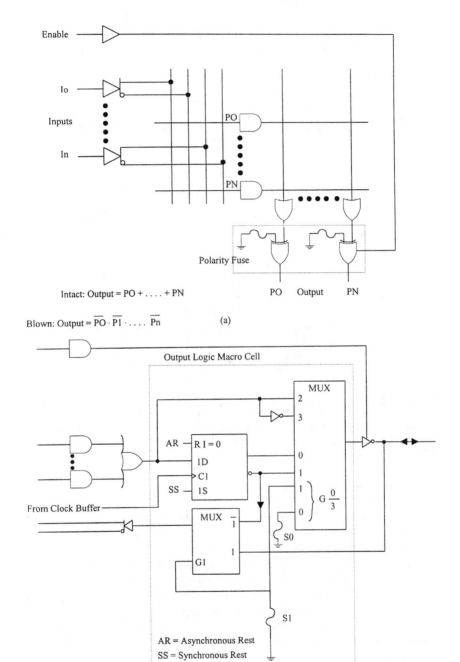

FIGURE 10-7 An output macro cell: (a) polarity selection, (b) output logic macro cell. (Source: Programmable Logic Databook, 1993, Texas Instruments Inc.)

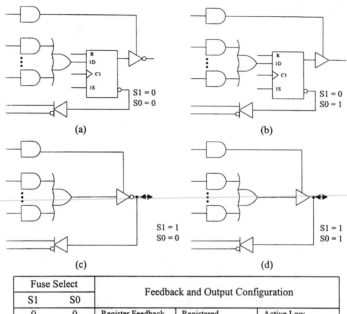

Fuse Select		Feedback and Output Configuration		
S1	S0			
0	0	Register Feedback	Registered	Active Low
0	1	Register Feedback	Registered	Active High
1	0	I/O Feedback	Combinational	Active Low
1	1	I/O Feedback	Combinational	Active High

FIGURE 10-8 User options, resulting macro cell feedback and output logic after programming: (a) register feedback, registered *active low* output; (b) register feedback, registered *active high* output; (c) I/O feedback, combinational *active low* output; (d) I/O feedback, combinational *active high* output. (Source: Programmable Logic Databook, 1993, Texas Instruments Inc.)

The user options are as follows:

1. *Clock polarity select.* The clock signal can be inverted via a clock polarity select fuse. This allows the transition of the register output to be on either the positive or negative edge of the clock pulse.
2. *Internal-state registers.* Several devices offer internal-state registers, often called *buried registers.* With the internal-state register, the output of the register is fed back into the AND array rather than to an output pin. This feature can be used for timing control sequences.
3. Variable product terms. Some PLD device architectures vary the number of product terms associated with each output pin to allow better utilization of the programmable array.

There may be variations of these in practical devices from different manufacturers (Dettmer, 1990; Hannington, 1990; Small, 1987, 1988).

10.2.3 FPLA Architecture

The field programmable logic array (FPLA) goes one step further in offering both a programmable AND array and a programmable OR array (Figure 10-9). This feature makes the FPLA the most versatile device of the three but impractical in most low-complexity applications. For applications in which complex timing

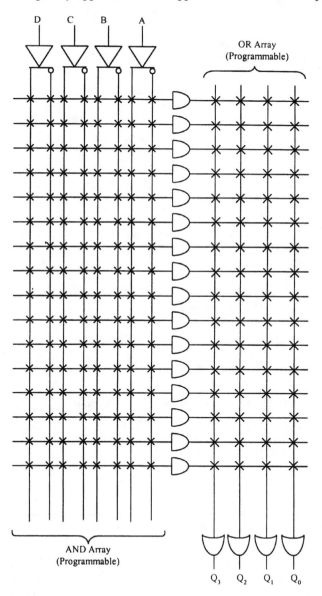

FIGURE 10-9 FPLA architecture

control is required, Texas Instruments (TI) offers several programmable state machines based on the FPLA architecture. Some of these devices incorporate internal state registers or on-chip binary counters to aid in generating complex timing sequences.

10.2.4 Erasable PLDs

Another type of programmable logic device is the erasable programmable logic device (EPLD). Based on traditional PLD architecture, these devices typically offer a higher level of flexibility in the input and output configuration, register selection, and clocking options. CMOS EPLDs provide a higher level of density over standard PLDs and have lower power dissipation than bipolar PLDs. All programmable logic approaches discussed have their own unique advantages and limitations. The best choice depends on the complexity of the function being implemented and the current cost of the devices themselves. It is important to realize that a circuit solution may exist for more than one of these logic families.

10.3 Advantages of Programmable Logic

Low-density PLDs offer many advantages to the system designer presently using several standard catalog small scale integrated circuit (SSI) and medium scale integrated circuit (MSI) functions. The following list mentions just few of the benefits achievable when using programmable logic:

- *Package count reduction.* Several MSI/SSI functions can be replaced with one PLD to reduce system power requirements.
- *PCB area reduction.* Fewer devices consume less PCB space.
- *Circuit flexibility.* Programmability allows for minor circuit changes without changing the PC board.
- *Improved reliability.* With fewer interconnections, overall system reliability increases.
- *Shorter design cycle.* When compared with standard-cell or gate-array approaches, custom functions can be implemented much more quickly.
- *Proprietary design protection* (fuse protection). Circuit can be protected by blowing the security fuse.
- *Minimization of design labor.* Advantages of PLD computer-aided design tools could help save time in logic implementation.

10.4 Designing with PLDs

Figure 10-10 shows a basic PLD design flowchart. The best way to demonstrate the unique capabilities of the PLD is through a design example. With the design example given later, the user will gain the basic understanding on how to design a PLD application using design tools.

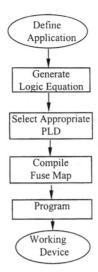

FIGURE 10-10 PLD process flowchart

10.5 Design Tools for PLDs

An important secondary issue to choosing and using PLDs is selecting the right design tools (Liebson, 1990; Conner, 1990; Schulze, 1991). Design tools for PLDs comprise compilers, simulation products, and test software. Compilers convert the design into a fuse map; simulation products allow testing the design before programming a PLD to make sure that the design is right. Test software, which includes fault graders and automatic test pattern generators, helps keep things right during production.

A number of software products are available to make logic designs easier and less cumbersome. When choosing a design tool, understanding the design process involving PLD designs and the key differences in design tools is necessary to help narrow down the choices. Figure 10-11 shows the typical design steps involved in designing with PLDs. The first step is to enter the design, and the common design entry methods are shown in Figure 10-11. Not all compilers support every design entry method. Schematic and Boolean entry methods remain popular among PLD designers, as they work well with 7400 TTL-style designing.

Many PLD compilers allow creating a design without regard to a particular device architecture. These compilers accept device specifications and allow postponing device selection until much later. Device specificity introduces one of the first criteria you must consider when selecting a PLD compiler. Another key factor involved in selecting a compiler is good optimization performance. Good optimization centers around the compiler's ability to generate device solutions that require minimal resources for a given logic function. Logic reduction algorithms eliminate redundant elements while preserving the required functionality.

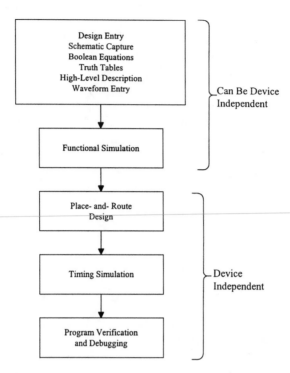

FIGURE 10-11 Steps involved in designing with PLD compilers

Different types of architecture often require different optimization techniques to maximize their use.

After optimization, a device must be selected with the physical resources necessary to implement the reduced equations. Most compilers can partition a design by trying single and multiple PLD solutions to create the "best fit." Some engineers appreciate a compiler's ability to partition. However, others know what PLDs they want to use and are not interested in the compiler's ability to partition. Even if familiar with all the devices a designer may wish to use, partitioning software can explore many more possibilities in far less time than by hand. Furthermore, some compilers will allow creating subset libraries that contain just the PLDs in the designer's lab stock or on a company's approved list, eliminating obviously useless choices when generating alternative solutions.

Most PLD compilers include some sort of simulation capability. However, sometimes simulators may have to be purchased separately to fulfill particular requirements. The simulators with the least capabilities accept input and output vectors created by the designer and simulate the designer's logic equations to ensure that the equations produce the results expected. More complex simulators use a model of a target PLD to create a simulation that reproduces the behavior specified. For this type of simulation, the designer specifies the input vectors, and the simulator generates output vectors or waveforms. The results can be checked to see if a design works as expected. Thus, the designer finds out how the circuit will behave instead of guessing and verifying that guess.

After obtaining satisfactory results from the functional simulation, the next step is to run the place and route software. This is a critical step to get accurate timing simulation. There will be numerous possible routings, and it is difficult to comment on the ability of the routers to successfully route designs that meet tough timing requirements. Although users prefer placing and routing to be completely automatic, most do not want to run into a brick wall if the automatic software cannot handle the job. Some vendors offer an interactive design editor which allows manually tidying up things. Once the routing is complete, the next step will be to verify the timing, by loading the timing information generated by the place and route software into a simulator and running the timing simulation. The final step is to verify that the program has been correctly loaded into the device and debug. Most PLD problems can be found and corrected with no significant cost or schedule impact.

To achieve satisfactory programming yields for PLDs, it is critical that the device programmers adhere to the programming algorithm specifications as defined by the device manufacturer. Some helpful hints for good programmability follow:

1. Follow accepted standards for electrostatic discharge (ESD) protection. Equipment, personnel and work surfaces should be grounded. Air ionization is recommended when handling static sensitive devices outside protective containers.
2. Misaligned connectors and worn sockets can contribute to poor programming yield. Be aware of the manufacturer's specification for number of insertions and be sure sockets are replaced frequently to ensure proper contact.
3. Be sure to use the latest update. Most program manufacturers offer updates and repair services to their users. The cost of service typically is not much more than the cost of a single update and the manufacturer may update four or more times per year.
4. Programming equipment should be calibrated. It is recommended to have no less than two calibrations per year.
5. Verify that the correct family pin-out codes or device entry codes are being used. It is important to understand that different algorithms may be needed for different speed versions of the same function.
6. Ensure that the programming equipment is evaluated and approved by the device manufacturer for highest possible programming yields and quality level.

In addition to the functional testing described earlier, static parametric testing such as input and I/O leakage currents, output high and low voltages under static loading, and power supply current will only improve the quality of the PLD going into the application. This will ensure that devices are functional and were not damaged due to ESD or electrical overstress (EOS) during the customization process.

The CUPL (Logical Devices Inc.), PL Designer (MINC Inc.), and ABEL (Data I/O) are universal computer-aided design (CAD) tools that support PLDs. With these software products, complex designs can be described using Boolean equations, truth tables, state machine diagrams, and schematic capture methods available on most CAD systems.

ABEL consists of a special purpose, high-level language used to describe logic designs and a language processor that converts logic descriptions to programmer load files or JEDEC files. These files contain the information necessary to program and test PLDs.

The ultimate function of these software products is to generate a JEDEC file of the original design and program the PLD. However, most software vendors provide more than a JEDEC file as output from the software. Here, we describe the attributes of a few of the popular logic design products. We recommend contacting specific manufacturers to obtain the latest and most comprehensive information available.

The CUPL (universal compiler for programmable logic), like ABEL, is a universal CAD tool that supports PLDs. It has utility files that facilitate conversion of designs done in other design software environments to the CUPL design environment. CUPL also produces a standard programmer load file in JEDEC format, making it compatible with logic programmers that accept JEDEC files. Some features of CUPL design language follow:

- *Flexible forms for design description.* The forms include Boolean equations, truth tables, and state diagrams.
- *Expression subsitutions and time-saving macros.* This involves the assignment of names to equations and having the software do the substitution any time the assigned name is encountered during the compiling process.
- *Shorthand features.* These features include list notation, bit fields, distributive properties, and De Morgan's theorem. List notation means the nested directive

```
[A4, A3, A2, A1, A0]
```

can be represented as

```
[A4 ..0]
```

Bit fields means that a group of bits may be assigned to a name as in

```
FIELD ADDR = [A4...0]
```

In the use of distributive property,

```
A & (B # C)
```

is replaced with

```
A & B # A & C
```

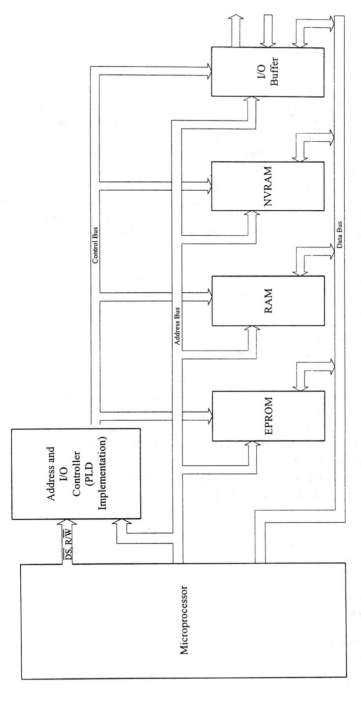

FIGURE 10-12 Address decoder for an Z8800-based embedded controller using a PLD/system block diagram

In De Morgan's theorem,

```
(A & B)
```

is replaced with

```
!A # !B
```

- *The CUPL language processor.* CUPL provides design templates that allow designers to just fill in the blanks when originating a design. Free form comments can be used throughout the design.
- *Error checking.* Error checking with detailed error messages directs designers to the source of problems during debugging.
- *Logic equation reduction.* The logic equation reduction capabilities on CUPL offers a choice of several minimization levels for a number of product terms. This can be from just fitting a design without minimization or reducing the product terms to an optimum, before finalizing the fuse map of the PLD.
- *Design simulation.* This is accomplished using the CSIM feature, which allows designers to check the workability of the design before a part is programmed. Functional simulation can be done by the programmer when test vectors are provided.

10.6 A Design Example

We take a simple example of an address decoder, as per Figure 10-12. This block illustrates the address decoding and generating the control signals in a memory-mapped Z8800-embedded application.

The PLD is expected to generate the chip select and control signals for the peripherals, as listed under output in the listing as per Listing 10.1. This is based on the package CUPL. Listing 10.2 is a comprehensive description of the logical expressions. CUPL creates expanded product terms, a variables list, a fuse plot, and the chip diagram (Figure 10-13) as per Listing 10.3.

Listing 10.1 PLD Design for an Address Decoder

```
Name        Address Decoder for Z8800 based Embedded
            Controller;
Partno      Uxx;
Date        16/06/98;
Revision    01;
Designer    Kaushalya;
Company     ACCMT;
Assembly    XXXXX;
Location    XXXXX;
```

```
/**********************************************/
/* This decodes the addresses for EPROM, RAM, NV RAM
(with RTC)                                    */
/* & an I/O space.                            */
/**********************************************/
/* Selected Target Device Types: GAL16V8/PALCE16V8*/
/**********************************************/

/**   Inputs   **/

Pin 2        =   !ds;
Pin 3        =   rw;
Pin [4..8]   =   [a15..a11];

/**   Outputs   **/

Pin 12       =   !mem_oe; /* memory output enable   */
Pin 13       =   !mem_we; /*memory write enable      */

Pin 14       =   !rom_cs; /* EPROM Chip select       */
Pin 15       =   !io_sel; /*Input-output select      */
Pin 16       =   !nvram_cs; /* Chip select of the
                                NVRAM with RTC    */
Pin 17       =   !ram_cs; /* Data Ram chip select    */

/** Declarations and Intermediate Variable
  Definitions **/

field    addr_map = [a15,a14,a13,a12,a11];

rom_addr    =    addr_map:[0000..7fff];
                      /* 32K PROGRAM MEMORY SPACE     */
io_addr     =    addr_map:[8000..83ff];
                      /*1K I/O SPACE                   */
nv_ram_addr =    addr_map:[c000..dfff];
                      /* 8K NV DATA MEMORY
                      SPACE & RTC                      */
ram_addr    =    addr_map:[e000..ffff];
                      /* 8K  DATA MEMORY SPACE         */

/**   Logic Equations   **/
mem_oe       =   ds & rw & (rom_addr # ram_addr
                      # nv_ram_addr);
mem_we       =   ds & !rw & (ram_addr # nv_ram_addr);

rom_cs       =   rom_addr;
io_sel       =   io_addr;
nvram_cs     =   nv_ram_addr;
ram_cs       =   ram_addr;
```

===
Chip Diagram
===

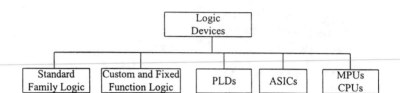

FIGURE 10-13 System-created pin configuration of the PLD

Listing 10.2 Listing for the Logic Description File

```
LISTING FOR LOGIC DESCRIPTION FILE:
pld_ex1.pld                        Page 1

CUPL: Universal Compiler for Programmable Logic
Version 4.2a Serial# MD-19111210
Copyright (c) 1983, 1991 Logical Devices, Inc.
Created Thu Oct 29 20:15:11 1998

   1:Name          Address Decoder for Z8800 based
                   Embedded Controller;
   2:Partno        Uxx;
   3:Date          16/06/98;
   4:Revision      01;
   5:Designer      Kaushalya;
   6:Company       ACCMT;
   7:Assembly      XXXXX;
   8:Location      XXXXX;
   9:
  10:/*******************************************************/
  11:/* This decodes the addresses for EPROM, RAM, NV
        RAM (with RTC)                                    */
  12:/* & an I/O space.                                   */
  13:/*******************************************************/
  14:/* Selected Target Device Types: GAL16V8 /
        PALCE16V8*/
  15:/*******************************************************/
```

```
16:
17:/**  Inputs  **/
18:
19:Pin 2        =  !ds;
20:Pin 3        =  rw;
21:Pin [4..8]   =  [a15..a11];
22:
23:/**  Outputs  **/
24:
25:Pin 12       =  !mem_oe;
26:Pin 13       =  !mem_we;
27:
28:Pin 14       =  !rom_cs;
29:Pin 15       =  !io_sel;
30:Pin 16       =  !nvram_cs;
31:Pin 17       =  !ram_cs;
32:
33:/** Declarations and Intermediate Variable
      Definitions **/
34:
35:field   addr_map = [a15,a14,a13,a12,a11];
36:
37:rom_addr     =  addr_map:[0000..7fff];
                     /* 32K PROGRAM MEMORY
                        SPACE                        */
38:io_addr      =  addr_map:[8000..83ff];
                     /* 1K I/O SPACE                 */
39:nv_ram_addr  =  addr_map:[c000..dfff];
                     /* 8K NV DATA MEMORY
                        SPACE & RTC                  */
40:ram_addr     =  addr_map:[e000..ffff];
                     /* 8K DATA MEMORY SPACE         */
41:
42:
43:/**  Logic Equations  **/
44:
45:mem_oe    =   ds & rw & (rom_addr # ram_addr
                       # nv_ram_addr);
46:mem_we    =   ds & !rw & (ram_addr # nv_ram_addr);
47:
48:rom_cs    =   rom_addr;
49:io_sel    =   io_addr;
50:nvram_cs  =   nv_ram_addr;
51:ram_cs    =   ram_addr;
52:
```

```
53:
Jedec Fuse Checksum        (23f1)
Jedec Transmit Checksum    (b4c4)

Listing 10.3 Documentation File

*********************************************************
                    Address
*********************************************************
CUPL           4.2a Serial# MD-19111210
Device         g16v8s  Library DLIB-h-28-9
Created        Thu Oct 29 20:15:16 1998
Name           Address Decoder for Z8800 based
               Embedded Controller
Partno         Uxx
Revision       01
Date           16/06/98
Designer       Kaushalya
Company        ACCMT
Assembly       XXXXX
Location       XXXXX

=====================================================
                Expanded Product Terms
=====================================================
addr_map =>
    a15,  a14,  a13,  a12,  a11

io_addr =>
    !a11 & !a12 & !a13 & !a14 & a15

io_sel =>
    !a11 & !a12 & !a13 & !a14 & a15

mem_oe =>
    a14 & ds & rw
  # !a15 & ds & rw

mem_we =>
    a14 & a15 & ds & !rw

nv_ram_addr =>
    !a13 & a14 & a15

nvram_cs =>
    !a13 & a14 & a15

ram_addr =>
    a13 & a14 & a15
```

```
ram_cs =>
    a13 & a14 & a15

rom_addr =>
    !a15

rom_cs =>
    !a15
```

```
=======================================================
                     Symbol Table
=======================================================
Pin  Variable                Pterms Max  Min
Pol    Name       Ext  Pin   Type  Used Pterms Level
---  --------     ---  ---   ----  ---- ------ -----
       a11         8    V     -     -     -
       a12         7    V     -     -     -
       a13         6    V     -     -     -
       a14         5    V     -     -     -
       a15         4    V     -     -     -
       addr_map    0    F     -     -     -
   !   ds          2    V     -     -     -
       io_addr     0    I     1     -     -
   !   io_sel      15   V     1     8     2
   !   mem_oe      12   V     2     8     2
   !   mem_we      13   V     1     8     2
       nv_ram_addr 0    I     1     -     -
   !   nvram_cs    16   V     1     8     2
       ram_addr    0    I     1     -     -
   !   ram_cs      17   V     1     8     2
       rom_addr    0    I     1     -     -
   !   rom_cs      14   V     1     8     2
       rw          3    V     -     -     -
LEGEND   D: default variable F: field G: group
         I: intermediate variable N: node M: extended
            node
         U: undefined V: variable X: extended
            variable
         T: function
=================================================
                   Fuse Plot
=================================================
Syn   02192 - Ac0   02193 x
Pin #19  02048  Pol x  02120  Ac1 -
00000 xxxxxxxxxxxxxxxxxxxxxxxxxxxxxxxxx
00032 xxxxxxxxxxxxxxxxxxxxxxxxxxxxxxxxx
```

```
00064  xxxxxxxxxxxxxxxxxxxxxxxxxxxxxxxxxx
00096  xxxxxxxxxxxxxxxxxxxxxxxxxxxxxxxxxx
00128  xxxxxxxxxxxxxxxxxxxxxxxxxxxxxxxxxx
00160  xxxxxxxxxxxxxxxxxxxxxxxxxxxxxxxxxx
00192  xxxxxxxxxxxxxxxxxxxxxxxxxxxxxxxxxx
00224  xxxxxxxxxxxxxxxxxxxxxxxxxxxxxxxxxx
Pin #18  02049  Pol x  02121  Ac1 -
00256  xxxxxxxxxxxxxxxxxxxxxxxxxxxxxxxxxx
00288  xxxxxxxxxxxxxxxxxxxxxxxxxxxxxxxxxx
00320  xxxxxxxxxxxxxxxxxxxxxxxxxxxxxxxxxx
00352  xxxxxxxxxxxxxxxxxxxxxxxxxxxxxxxxxx
00384  xxxxxxxxxxxxxxxxxxxxxxxxxxxxxxxxxx
00416  xxxxxxxxxxxxxxxxxxxxxxxxxxxxxxxxxx
00448  xxxxxxxxxxxxxxxxxxxxxxxxxxxxxxxxxx
00480  xxxxxxxxxxxxxxxxxxxxxxxxxxxxxxxxxx
Pin #17  02050  Pol x  02122  Ac1 x
00512  --------x---x---x--------------
00544  xxxxxxxxxxxxxxxxxxxxxxxxxxxxxxxxxx
00576  xxxxxxxxxxxxxxxxxxxxxxxxxxxxxxxxxx
00608  xxxxxxxxxxxxxxxxxxxxxxxxxxxxxxxxxx
00640  xxxxxxxxxxxxxxxxxxxxxxxxxxxxxxxxxx
00672  xxxxxxxxxxxxxxxxxxxxxxxxxxxxxxxxxx
00704  xxxxxxxxxxxxxxxxxxxxxxxxxxxxxxxxxx
00736  xxxxxxxxxxxxxxxxxxxxxxxxxxxxxxxxxx
Pin #16  02051  Pol x  02123  Ac1 x
00768  --------x---x----x-------------
00800  xxxxxxxxxxxxxxxxxxxxxxxxxxxxxxxxxx
00832  xxxxxxxxxxxxxxxxxxxxxxxxxxxxxxxxxx
00864  xxxxxxxxxxxxxxxxxxxxxxxxxxxxxxxxxx
00896  xxxxxxxxxxxxxxxxxxxxxxxxxxxxxxxxxx
00928  xxxxxxxxxxxxxxxxxxxxxxxxxxxxxxxxxx
00960  xxxxxxxxxxxxxxxxxxxxxxxxxxxxxxxxxx
00992  xxxxxxxxxxxxxxxxxxxxxxxxxxxxxxxxxx
Pin #15  02052  Pol x  02124  Ac1 x
01024  --------x----x---x---x---x------
01056  xxxxxxxxxxxxxxxxxxxxxxxxxxxxxxxxxx
01088  xxxxxxxxxxxxxxxxxxxxxxxxxxxxxxxxxx
01120  xxxxxxxxxxxxxxxxxxxxxxxxxxxxxxxxxx
01152  xxxxxxxxxxxxxxxxxxxxxxxxxxxxxxxxxx
01184  xxxxxxxxxxxxxxxxxxxxxxxxxxxxxxxxxx
01216  xxxxxxxxxxxxxxxxxxxxxxxxxxxxxxxxxx
01248  xxxxxxxxxxxxxxxxxxxxxxxxxxxxxxxxxx
Pin #14  02053  Pol x  02125  Ac1 x
01280  ---------x--------------------
01312  xxxxxxxxxxxxxxxxxxxxxxxxxxxxxxxxxx
```

```
01344 xxxxxxxxxxxxxxxxxxxxxxxxxxxxxxxx
01376 xxxxxxxxxxxxxxxxxxxxxxxxxxxxxxxx
01408 xxxxxxxxxxxxxxxxxxxxxxxxxxxxxxxx
01440 xxxxxxxxxxxxxxxxxxxxxxxxxxxxxxxx
01472 xxxxxxxxxxxxxxxxxxxxxxxxxxxxxxxx
01504 xxxxxxxxxxxxxxxxxxxxxxxxxxxxxxxx
Pin #13  02054  Pol x  02126  Ac1 x
01536 -x---x--x---x------------------
01568 xxxxxxxxxxxxxxxxxxxxxxxxxxxxxxxx
01600 xxxxxxxxxxxxxxxxxxxxxxxxxxxxxxxx
01632 xxxxxxxxxxxxxxxxxxxxxxxxxxxxxxxx
01664 xxxxxxxxxxxxxxxxxxxxxxxxxxxxxxxx
01696 xxxxxxxxxxxxxxxxxxxxxxxxxxxxxxxx
01728 xxxxxxxxxxxxxxxxxxxxxxxxxxxxxxxx
01760 xxxxxxxxxxxxxxxxxxxxxxxxxxxxxxxx
Pin #12  02055  Pol x  02127  Ac1 x
01792 -x--x-------x------------------
01824 -x--x----x--------------------
01856 xxxxxxxxxxxxxxxxxxxxxxxxxxxxxxxx
01888 xxxxxxxxxxxxxxxxxxxxxxxxxxxxxxxx
01920 xxxxxxxxxxxxxxxxxxxxxxxxxxxxxxxx
01952 xxxxxxxxxxxxxxxxxxxxxxxxxxxxxxxx
01984 xxxxxxxxxxxxxxxxxxxxxxxxxxxxxxxx
02016 xxxxxxxxxxxxxxxxxxxxxxxxxxxxxxxx
LEGEND    X: fuse not blown
          -: fuse blown
```

Listing 10.4 Simulation Input File for the Address
Decoder

```
Name        Address Decoder for Z8800 based Embedded
            Controller;
Partno      Uxx;
Date        16/06/98;
Revision    01;
Designer    Kaushalya;
Company     ACCMT;
Assembly    XXXXX;
Location    XXXXX;
/**************************************************/
/* This decodes the addresses for EPROM, RAM, NV RAM
   (with RTC)    */
/* & an I/O space.                              */
/**************************************************/
/* Selected Target Device Types: GAL16V8/PALCE16V8*/
/**************************************************/
```

```
order: !ds,rw,a15,a14,a13,a12,a11,%2,!mem_oe,
!mem_we, !rom_cs, !io_sel,
!nvram_cs, !ram_cs;

vectors:    1000000 HHLHHH  /* no !ds asserted,
                                  idle condition    */
            0100000 LHLHHH  /* ROM Read Cycle      */
            0110000 HHHLHH  /* I/O select Cycle    */
            0111000 LHHHLH  /* NV RAM Read Cycle   */
            0011000 HLHHLH  /* NV RAM Write Cycle  */
            0111100 LHHHHL  /* RAM Read Cycle      */
            0011100 HLHHHL  /* RAM Write Cycle     */
```

Listing 10.5 Simulation Result of the Address Decoder

```
CSIM: CUPL Simulation Program
Version 4.2a Serial# MD-19111210
Copyright (c) 1983, 1991 Logical Devices, Inc.
CREATED Fri Oct 30 18:52:24 1998

LISTING FOR SIMULATION FILE: C:\CUPL\WORK\PLD_EX1.si

  1: Name        Address Decoder for Z8800 based
                 Embedded Controller;
  2: Partno      Uxx;
  3: Date        16/06/98;
  4: Revision    01;
  5: Designer    Kaushalya;
  6: Company     ACCMT;
  7: Assembly    XXXXX;
  8: Location    XXXXX;
  9:
 10: /************************************************/
 11: /* This decodes the addresses for EPROM, RAM,
         NV RAM (with RTC)                          */
 12: /* & an I/O space.                             */
 13: /************************************************/
 14: /*  Selected Target Device Types: GAL16V8 /
         PALCE16V8                                  */
 15: /************************************************/
 16:
 17: order: !ds,rw,a15,a14,a13,a12,a11,%2,!mem_oe,
     !mem_we, !rom_cs,
 !io_sel, !nvram_cs, !ram_cs;
 18:
```

```
===================================================
                 Simulation Results

0001: 1000000   HHLHHH
0002: 0100000   LHLHHH
0003: 0110000   HHHLHH
0004: 0111000   LHHHLH
0005: 0011000   HLHHLH
0006: 0111100   LHHHHL
0007: 0011100   HLHHHL
```

10.7 High-Density PLDs

Even the largest PLAs and PALs can achieve integration levels equivalent to only a few hundred gates. The limitations on low-density PLDs are attributable to the rigidity of the AND-OR plane organization and its inability to perform any sequential or counter register functions at a rudimentary level.

As the progress of semiconductor technology made higher integration levels possible, high-density PLDs such as CPLDs and FPGAs emerged. The present marketplace of PLDs can be further classified into the following families (Figure 10-14):

1. Programmable array logic (PAL).
2. Simple PLDs (SPLDs).
3. Complex PLDs (CPLDs).
4. Field programmable gate arrays (FPGAs).

CPLDs (sometimes called *super PALs*; Motorola Inc., Application note AN-1615) essentially are a large collection of PAL-like structures on one chip, desirable for designs that have tight speed requirements. FPGAs, on the other hand, consist of arrays of logic blocks connected by rows and columns of programmable interconnection lines and excel at designs that require large numbers of registers.

Choosing CPLDs or FPGAs or both for a design requires understanding the strengths of each (Swager, 1992; Miller, 1993). CPLDs usually are a better

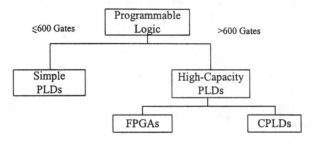

FIGURE 10-14 Programmable logic marketplace. (Reproduced by permission of Motorola Inc.)

choice for designs that have very tight speed requirements, simply because of their deterministic timing. This is not to say that CPLDs will be the fastest implementation of a design but simply that the final speed will be fairly predictable at the beginning of the design cycle. Creative FPGA design can result in very fast logic, but it also can require time-consuming attention to the placement and routing details of internal logic blocks. Generally, CPLDs excel at implementing large amounts of combinatorial logic, and FPGAs excel at designs that require large numbers of registers. For example, implementing control functions such as state machines depends on taking advantage of the architectural features of the device being used. If encoding is used to develop a state machine, the resulting logic tends to work best in logic devices having lots of combinatorial gates and few registers, such as CPLDs. On the other hand, if a bit-per-state mapping method is used to develop a state machine, the design will require more registers but few logic gates, which can be best implemented using FPGAs. A comprehensive description of these devices and designing with CPLDs and FPGAs are beyond the scope of the chapter.

If a design is dense, requiring more than 20,000 gates and high-volume production is expected, the best choice for implementation would be full-custom ASICs or standard cell-based designs (Neigh and Bishop, 1988; Napier, 1992). In full-custom ASICs, including cell-based systems and optimized arrays, there is no preprocessing of the silicon and the designer can place any structure onto the design. In cell-based designs, circuit elements such as D-types and full addresses are predesigned. The designer places these in position and forms the connections between them. This generally is done manually and requires a detailed knowledge of silicon technology.

Recently, PLDs, particularly FPGAs, have been gaining momentum over traditional ASICs as the logic device of choice for system designs with fewer than 20,000 gates. High-density PLDs are starting to break the price barrier that kept them from volume applications. Keeping in mind that mask-programmed gate arrays are faster and more capacious than FPGAs, logic designers have several product development paths they could embark on (Small, 1992):

- Do an FPGA design, reserving the option to convert the FPGA to a mask-programmed gate array later.
- Do a multiple FPGA design, reserving the option to convert the FPGA to a single mask-programmed gate array later.
- Do an FPGA/multiple FPGA and mask-programmed gate-array design in parallel, using FPGAs for prototypes or initial production runs.

Several companies offer FPGA-to-ASIC conversion to achieve these goals. Historically, designers have viewed PLDs as prototypes and used gate arrays for more cost-effective high-volume applications. However, high-density PLDs are making inroads on territory that traditional ASICs and gate arrays once dominated.

References

Conner, Doug. "Design Tools Smooth FPGA Configuration." *EDN* (June 7, 1990), pp. 49–58.

Dettmer, Roger. "User Programmable Logic — Chasing the Gate Array." *IEE Review* (May 1990).

Hannington, Steven. "Choosing and Using Programmable Logic." *Electronics and Wireless World* (July 1990), pp. 619–625.

Liebson, Steven. "PLD Development Software." *EDN* (August 1990), pp. 100–116.

Maxfield, Clive. "Field Programmable Devices." *EDN* (October 10, 1996), pp. 201–206.

Miller, Warren. "Design Efficient State Machines for PLDs or FPGAs." *EDN* (March 18, 1993), pp. 225–232.

Motorola Inc. "An FPGA Primer for PLD Users." Application note 1615. Motorola Inc., Phoenix, AZ, USA ●●●.

Napier, John C. "Bringing IC Layout In-House." *EDN* (September 3, 1992), pp. 108–118.

Neish, P., and P. Bishop, *Designing ASICs*. Ellis Howard Publications, England: 1988.

Pellerin, David, and Michael Holley. *Practical Design Using Programmable Logic*. Englewood Cliffs, NJ: Prentice-Hall, 1991.

Schulze, Bill. "Avoid Pitfalls in Selecting the Right Programmable-Logic Design Tool." Electronic Design (November 7, 1991), pp. 71–81.

Small, Charles S. "Programmable Logic Devices." *EDN* (February 5, 1987), pp. 112–127.

Small, Charles S. "Programmable Logic Devices." *EDN* (November 10, 1988), pp. 142–156.

Small, Charles S. "Programmable Gate Arrays." *EDN* (April 27, 1989), pp. 146–154.

Small, Charles S. "FPGA Conversion." *EDN* (June 4, 1992), pp. 107–116.

Swager, Anne Watson. "Choosing Complex." *EDN* (September 17, 1992), pp. 74–84.

Texas Instruments. *Programmable Logic Databook — 1993*. Texas Instruments, ●●●.

Index

443